Weather Radar

Springer
Berlin
Heidelberg
New York
Hong Kong
London
Milan
Paris
Tokyo

Physics and Astronomy ONLINE LIBRARY

springeronline.com

Physics of Earth and Space Environments

springeronline.com

The series *Physics of Earth and Space Environments* is devoted to monograph texts dealing with all aspects of atmospheric, hydrospheric and space science research and advanced teaching. The presentations will be both qualitative as well as quantitative, with strong emphasis on the underlying (geo)physical sciences.
Of particular interest are

- contributions which relate fundamental research in the aforementioned fields to present and developing environmental issues viewed broadly

- concise accounts of newly emerging important topics that are embedded in a broader framework in order to provide quick but readable access of new material to a larger audience

The books forming this collection will be of importance for graduate students and active researchers alike.

Series Editors:

Professor Dr. Rodolfo Guzzi
Responsabile di Scienze della Terra
Head of Earth Sciences
Via di Villa Grazioli, 23
00198 Roma, Italy

Professor Dr. Dieter Imboden
ETH Zürich
ETH Zentrum
8092 Zürich, Switzerland

Dr. Louis J. Lanzerotti
Bell Laboratories, Lucent Technologies
700 Mountain Avenue
Murray Hill, NJ 07974, USA

Professor Dr. Ulrich Platt
Ruprecht-Karls-Universität Heidelberg
Institut für Umweltphysik
Im Neuenheimer Feld 366
69120 Heidelberg, Germany

Peter Meischner (Ed.)

Weather Radar

Principles and Advanced Applications

Springer

Dr. Peter Meischner
DLR, Institut für Physik der Atmosphäre
Oberpfaffenhofen
82234 Wessling
Germany

Library of Congress Cataloging-in-Publication Data
Weather Radar: Principles and Advanced Applications / P. Meischner (ed.). p.cm. – (Physics of Earth and Space Environments, ISSN 1610-1677). Includes bibliographical references and index. ISBN 3-540-00328-2 (acid-free paper) 1. Radar meteorology. I. Meischner, P. (Peter), 1973. - II. Series.
QC973.5.W43 2003

ISSN 1610-1677

ISBN 3-540-000328-2 Springer-Verlag Berlin Heidelberg New York

This work is subject to copyright. All rights are reserved, whether the whole or part of the material is concerned, specifically the rights of translation, reprinting, reuse of illustrations, recitation, broadcasting, reproduction on microfilm or in any other way, and storage in data banks. Duplication of this publication or parts thereof is permitted only under the provisions of the German Copyright Law of September 9, 1965, in its current version, and permission for use must always be obtained from Springer-Verlag. Violations are liable for prosecution under the German Copyright Law.

Springer-Verlag is a part of Springer Science+Business Media
springeronline.com

© Springer-Verlag Berlin Heidelberg 2004
Printed in Germany

The use of general descriptive names, registered names, trademarks, etc. in this publication does not imply, even in the absence of a specific statement, that such names are exempt from the relevant protective laws and regulations and therefore free for general use.

Typesetting by the authors
Data conversion: Frank Herweg, Leutershausen
Cover design: Erich Kirchner, Heidelberg

Printed on acid-free paper   54/3141/ts - 5 4 3 2 1 0

Introduction

All aspects of our lives are influenced by the weather and especially severe weather events – such considerations tend to become more pertinent as such weather events may become more frequent with global warming. It is evident that better weather forecasts for the next few hours, days or even weeks have a high economic value for our developed and densely populated world. Ground and air traffic management benefits, as does the construction industry, tourism, agriculture and big outdoor events such as the Olympic games. Weather forecasting nowadays relies on numerical weather prediction models. The reliability and accuracy of forecasts however depends not only on the quality and resolution of these models, but to a high degree on the density and quality of input data from observations. Weather radar measurements complement current data such as ground observations and soundings at smaller scales and satellite observations on larger scales.

Targeted weather radar measurements additionally give detailed insight into the processes governing the developments of, for example, frontal systems and thunderstorms, which is needed for improving the model quality by comparing the observations with model forecasts. Weather radar observations on their own can be used directly for short-term forecasting, called nowcasting, just by observing the developments of weather systems some 100 km around and extrapolating the observations with knowledge of their typical behaviour. Finally, with reliable quantitative measurements of area precipitation by radar, flood forecasts will be improved. Reliable quantitative measurements of precipitation over large areas are essential, too, for the verification of upcoming global precipitation data sets from satellites, an urgent topic in climate research for understanding the hydrological cycle.

Technical advances in weather radar during the last decade have enabled traditional applications to be implemented much more precisely and reliably and in addition, enabled new applications to be tackled.

This growing interest in using advanced weather radar technologies is demonstrated by the numerous recent presentations at international conferences on radar meteorology such as the traditional biennial conference of the American Meteorological Society and the biennial European conference on radar meteorology, ERAD, a new series started in 2000. Besides the improvements in the quality of the measured parameters themselves, more versatile applications have emerged together with improved complementary observations, for example, from satellite and LIDAR systems. The data are used for better understanding of the characteristics of weather systems, for improved forecasts, especially of severe weather, for hydrological applications and flood warnings, as well as for climate research where ground verification is needed for global precipitation measurements by satellites.

This book presents a variety of advanced methods, applications and research results mainly from the European research arena. The contributions cover weather forecasting, hydrology and atmospheric and climate research. Much has been developed and discussed within the European COST-actions on weather radar technology and applications operating mostly with 3 year periods during the last 15 years. Every discipline benefits from the advances in other disciplines; therefore, the topics covered are related and some overlap is natural; however, common to all is the need for the highest possible data quality, which can only be assured by well designed, well calibrated and maintained radar systems together with adequate measurement methods and proper correction for errors caused by the varying atmospheric conditions themselves. The intention of this volume is not to present a general textbook on radar meteorology, since excellent monographs on Doppler and polarimetric radar are available and will be cited. An introductory Chap. 1, however, gives an overview of the potential of modern radar systems including Doppler and polarisation techniques, data processing, product generation and error correction in order to demonstrate the data quality and resolution achievable today. Because high quality data are needed for much larger areas than are covered by just one radar, weather radar networks are presented, too, especially for demonstrating the improvements in data quality obtinaed by cross-checking the performance of the individual systems in overlapping observation areas.

Quantitative precipitation measurement remains the topic of widest interest in radar meteorology. On the local scale of urban hydrology, the very high time and space resolution required over wide areas can only be achieved with weather radar. The accuracy required, however, is a challenge. On the global scale, sufficiently dense space and time sampling for monitoring elements of the hydrological cycle will only be possible by satellite observation. The accuracy of satellite retrievals of precipitation is presently open to question; ground validation is needed in different climate regimes (e.g. warm and cold regions) where different cloud and precipitation microphysical processes may be operating. Radar networks and polarimetric weather radars providing the best precipitation estimates will play a key role in such validation programmes.

Chapters 2 and 3 cover the most advanced strategies of quantitative precipitation measurement with operational Doppler radar networks in particular environments. In Chap. 2 precipitation measurement in mountaneous terrain such as the Alps is discussed, where the precipitation itself is strongly influenced by the orography – a process which should be understood – and where radar measurements frequently are degraded by beam blockage and ground clutter. Strategies are discussed for the avoidance and efficient elimination of ground clutter from mountains, and how to extrapolate measurements from above the Alps down to the ground level, and how to incorporate gauge data.

Chapter 3, closely related to the topic of Chap. 2, presents strategies for quantitative precipitation measurements in cold climates, where sleet and snow are always present and where radar beam ducting over cold surfaces is a major problem. It further impressively shows the tremendous improvements of data quality when care is taken to use the full potential of a radar network.

Chapter 4 gives a condensed overview of the use of radar measurements in hydrometeorology, flood forecasting and water resources management. Data quality again is a key issue, as is the long time reliability of precipitation data for the design of dams and flood defences.

Polarimetric methods have been the subject of much research and are now being considered for more extended operational use. The new parameters essentially measure hydrometeor shape and fall mode and can be used for classification of precipitation type. This allows precipitation growth to be followed and conceptual models to be formulated, as shown in Chap. 6. Raindrops are oblate to a degree which depends upon their size, so this additional size information can be used for improving quantitative rainfall estimates when compared to those available from conventional measurements of radar reflectivity. The new parameters also provide better quality control and error estimation. Both better rainfall estimates and better overall quality control are urgently needed for forthcoming local and global quantitative precipitation measurements. However, the precise choice of the combination of parameters to achieve these goals is still not clear. Chapter 5 reviews these aspects in the context of the requirements of an operational system.

Chapter 6 illustrates the improved understanding of the dynamics and precipitation production in midlatitude weather systems such as fronts and in the mesoscale organisation of severe convection and accompanying hazardous weather, including tornadoes, by making full use of Doppler and polarimetric measurements. Conceptual models that help in forecasting severe events have been developed from Doppler measurements that indicate air flow patterns, together with polarimetric measurements that simultaneously identify the different precipitation particle types, their location and growth. Well-documented cases are the basis for improving and verifying cloud resolving prediction models, and for the development of parameterisations of the relevant subgrid scale processes in larger scale weather prediction models.

Global observations of precipitation, as mentioned, are urgently needed for climate research and monitoring, and for more local applications, too, if the required resolution can be provided. Global coverage with such measurements can only be performed with satellites, and steps towards such missions include airborne instrumentation. Chapter 7 presents the latest strategies for measurement of precipitation by spaceborne and airborne radar. Because payload capacity is limited, such systems need to operate with smaller antennas than ground based instruments, implying smaller wavelengths to maintain sufficient spatial resolution. Radar radiation of about 1 cm wavelength rather than 5 or 10 cm, however, is much more prone to attenuation. Therefore, special retrieval algorithms and correction schemes for attenuation have been developed which can now also be applied to ground based radars. This, together with the polarimetric methods presented in Chap. 5, will further improve data quality.

Airborne weather radars not only are steps toward spaceborne systems, but are needed, for example, for measurements in remote areas, or – most importantly – to follow moving cloud systems thus maintaining optimised measurement conditions. Wind field retrievals by airborne Doppler radar, described here also, help to understand and forecast weather developments, thus complementing the measurement techniques presented in Chap. 6.

Conventional operational weather radars mainly operate in the S-band (10 cm wavelength) and C-band (5 cm wavelength). They are not sensitive enough to detect cloud particles, droplets or pristine ice crystals, some 100 µm in diameter or smaller. So for the study of nonprecipitating clouds, which contribute much to the global radiation balance, shorter radar wavelengths of 1 cm and less must be used. The combination with Lidar measurements further extends the ability to follow condensation processes and the interaction with aerosol particles. Chapter 8 deals with the combination of cloud radars with lidar and microwave radiometers, and case studies demonstrate the synergistic effect of combining such instrumentation.

The use of radar measurements for assimilation in numerical weather prediction models has already begun to improve forecast quality and will contribute more in the future. Radar estimated precipitation and radar estimated wind profiles, radial Doppler winds or 3-D wind fields from bistatic radar systems can all be assimilated. Realising the optimum benefit from these data requires that they be quality controlled and accompanied by error estimates. In Chap. 9, requirements and promising assimilation strategies are presented, and future prospects discussed.

This book closes with Chap. 10, which is a glossary explaining radar meteorology terms used here, and which might be of general use. Abbreviations used throughout the book are also defined in this chapter.

Acknowledgements

This book has become a reality through a coordinated effort of many internationally renowned European scientists in the field of radar meteorology. They represent research, academia, operations and industry, and such assure that all aspects from data acquisition, data quality and all the advanced applications in hydrology, meteorology and weather forecast are covered.

I want to express my deep gratitude to all who contributed, knowing only too well that this work had to be done in addition to the normal burden of duties. This holds for all institutions involved but to a special degree for the DLR, Institut of Atmosphic Physics, where resources generously could be used. Special thanks go to Brigitte Ziegele and Thomas Jank who were heavily involved in typing and installing and maintaining the special LaTeX tools. Their personal commitment was fundamental for the final technical success.

Thanks to George Craig some of the nonnative formulations have been smoothed over for easier reading.

We all now hope that this book will be helpful for all those involved in the future use of modern weather radar systems. Enjoy it.

Oberpfaffenhofen, March 2003 *Peter Meischner*

Contents

1 The State of Weather Radar Operations, Networks and Products
Frank Gekat, Peter Meischner, Katja Friedrich, Martin Hagen, Jarmo Koistinen, Daniel B. Michelson, and Asko Huuskonen 1
1.1 Introduction 1
1.2 Doppler Radars 3
1.3 Polarimetric Radars 13
1.4 Weather Radar Networking 17
1.5 Data Quality 19
1.6 Products of Modern Weather Radars and Networks 35

2 Operational Measurement of Precipitation in Mountainous Terrain
Urs Germann and Jürg Joss 52
2.1 Introduction with a Focus on the Variability of Precipitation in the Alps 52
2.2 Clutter Elimination 58
2.3 Correction for Visibility 63
2.4 Profile Correction 67
2.5 Adjustments by Gauges 71
2.6 What Next? 74
References 75

3 Operational Measurement of Precipitation in Cold Climates
Jarmo Koistinen, Daniel B. Michelson, Harri Hohti, and Markus Peura .. 78
3.1 Introduction 78
3.2 Problems Due to Ducting over Cold Surfaces 80
3.3 Precipitation Phase 90
3.4 Shallow Precipitation: the Main Limiting Factor 94
3.5 Nowcasting 104
3.6 Future Outlook 109
3.7 Acknowledgements 110
References 110

4 Using Radar in Hydrometeorology
Christopher Collier, Paul Hardaker 115
4.1 The Role of Radar in Flood Forecasting
    and Water Resources Management 115
4.2 Radar Data Quality Control 116
4.3 Flood Forecasting 118
4.4 Coupled Atmospheric and Hydrologic Numerical Models 123
4.5 Engineering Design 125
4.6 Future Prospects 127
References 127

5 Improved Precipitation Rates and Data Quality by Using Polarimetric Measurements
Anthony Illingworth 130
5.1 Introduction 130
5.2 The Polarisation Parameters 131
5.3 Raindrop Shapes and Size Spectra 139
5.4 Identification of Ground Clutter and Anomalous Propagation . 144
5.5 Improved Rainfall Rates using Polarisation Parameters 145
5.6 Improved Rainfall Rate Using Integrated Polarisation Parameters 150
5.7 Improved Rainfall Rates When Ice May Be Present 157
5.8 Correction for Attenuation 158
5.9 Identification of Hydrometeors 159
5.10 Conclusion 160
References 163

6 Understanding Severe Weather Systems by Advanced Weather Radar Observations
Peter Meischner, Nikolai Dotzek, Martin Hagen, Hartmut Höller 167
6.1 Introduction 167
6.2 Frontal Systems 168
6.3 Deep Convective Systems 172
6.4 The Future 195
References 196

7 Precipitation Measurements from Space
Jacques Testud 199
7.1 Correcting for Attenuation:
    A Major Problem with a Space Borne or an Airborne Radar 199
7.2 Statistical Properties of the Drop-Size Distribution
    and Parameterisation of the Rain Relations 204
7.3 Algorithms for Rain Retrieval with a Spaceborne Weather Radar 215
7.4 Airborne Dual Beam Doppler Radar 218
7.5 Validation of the Algorithm Product
    with Airborne or Space Borne Radars 222

7.6 The Spaceborne Technology Applied
to Ground Based Polarimetric Radar: Algorithm ZPHI 226
7.7 The Rain Profiling Algorithm in Future Operational Applications 232
References 232

8 Radar Sensor Synergy for Cloud Studies; Case Study of Water Clouds
Herman Russchenberg, Reinout Boers 235
8.1 Introduction 235
8.2 Particle Scattering 236
8.3 Sensor Synergy 240
8.4 Case Study: Retrieval of the Microstructure of Water Clouds 243
8.5 Acknowledgements 252

9 Assimilation of Radar Data in Numerical Weather Prediction (NWP) Models
Bruce Macpherson, Magnus Lindskog, Véronique Ducrocq, Mathieu Nuret, Gregor Gregorič, Andrea Rossa, Günther Haase, Iwan Holleman, and Pier Paolo Alberoni 255
9.1 Introduction 255
9.2 Data Assimilation 256
9.3 Assimilation of Radar Precipitation Data in NWP Models 259
9.4 Assimilation of Radar Wind Data 268
9.5 Quality Control of Radar Data for NWP 273
9.6 Treatment of Radar Data Errors in Assimilation 274
9.7 Future Prospects 275
References 276

Glossary 281

Color Plates 313

Index 333

List of Contributors

Pier Paolo Alberoni
Arpa - Servizio Meteorologico
Regionale
Viale Silvani 6
40122 Bologna
Italy
palberoni@smr.arpa.emr.it

Reinout Boers
Royal Netherlands
Meteorological Institute
Wilhelminalaan 10
3730 AE De Bilt
Netherlands
boers@knmi.nl

Christopher Collier
Telford Institute of
Environmental Systems
University of Salford
Salford M5 4WT
United Kingdom
c.g.collier@civils.salford.ac.uk

Veronique Ducrocq
Météo-France
CNRM-GMME
42, avenue G. Coriolis
31057 Toulouse
France
veronique.ducrocq@meteo.fr

Nikolai Dotzek
DLR
Institut für Physik der Atmosphäre
Oberpfaffenhofen

82234 Wessling
Germany
nikolai.dotzek@dlr.de

Katja Friedrich
DLR
Institut für Physik der Atmosphäre
Oberpfaffenhofen
82234 Wessling
Germany
katja.friedrich@dlr.de

Frank Gekat
Gematronik GmbH
Raiffeisenstr. 10
41470 Neuss
Germany
f.gekat@gematronik.com

Urs Germann
Meteo-Swiss
Via ai Monti 146
6605 Locarno Monti
Switzerland
uge@meteoswiss.ch

Gregor Gregorič
University of Ljubljana
Department of Physics
Jadranska 19
1000 Ljubljana
Slowenia
gregor.gregoric@uni-lj.si

List of Contributors

Günther Haase
Swedish Meteorological and
Hydrological Institute (SMHI)
60176 Norrköping
Sweden
gunter.haase@smhi.se

Martin Hagen
DLR
Institut für Physik der Atmosphäre
Oberpfaffenhofen
82234 Wessling
Germany
martin.hagen@dlr.de

Paul Hardaker
Met Office
Beaufort Park
Easthamstead Wokingham Berkshire
RG40 3DN
United Kingdom
paul.hardaker@metoffice.com

Harri Hohti
Finnish Meteorological Institute
(FMI)
Product Development
P.O. Box 503
00101 Helsinki
Finland
Harri.Hohti@fmi.fi

Iwan Holleman
KNMI
Department of Satellite Data
Wilhelminalaan 10
3730 AE De Bilt
Netherlands
holleman@knmi.nl

Hartmut Höller
DLR
Institut für Physik der Atmosphäre
Oberpfaffenhofen
82234 Wessling
Germany
hartmut.hoeller@dlr.de

Asko Huuskonen
Finnish Meteorological Institute
(FMI)
Observational Services
P.O. Box 503
00101 Helsinki
Finland
Asko.Huuskonen@fmi.fi

Anthony Illingworth
Dept. of Meteorology
University of Reading
Reading RG6 6BB
United Kingdom
a.j.illingworth@reading.ac.uk

Jürg Joss
Meteo-Swiss
Via ai Monti 146
6605 Locarno Monti
Switzerland
jjo@freesurf.ch

Jarmo Koistinen
Finnish Meteorological Institute
(FMI)
P.O. Box 503
00101 Helsinki
Finland
Jarmo.Koistinen@fmi.fi

Magnus Lindskog
Swedish Meteorological and
Hydrological Institute (SMHI)
60176 Norrkøping
Sweden
magnus.lindskog@smhi.se

Bruce Macpherson
Met Office
London Road
Bracknell RG12 2SZ
United Kingdom
bruce.macpherson@metoffice.com

Peter Meischner
DLR
Institut für Physik der Atmosphäre
Oberpfaffenhofen
82234 Wessling
Germany
peter.meischner@dlr.de

Daniel Michelson
Swedish Meteorological and
Hydrological Institute (SMHI)
60176 Norrkøping
Sweden
daniel.michelson@smhi.se

Mathieu Nuret
Météo-France
CNRM-GMME
42, avenue G. Coriolis
31057 Toulouse
France
mathieu.nuret@meteo.fr

Markus Peura
Finnish Meteorological Institute
(FMI)
Product Development
P.O. Box 503
00101 Helsinki
Finland
Markus.Peura@fmi.fi

Andrea Rossa
Météo-Swiss
Krähbühlstrasse 58
8044 Zürich
Switzerland
andrea.rossa@meteoswiss.ch

Herman Russchenberg
Delft University of Technology
Mekelweg 4
2600 GA Delft
Netherlands
h.w.j.russchenberg@et.tudelft.nl

Jacques Testud
CETP
10-12 Avenue de l'Europe
78140 Velizy
France
jacques.testud@cetp.ipsl.fr

1 The State of Weather Radar Operations, Networks and Products

Frank Gekat,[1] Peter Meischner,[2] Katja Friedrich,[2] Martin Hagen,[2] Jarmo Koistinen,[3] Daniel B. Michelson,[4] and Asko Huuskonen[3]

[1] Gematronik GmbH, Raiffeisenstr. 10, D-41470 Neuss
[2] DLR, Institut für Physik der Atmosphäre, Oberpfaffenhofen, D-82234 Wessling
[3] Finnish Meteorological Institute (FMI) Helsinki, Finland
[4] Swedish Meteorological und Hydrological Institute (SMHI), Norrköping, Sweden

1.1 Introduction

Do you remember the famous cars of the early 1950s, the 2CV and the Volkswagen "Beetle"? They were simple tin boxes with four wheels, four seats, a motor, a steering wheel, and a speedometer. When opening the bonnet, you could recognise nearly every part, you understood all the functions and could handle most shortcomings yourself. How does a modern upper class car look nowadays in comparison? The front panel is like the cockpit of an aircraft and when opening the bonnet, you just may recognise where to fill the water for cleaning your windows. Similarly, weather radar systems developed from microwave devices transmitting radiation pulses and detecting the location and distance of a target by its reflection – RADAR means "RAdiation Detection And Ranging" – to "high tech" systems delivering a number of products of interest for the user. The "targets" we now have in mind are cloud and precipitation particles.

Fig. 1.1. Enjoy driving a modern car!

Modern weather radars, Doppler radars, or especially polarimetric Doppler radars use the latest technological developments. Microwave devices and components are developed by electrical engineers, the sophisticated data processing algorithms use mathematics and applied computer science, the latest colour display techniques are used, including image processing methods, and data must be exchanged by powerful telecommunication links within weather radar networks and for remote operation and maintenance.

You, the user, now have the choice. If you just want to drive a nice car to arrive in time, enjoying the landscape and listening to Mozart, you will fully rely on your manufacturer and your workshop. If, however, you, as the pilot, want to win a race, you need to optimise the performance of your car specially for the actual race track and the actual conditions. For that you need to know the components of your car more precisely, and you must be able to advise your specialised racing team.

An ideal radar meteorologist therefore should be an electrical engineer, a mathematician, a computer scientist, a meteorologist, a cloud physicist, and hopefully a hydrologist in one person. Unfortunately, he mostly is not! Instead, he will be more experienced in applications. But to make best use of their system, users from operations and research as well need to know the principles in order to be aware of limitations and to locate errors and uncertainties. Do those originate from the radar system not working correctly or from changes in the performance of subsystems, are they due to the measurement method, for example, the scan strategy, or are the assumptions on atmospheric conditions – for example, that within the resolution volume we have only liquid raindrops of spherical shapes of a known size distribution – not fulfilled? Or is all the data processing not working correctly? Usually, we have contributions from all such error sources and need to understand the different weightings.

Every user, the holiday driver, and the racing pilot are affected by such errors. No wonder that all chapters of this book touch on this issue in some depth, especially those dealing with quantitative precipitation measurements. Within radar networks, which are combined from different radar systems, differences in performance emerge clearly. Therefore, in this chapter, we already gives examples and discuss experiences gained with NORDRAD, the Nordic weather radar network. They show most dramatically that you never should rely on the performance of a weather radar system without doing your own testing.

It is the aim of this first introductory chapter to give a general overview of advanced weather radar systems and their subsystems, their performance, different error sources as well as on basic and some advanced products. Readers may have a short glance at this chapter and then concentrate on applications of interest. When, however, reading this chapter, they may ask for much more detail. These are available in a number of special publications, reports and manuals cited within the text.

In any case, we hope it helps you to become an excellent driver!

1.2 Doppler Radars

A modern weather radar consists, in principle, of five main subsystems, as displayed in Fig. 1.2. Controlled by the radar signal processor, the transmitter generates microwave pulses, which are guided to the duplexer. The duplexer is a nonlinear microwave circuit which routes the signal to the antenna and the backscattered received signals to the receiver, which amplifies them and removes the microwave carrier frequency. The output signal then is digitised and processed by the signal processor, which also controls the timing of the transmitter, the transmitted pulse width, and the pulse repetition frequency, PRF. The output data are send to the radar product generator, the workstation of the user. It generates three-dimensional meteorological data sets. Depending on the product of interest, the operator needs to control the transmitter timing, the scan strategy, and if applicable, the polarisation settings. **The range** r, the distance between radar and target, is determined by the delay time t_d the signal needs to travel to and back from the target at c, the speed of light:

$$r = t_d c / 2 \qquad (1.1)$$

The range resolution $\Delta r = \tau c / 2$ is given by the duration of the transmitted radiation, the pulse length τ. A typical pulse length is 1µs, corresponding to 300 m in space. See Fig. 1.3.

The range $r_{\max}$ up to which measurements can be performed **unambiguously** depends on the pulse repetition frequency, PRF, because a signal return must be received before the next pulse is transmitted.

$$r_{\max} = t_{d,\max} c / 2 = c / 2(PRF). \qquad (1.2)$$

Figure 1.4 shows a radar operating with a PRF of 1000 Hz. The maximum unambiguous range then is 150 km. If the receiving channel is open longer than $t_{t,\max}$, a target at a distance of 189 km would be displayed at 39 km.

Modern Doppler weather radars may extend the unambiguous range by the so-called second trip recovery technique. Subsequent pulses are coded differently, so that the received return signals can be related unambiguously to the transmitted pulses (Siggia, 1993).

The reflectivity Z of a target is estimated from the received power P_R. This power depends on the technical characteristics of the radar, the propagation conditions, the distance to the target and its reflectivity Z. The meteorological radar equation, as given by Doviak and Zrnic (1984), is

$$P_R(r) = \frac{\pi^3 P_T G^2 G_R \Theta^2 c\tau |K|^2 Z}{2^{10} \ln(2) \lambda^2 r^2 L_{\mathrm{atm}}^2 L_{\mathrm{MF}}}, \qquad (1.3)$$

with

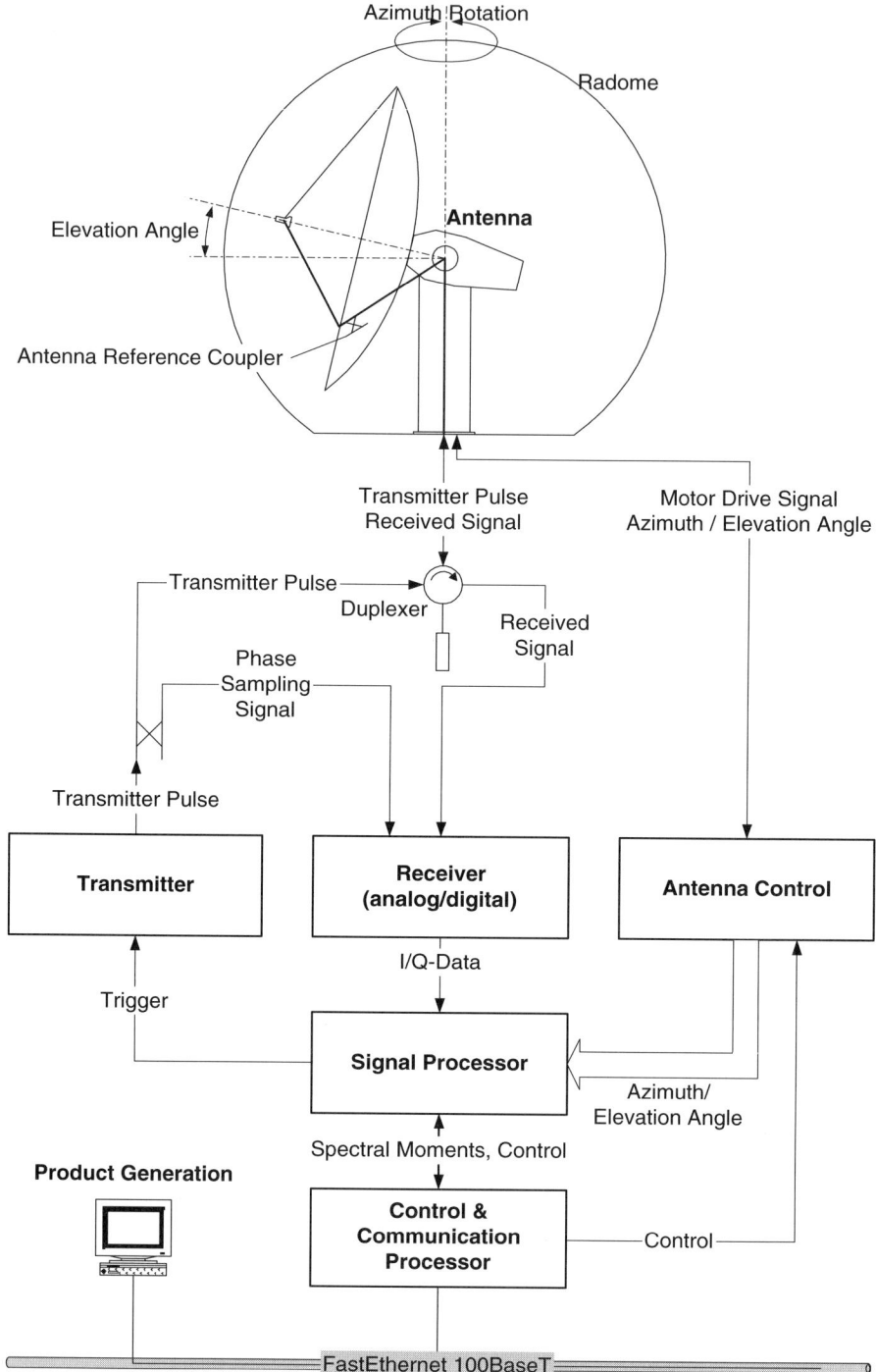

Fig. 1.2. Basic components of a Doppler radar system

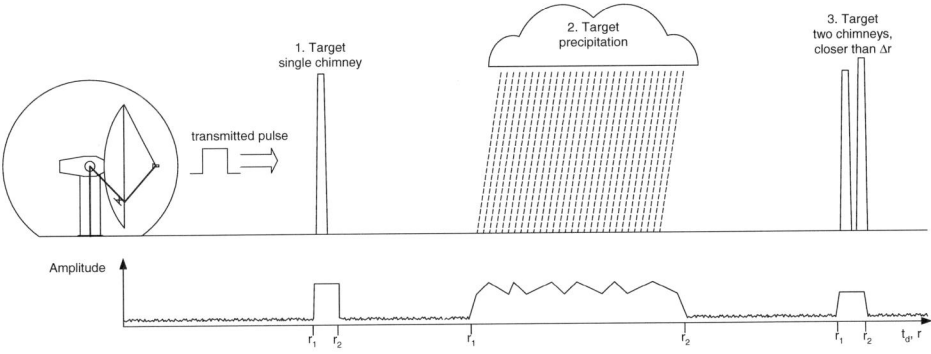

Fig. 1.3. Range resolution. The horizontal axis is the time axis, equivalent to range

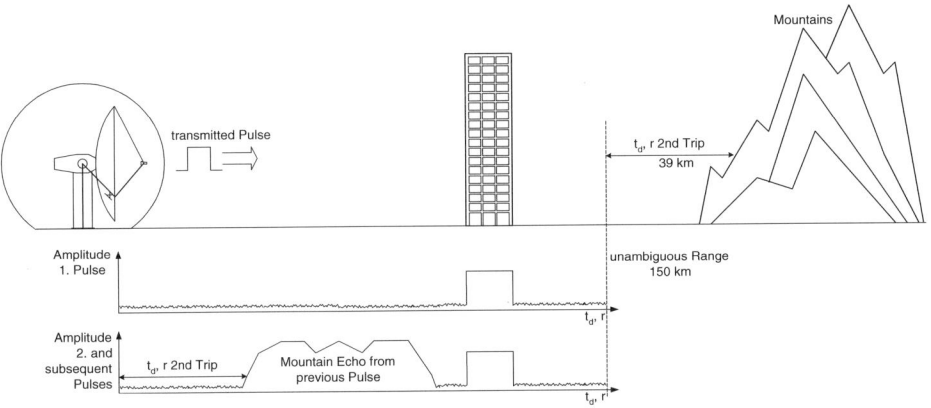

Fig. 1.4. Unambiguous range

P_T Transmitted peak power at the antenna
G Total antenna gain
G_R Total receiver gain, considering also path losses between antenna and electronics
Θ Antenna 3 dB beam width
$|K|^2$ Constant factor, 0.93 for rain and 0.2 for ice and snow
λ Wavelength of transmitted radiation
L_{atm} Atmospheric attenuation between antenna and target (one way)
L_{MF} Matched filter losses.

This equation is valid under the assumption that the beam formed by the antenna has a circularly symmetric shape and the resolution volume is completely filled with precipitation.

With $c = \lambda f$, where f is frequency, (1.3) can be rearranged to calculate the reflectivity Z, with terms grouped according to the different radar subsystems:

$$Z = \underbrace{\frac{2^{10}\ln(2)\,c}{\pi^3 |K|^2}}_{\text{constant}} \cdot \underbrace{\frac{1}{f^2 P_T \tau}}_{\text{transmitter}} \cdot \underbrace{\frac{1}{G^2 \Theta^2}}_{\text{antenna}} \cdot \underbrace{\frac{L_{\text{MF}}}{G_R}}_{\text{receiver}} \cdot r^2 \cdot L_{\text{atm}}^2 \cdot P_R \,. \tag{1.4}$$

If the radar system is well calibrated and stable, we can combine all constants and use C_R for that system. It includes the pulse length used and must further use the right $|K|^2$ which is quite different for rain and snow. This important point will be further discussed in this book by Koistinen et al. (2003). The measured reflectivity then is given by

$$Z = C_R \cdot r^2 \cdot L_{\text{atm}}^2 \cdot P_R \,. \tag{1.5}$$

L_{atm}^2 must be estimated by actual measurement. For the most part, such a procedure is already provided by modern systems. Z is measured in mm^6/m^3 representing the scattering cross section of all hydrometeors within one cubic meter. Because of the large dynamic range of observed reflectivities, logarithmic units are used

$$dBZ = 10 \log \left[Z/(\text{mm}^6/\text{m}^3) \right] \,. \tag{1.6}$$

A Doppler or coherent weather radar, in addition to reflectivity measurements, is capable of controlling and measuring **the phases** of the transmitted and received signals. The measured phase shift of the received backscattered signal, compared to the phase of the transmitted signal, allows the estimation of the mean radial, so-called Doppler velocity of the particles within the resolution volume. The principles of such a radar are given in Fig. 1.5.

The steps of signal generation and processing are

The *RF* signal is generated by the Stable Local Oscillator, STALO, and fed to the exciter and to the downconversion mixers.	$\exp(j\omega t)$
The signal is modulated by the exciter and routed to the transmitter.	$M(\tau, t) \exp(j\omega t)$
After amplification, the transmitter pulse passes the duplexer and is applied to the antenna. Z_L is the line impedance.	$\sqrt{P Z_L} M(\tau, t) \exp(j\omega t)$
The antenna emits an electromagnetic wave described by the amplitude E_T of the emitted pulse, and k is the wave number.	$E_T(r) M(\tau, t) \exp[j(\omega t - kr)]$

The antenna signal from backscattering includes the factor 2 because the signal has traveled the range r twice.

$$E_{\text{Refl.}}(r)\,M(\tau,t)\exp\left[j(\omega t - 2kr)\right]$$

After amplification by the low noise amplifier (LNA) and matched filtering, the amplitude is calculated with the radar equation (3), accounting for the line impedance. The received power is multiplied by 2 for normalisation.

$$\sqrt{2P_{\text{R}}(r)\,Z_{\text{L}}}\,M(\tau,t)\exp\left[j(\omega t - 2kr)\right]$$

To convert the rf signal to the baseband, the received rf signal s_{rf} is transformed to the time domain:

$$\begin{aligned}
s_{\text{RF}} &= \text{Re}\left\{\sqrt{2P_{\text{RX}}(r)\,Z_{\text{L}}}\,M(\tau,t)\exp\left[j(\omega t - 2kr)\right]\right\} \\
&= \sqrt{2P_{\text{RX}}(r)\,Z_{\text{L}}}\,M(\tau,t)\cos(\omega t - 2kr).
\end{aligned} \quad (1.7)$$

In order to extract the phase information $2kr$, the signal is split into its so-called in-phase component, s_{I}, and its quadrature component, s_{Q}, by splitting and mixing one component with the STALO signal and the other with the STALO

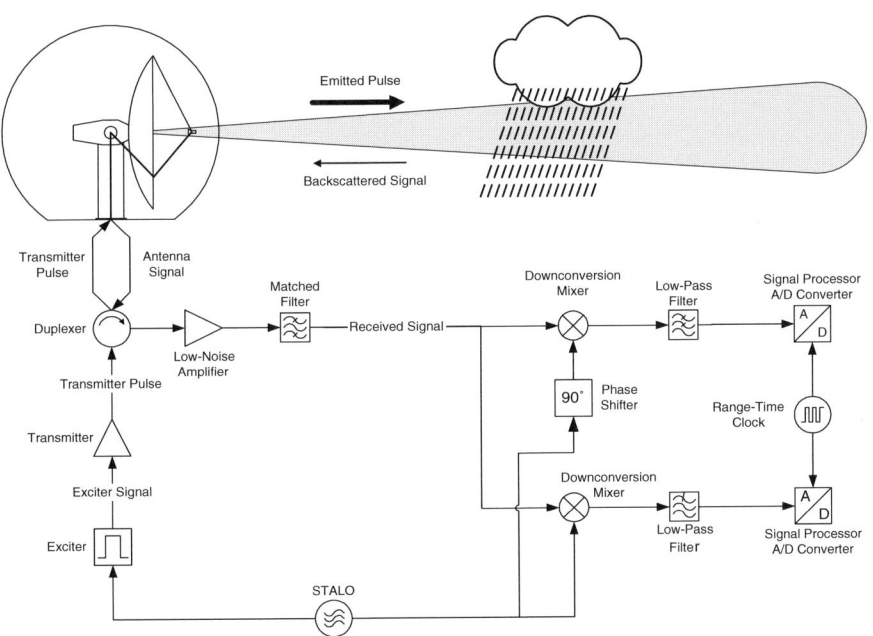

Fig. 1.5. Signal flow of a Doppler weather radar with transmitter, left, and receiver, right

signal shifted by 90°. This enables the distinction between positive and negative phase shifts given by targets moving toward or away from the radar site. The signal power P_R is calculated as

$$P_R = (s_I^2 + s_Q^2)/Z_L \tag{1.8}$$

with Z_L the line impedance.

For simplicity, the operation with a homodyne receiver was described, which requires only one oscillator. Modern weather radars, however, work with heterodyne receivers, applying at least two oscillators, the STALO and an intermediate frequency (if) oscillator for the frequency transformation between rf and baseband. In those systems, the exciter not only provides the rf pulse modulation but also the upconversion of the if signal to the rf. Each stage features its own amplifiers, and matched filtering is usually applied to the if signal, not to the rf signal.

The signal flow described so far applies to klystron transmitters, which are high-power amplifiers and therefore stable in phase. For systems with magnetrons, which are pulsed high-power oscillators, the phase of each magnetron pulse with respect to the STALO needs to be measured. The phase offset is added to the phase of the backscattered received signal as indicated in Fig. 1.5, where a phase sampling signal is coupled from the transmitter to the receiver.

In order to optimise the signal-to-noise ratio, SNR, of the system, the noise figure N,

$$N = k_B T_{SYS} B \tag{1.9}$$

which depends on the system temperature T_{sys} and the receiver bandwidth B (k_B is the Boltzmann constant) needs to be minimised. This is performed by a **filter matched** to the transmitter bandwith. The matched filter loss is a pure spectral loss. Figure 1.6 shows the spectrum of the transmitted pulse. Only spectral lines which fall within the matched filter bandwidth are processed, all others are rejected. The nature of this loss has to be considered when the receiver is calibrated with a test signal generator.

When the transmitted microwave pulse travels outward, backscattered signals arrive at the antenna continuously. The **range sampling** is performed by a range time clock which triggers the analog-to-digital converters (ADC) of the radar signal processor. The clock frequency determines the range resolution. In order to acquire one sample per range gate, the frequency of the range time clock f_{rtc} must be the reciprocal of the pulse width τ.

For each range gate, a complex signal $\underline{s}$ is sampled. For the nth range gate, it is

$$\underline{s}_n = \frac{\sqrt{2 P_R(r_n) Z_L}}{2} \left[\cos(2kr_n) + j \sin(2kr_n) \right]. \tag{1.10}$$

The number, M, of samples which can be taken for each range gate depends on the antenna rotation rate, ω_{az}, the pulse repetition frequency, PRF, and the beamwidth Θ:

$$M = \text{PRF } \Theta/\omega_{az} = PRF\, T_D. \tag{1.11}$$

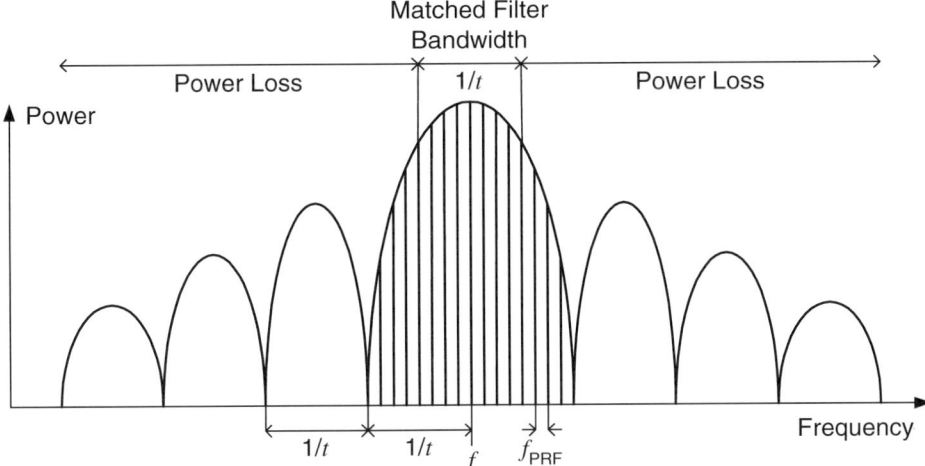

Fig. 1.6. Pulse spectrum

The time T_D, that the antenna beam stays on the target is called the integration time, dwell time or coherent processing interval (CPI). The sampling process, as illustrated in Fig. 1.7, provides complex time series for each range gate. Typically, 20 to 200 samples are collected within less than 0.1 s depending on the PRF. These time series are the basis for further processing steps.

The further **signal flow** with a modern Doppler weather radar signal processor is shown in Fig. 1.8. It shows the most important steps to providing basic radar products of use to the customer.

First the analog signals are digitised. The noise floor of the system is composed of the noise figure of the analog components of the receiver chain and the digitisation noise of the ADCs. The upper limit of the dynamic reception range of the radar system is thus determined by the resolution of the ADCs and the saturation limits of the analog components of the receiver.

In order to remove returns from mountains, buildings or just ground without removing the power backscattered from precipitation, **clutter filtering** is performed. Different options commonly are available with recent Doppler radar systems. The often applied "time-domain filtering" is illustrated in Fig. 1.9. The major disadvantage of this filtering is the unavoidable distortion of the weather signal, especially from slowly moving weather targets. This bias is avoided with frequency-domain filtering (Fig. 1.10). Here the signal data of each range gate are transformed by a Fast Fourier Transform (FFT) algorithm into the frequency domain, and bins representing zero velocity are discarded. Then the likely signal amplitudes for these frequency bins are reconstructed by interpolation. Finally, the signal is transformed back into the time domain for moment estimation, see SIGMET (2000). Although this technique needs higher digital signal processing resources than time domain filtering, it will be standard for future systems.

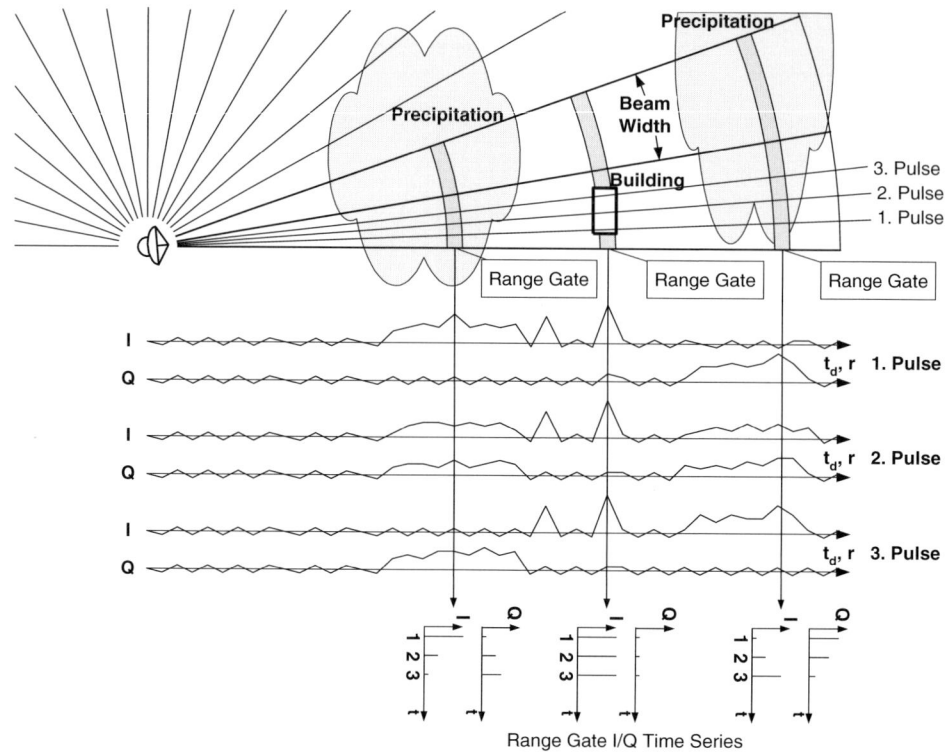

Fig. 1.7. Coherent sampling process

Careful clutter removal is essential when quantitative precipitation measurement is the goal. Doppler techniques however are only one approach; they generally are combined with further methods as discussed below, and more comprehensively by Germann and Joss (2003) in this book.

Clutter filtering is followed by covariance processing of the time series of each range gate. The first lags of the autocorrelation function of the complex signal series are estimated, as described by Keeler and Passarelli (1990), followed by time integration and, possibly, range integration. The autocorrelation function for the time lag T–L can be calculated from the complex signal series as

$$\hat{R}(T_\mathrm{L}) = \frac{1}{M-L} \sum_{i=1}^{M-L} \underline{s}_i \underline{s}^*_{i+L}, \qquad (1.12)$$

with M the number of signal samples from one range gate and L the number of lags between the two signal series. The asterisk* denotes the complex conjugate. $T - L$ is the time lag between the signal series, equal to the pulse repetition time $T_\mathrm{L} = 1/\mathrm{PRF}$.

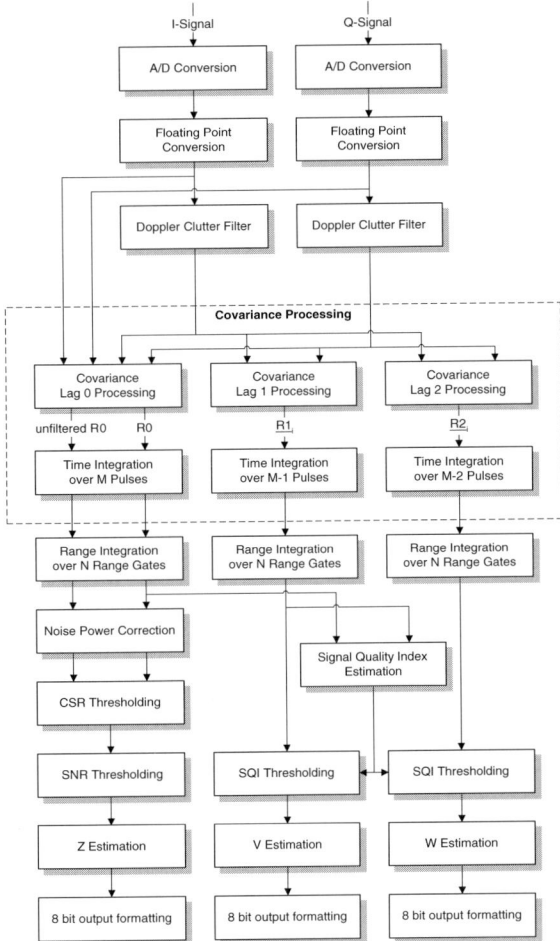

Fig. 1.8. Signal processing data flow

Finally, we receive the parameters provided by a Doppler weather radar. The **signal power** then is estimated by

$$\hat{R}_0 = \frac{1}{M} \sum_{i=1}^{M} \underline{s}_i \underline{s}_i^*, \tag{1.13}$$

the **mean radial velocity** by

$$\hat{R}_1 = \frac{1}{M-1} \sum_{i=1}^{M-1} \underline{s}_i \underline{s}_{i+1}^*, \tag{1.14}$$

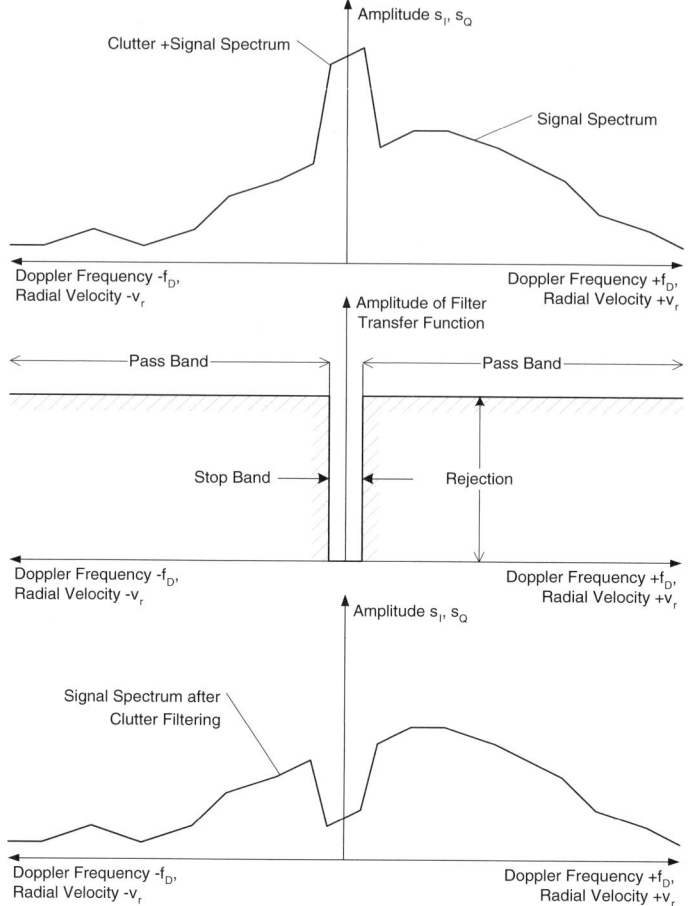

Fig. 1.9. Time-domain clutter filtering, described in the frequency domain

and the **spectral width**, related to the variance of the mean Doppler velocity within the resolution volume, by

$$\hat{R}_2 = \frac{1}{M-2} \sum_{i=1}^{M-2} \underline{s}_i \underline{s}_{i+2}^*. \tag{1.15}$$

In order to reduce the variance of these estimates, averaging of several contiguous range gates of the same dwell time may be applied, thus, however, reducing the range resolution. These three quality controlled parameters, received power and from that the **reflectivity factor**, the **Doppler velocity** and the **Doppler spectral width** are of primary interest to the user. They are displayed on the radar screen according to the scan strategy as PPI (horizontal overview) or RHI (vertical cross section) formats. There are numerous examples throughout this book.

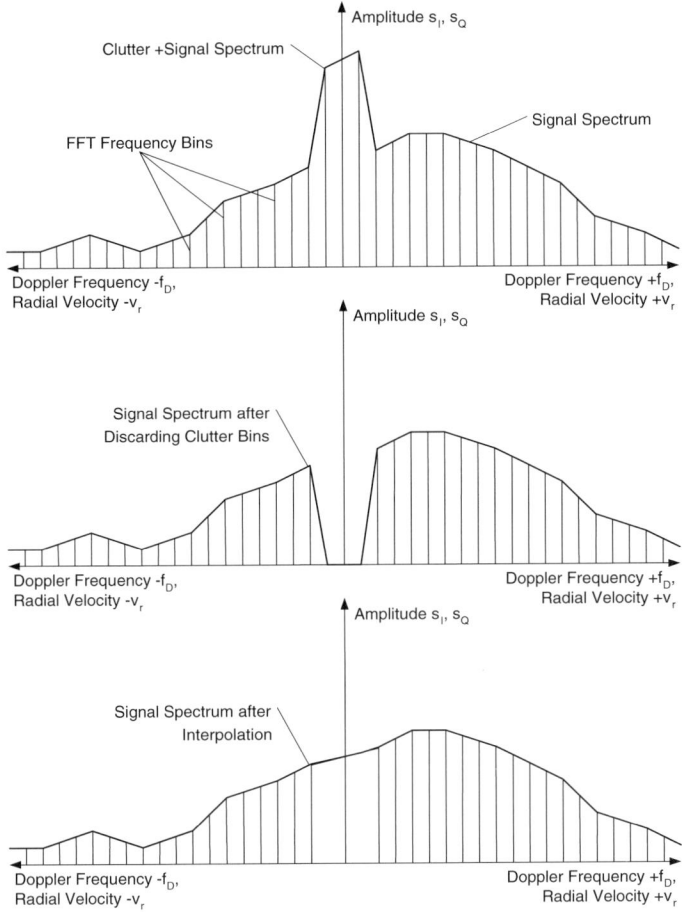

Fig. 1.10. Frequency-domain clutter filtering

1.3 Polarimetric Radars

Hydrometeors in the form of raindrops and ice particles are characterised by different shapes, different orientations during fall and different dielectric constants. Differently polarised radiation, therfore, will be backscattered differently, and polarimetric radar measurements provide additional parameters for recognising and classifying different precipitation types such as rain, hail, graupel, and snow, and for improving the quality of quantitative rain estimates. Such applications will be discussed further by Illingworth (2003), this book, and particle classification is used by Meischner et al. (2003), this book.

A polarimetric radar is capable of controlling the polarisation of the transmitted wave for each pulse as well as receiving selected polarisation states for the return of each pulse. Most polarimetric radars use only two selected orthogonal polarisations, usually horizontal, H, and vertical, V, to transmit and receive. Because the recent book of Bringi and Chandrasekar (2001) very comprehensively describes theory, system designs and applications, only we touch general principles here.

Common technical approaches are:

1) The switched dual-polarisation configuration
2) The dual-channel dual-polarisation configuration
3) The switched dual-channel, polarisation-agile radar

Special care is needed for the antenna design; it should be symmetrical for both polarisations. Sometimes this requirement must be compromised, for example, to minimise side lobes. A critical issue is the radome, made of several layers of electrically matched dielectric material.

A so-called pseudoradom panel cut is suggested for polarimetric radars because the orientation of the cuts is randomised, yielding hexagonal and pentagonal sphere sectors (Manz et al., 1998). Water layers, especially as rivulets on the hydrophobic coatings, however, cause attenuation and disturb polarisation purity. Therefore up to now, most polarimetric weather radars are operated without radomes.

A **switched-dual polarisation** radar switches the polarisation state from pulse to pulse. A block diagram is given in Fig. 1.11. The fast ferrite waveguide switch consists of two ferrite loaded waveguides in parallel. When the ferrites are magnetised by external magnetic fields, the propagation constant of the waveguide changes, and the phase of the wave passing the waveguide is shifted. By proper shifting and combination, the desired polarisation is created and fed to the antenna.

A **dual-channel dual-polarisation** radar transmits a selected polarisation and receives two polarisations simultaneously, Fig. 1.12. These two commonly are the co- and cross-polarisations of the transmitted polarisation state. Both signals are processed by individual receivers, before they are routed to the radar signal processor. For magnetron transmitters, phase sampling is required to establish coherent on-receive operation for both channels.

A **switched dual-channel, polarisation-agile radar**, then, is the most versatile system.

Because modern polarisation radars include Doppler capability, Doppler processing, as described above, can be done for each polarisation channel. The most important and widely used additional parameters from polarimetric Doppler weather radars are

The **differential reflectivity** $Z_{\mathrm{DR}} = 10\log(Z_\mathrm{H}/Z_\mathrm{V})$, a measure of mean particle shape. Because, for example, falling raindrops have oblate shapes with the axial ratio depending on size, this is a valuable additional parameter for estimation of rain intensity.

The **linear depolarisation ratio** LDR $= 10\log(Z_{\mathrm{VH}}/Z_{\mathrm{HH}})$, with Z_VH the vertically received return for horizontal transmission, is indicative of melting and tumbling snowflakes and is useful for identifying ground clutter.

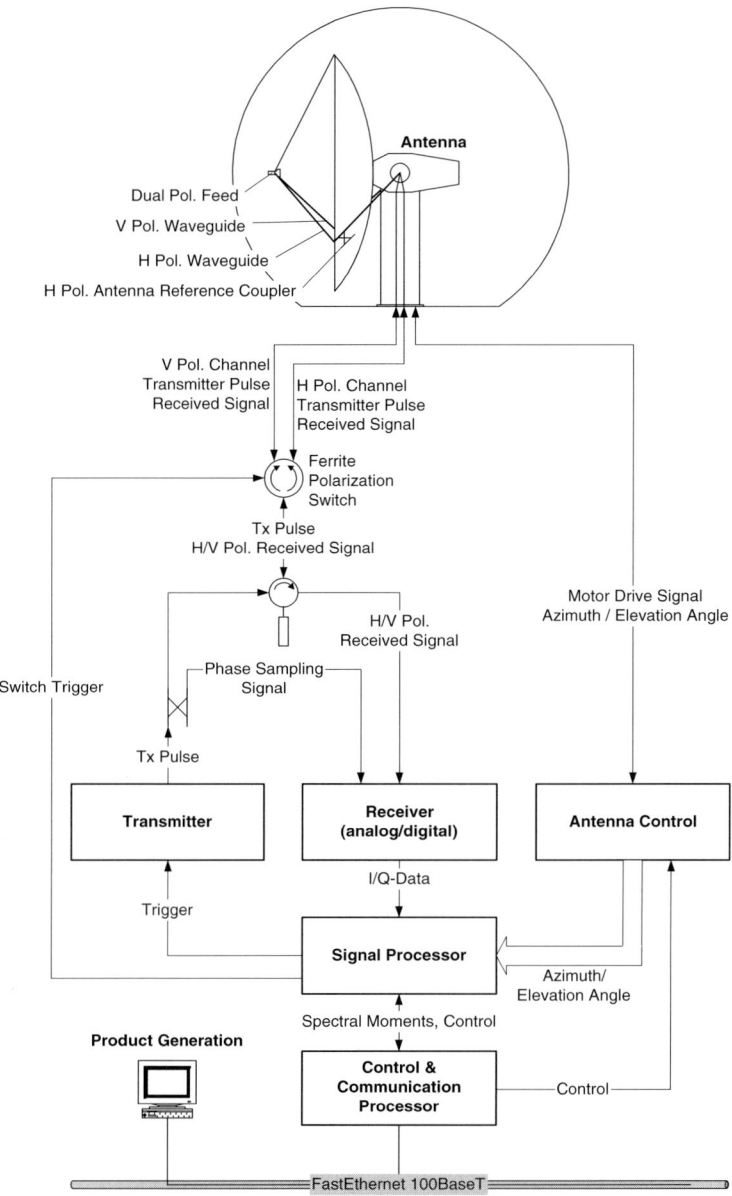

Fig. 1.11. Switched dual-polarisation radar

The **specific differential phase** $K_{\rm DP}$ is the difference of phase shifts for horizontally and vertically polarised radiation on propagation over 1 km and measured in $°{\rm km}^{-1}$. This parameter is of high value for rain measurements with increased accuracy.

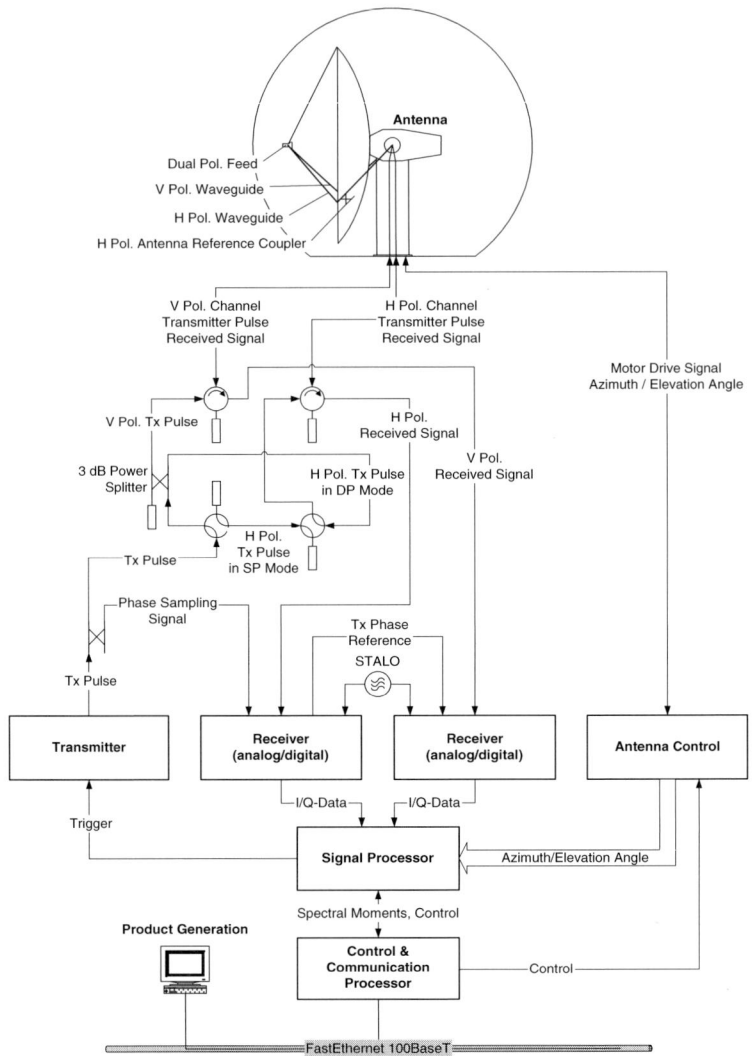

Fig. 1.12. Dual-channel dual-polarisation radar

The co-polar **correlation coeffient** ρ_{HV} of the time series of Z_H and Z_V estimates, ρ_{HV} indicates the variability of scatterers in shape, size, and thermodynamic phase.

The added value of these polarimetric parameters for rain rate estimates and for data quality is demonstrated by Illingworth (2003), this book, and for particle classification by Meischner et al. (2003), this book.

1.4 Weather Radar Networking

Since the 1990s, with technological development, networks of weather radars came into operation, which not only cover individual countries but geographical regions across national borders. Figure 1.14 shows actual European networks. The latest status of operational weather radars in Europe can be found at http://www.chmi.cz/OPERA/.

A radar network essentially consists of two or more weather radars which send data to a common center, as shown in Fig. 1.13.

The center merges the data from the individual radars and generates quality proofed composite products. Modern centers, very importantly, not only provide the data composites but also control the functions of unattended radars remotely.

In further processing steps, the radar data composites may be complemented with data from sensors like rain gauges or satellites, increasing the value for end users. A modern sensor network design is described by Weipert and Pierce (2000).

Radar networks became feasible with the availability of high rate data exchange. In a modern network, the radar stations and the data center are commonly equipped with a fast local area network (LAN). The wide area network (WAN) linking the different LANs, however, is still a bottleneck. This is underlined by estimating the required raw data rate, for example, from spectral and polarimetric moments as reflectivity, velocity, differential reflectivity, and

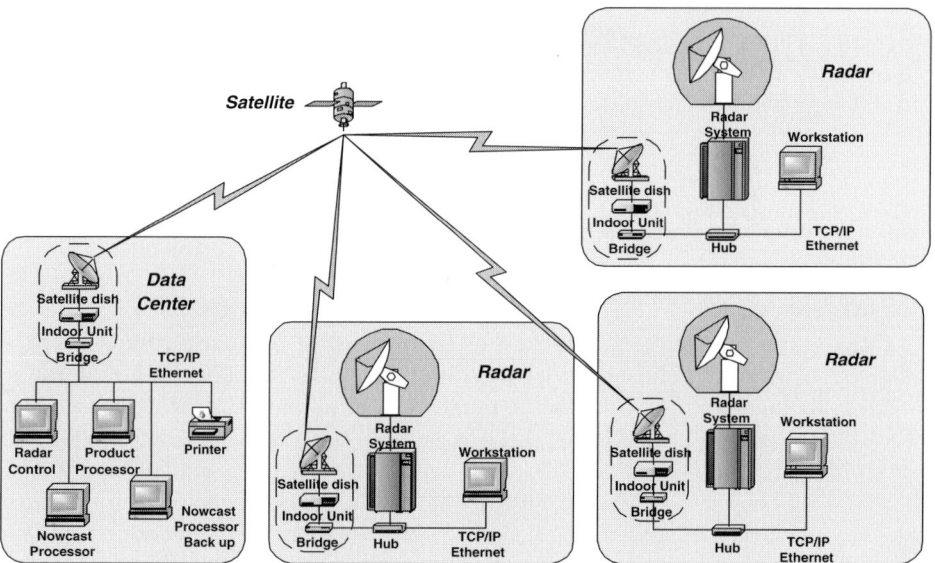

Fig. 1.13. Weather radar network, consisting of different weather radars, a data center for operating control, and data processing and distribution

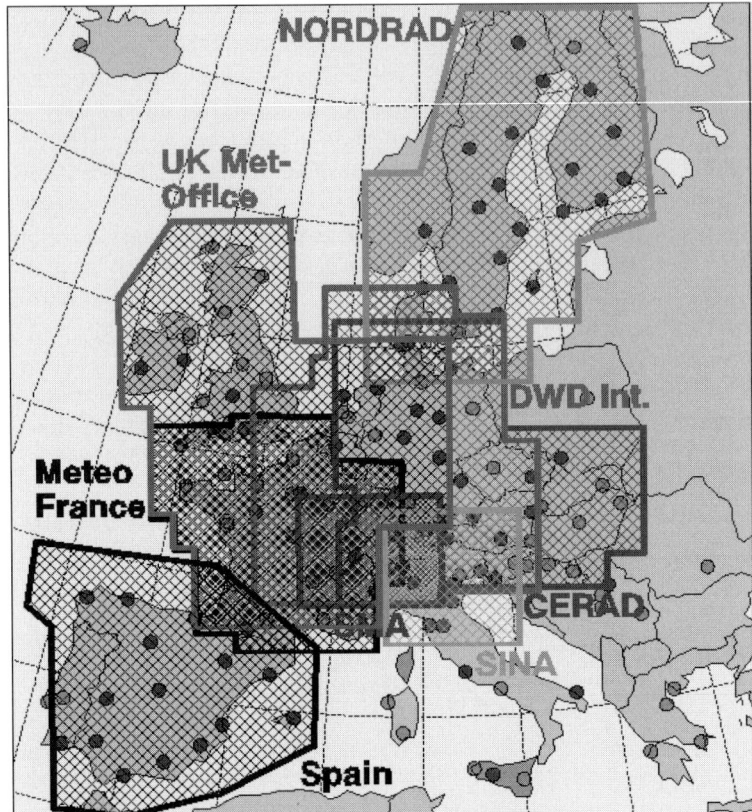

Fig. 1.14. European weather radar networks. The color Fig. 1 on page 313 shows the overlapping networks more clearly

differential phase. The raw data rate R_{RAW} can be calculated as

$$R_{\mathrm{RAW}} = \frac{r}{\Delta r} \cdot \omega_{\mathrm{AZ}} \cdot C_{\mathrm{RES}} \cdot n \,. \tag{1.16}$$

C_{RES} is the digital resolution and n the number of moments which are provided by the radar. The other parameters have been explained. For a radar covering a range r of 150 km with a range resolution of 150 m and an azimuth rotation rate of 18°/s, providing three moments, the resulting raw data rate is 0.4 Mbit/s (1 Mbit/s = 2^{20} bit/s). Such a data rate is no problem for a LAN, but could be one for a WAN, i.e. if only analog phone lines or expensive satellite links are available. In such cases, it is necessary to generate the products at the radar sites and send products instead of raw data to the center. This, however, reduces the flexibility of the whole data generation and utilization process.

Modern weather radar networking solutions, therefore, use efficient data compression algorithms in order to minimize the required bandwidth. For example, when sending via satellite, the compression ratio depends on the weather situation with a larger variance in severe weather situations. Equation (16), therefore, represents the worst case regarding the required data bandwidth.

Asynchronous digital subscriber lines (ADSL) and other broadband WAN techniques, however, will support the distribution of raw data in the future.

The benefit of networks is obvious. Larger scales than those covered by single radars are necessary for weather observation and weather forecasting, and the data are then available for verification and for assimilation in numerical weather prediction models, a topic discussed by Macpherson (2003) in this book. These applications, however, need high quality data and knowledge of error structures.

Radar networks with overlapping measurement areas further give the important chance for cross-checking and cross calibration of the individual systems, as will be shown impressively in Sect. 5.2.

1.5 Data Quality

In order to assure high data quality, three kinds of errors sources need to be understood and quantified. These are the measurement errors of the "weather radar" instrument, including all the data processing, errors inherent in the measurement method such as the scan strategy, and those caused by varying meteorological conditions such as drop size distributions, attenuation, ground clutter, anomalous propagation, and incomplete beam filling. Many of these errors together with reduction and correction strategies are individually discussed in several chapters throughout this book. We, therefore, here only want to stress the differences, to focus on instrumental errors to be minimised by proper calibration and maintenance, and touch on the principles of methodical errors.

1.5.1 Instrumental Accuracy

The radar manufacturers provide advance testing and build in calibration procedures as well as a choice of different quality thresholds. The users should be familiar with those in order to make best use of their system.

The advance testing includes

- the estimation of system parameters and system losses including matched filter losses.
- the range calibration which assures the correct distance for the range gates. The accuracy is determined by the trigger clock. A resolution of 100 ns is typical, corresponding to 15 m in range.
- the pointing of the radar antenna which is measured with precise position encoders mounted on the rotation axis of the antenna. A careful adjustment, for example, by suntracking provides total accuracies of the order of hundredths of a degree.
- the velocity measurement, based on the difference of two subsequent phase measurements. The typical phase stability from pulse to pulse is $\pm 1°$ for coherent on receive radar systems and $\pm 0.2°$ for fully coherent radars. A

system with a frequency $f = 3000$ MHz and a pulse repetition frequency of 1000 Hz measures a phase shift $\Delta\varphi$ of 7.2° for a radial velocity of 1 m/s. Therefore a *rms* phase noise of 1° will bias the velocity measurement by some 10 cm/s.
- the reflectivity calibration. A way to calibrate the complete radar is by remote calibrated radar targets. An active source is a beacon or transponder which receives the transmitted signal from the radar and sends it back with calibrated amplification. A passive source is a reflector with a calibrated radar cross section. Any clutter contamination, however, will distort the calibration with remote sources. A transponder is a versatile testing device. Because it amplifies the received signal, it can cover the complete dynamic range of the radar. It is further possible to modulate the phase of the returned signal to test the Doppler response. The clear air transponder response can be compared with a precipitation situation thus estimating actual attenuation. Similarly, the attenuation of a wet radome can be estimated. If this is done regularly the aging of the radome surface can be monitored, and the necessity for recoating it can be determined. The transponder further may be utilised to determine the antenna beam width and, if the position of the transponder is known precisely, the antenna pointing. The accuracy of the transponder calibration is solely dependent on the calibration of the transponder antenna gain and amplification, which can be maintained to about ±0.8 dB.

Passive calibration is performed with spheres, either tethered or on a mast. Because the dimension – giving the radar cross section – can be controlled very accurately, the calibration is also quite accurate. It, however, is necessary to keep the sphere stable within the range gate during the measurement which is not an easy task. Any movements or precipitation will distort the calibration. Under ideal conditions, an accuracy of ±0.5 dB should be achieved; see below.

Beside the overall system calibration, subsystems may be calibrated separately:

- The antenna gain and beam width can be calibrated with a test transmitter with calibrated output power and antenna gain, placed in a clutter free range gate. With a careful setup, accuracy in the gain calibration of ±0.5 dB and in the beam width of ±0.2° is possible.
- The radar transmitter is calibrated by coupling a fraction of its power to a measurement port and measuring this fraction with a microwave power meter. With precise couplers and a proper procedure, a total accuracy of ±0.2–0.5 dB can be achieved.

Modern Doppler weather radars with digital receivers have a dedicated channel for sampling the transmitter pulse. These data are used for phase referencing and also for tracking power drifts of the transmitter. The reflectivity measurement then is corrected for each pulse with respect to the transmitter power.
- The receiver, the most complex subsystem, needs to be controlled for gain, linearity, dynamic range and noise figure. Modern Doppler weather radars

with digital receivers do not require logarithmic receiver channels anymore because the dynamic range of their linear channels already exceeds the figures of logarithmic receivers. A linear receiver only requires a calibration for gain at one point within its dynamic range, whereas logarithmic receivers need gain calibration for the complete dynamic range. The receiver gain can be calibrated with an accuracy of ±0.2–0.5 dB.

During operation, modern systems automatically

- measure the transmitter power of each pulse and insert the measured figure in real-time in the radar equation
- correct for system noise by frequently taking noise samples during measurements
- provide tools for calibration of the linear receiver and testing its response curve
- provide range normalisation $1/r^2$
- provide gaseous attenuation correction
- provide clutter correction alternatives

With a carefully designed, calibrated and maintained radar, the overall uncertainty for the reflectivity factor has been estimated at 1.7 dB (Paul and Smith, 2001).

Important quality thresholds to be set by the operator are

- clutter-to-signal ratio, CSR. The noise power, as discussed above, depends on elevation as well as on the actual weather conditions. In order to remove this bias, the noise power must be estimated frequently, for example, for every elevation scan with a separate noise sampling in clear air. The noise power estimate will be stored and the reflectivity estimate corrected accordingly.
- the signal quality index SQI

$$\text{SQI} = |R_1|^2 \big/ R_0^2, \qquad (1.17)$$

which eliminates signals either too weak to be useful or with too large a spectral width (caused by too low a number of samples or by decorrelation because of too low a PRF, by too fast a moving antenna, or just by phase noise) to justify further analysis. Its value varies between 0 for an uncorrelated signal (white noise) to 1 for a noise-free zero-width signal. This threshold is typically set to values of 0.3 to 0.5. However, it should be used with care because the velocity variance is not only given by the radar phase noise and the number of time samples, but also by atmospheric turbulence. If a user is interested in turbulence measurements, in particular, a higher threshold might be adequate.

During operation, calibration should be performed from time to time by monitoring all critical subsystems. It is unlikely that the antenna gain or beam width will change with time, but electronic components such as the receiver might drift, or a radome might change its characteristics due to aging and environmental influences.

Several calibration strategies applied in Europe have been reported in Collier (2001), and the important potential for further improving data quality by using polarimetric measurements is shown by Illingworth (2003), this book.

As an example, the practice and the most valuable experience of data quality assurance with the Nordic weather radar network NORDRAD will now be presented in some more detail.

1.5.2 Data Quality Assurance for the NORDRAD Network

Inhomogeneities in data due to differences in calibration, signal processing algorithm errors, and siting problems can more easily be seen in radar products of networks in a cold climate because the sensitivity required to achieve accurate and homogeneous measurements is very high.

In 1990, the common real-time Nordic weather radar network (Carlson, 1995) and its software system, called NORDRAD, became fully operational during 1993, managing data from the radar systems in Norway, Sweden, and Finland.

Within NORDRAD, the focus has been largely on the operational exchange of 8-bit reflectivity data in the form of 2-km horizontal resolution, 500 m height pseudo-CAPPI images generated every 15 minutes. The common software includes the ability to generate composites based on these single-site products using an algorithm which is largely based on the altitude of input data but combines this with shielding information for individual radars where available (Andersson, 1992). Higher level CAPPI products, 3-hour accumulated precipitation, echo top, and vertical profiles of horizontal winds, using VAD and VVP techniques, see below, have also been on the list of agreed-upon products for operational exchange.

The present NORDRAD network, illustrated in Fig. 1.15, is by no means static. Estonia became the fourth NORDRAD country in 2002, and a second-generation software system was phased in early in 2003, which implies a migration to contemporary network standards and cutting-edge software technology. Operational exchange of polar volume data is on the scope, which opens up new potential for product algorithm and quality control developments.

NORDRAD cooperation has shown that it is relatively easy to establish a radar network but not so easy to reach high availability and quality of the data from it. Real-time calibration and monitoring tools are necessary in any radar network, and much work must be conducted still to reach a fully satisfactory level. Based on experience gained from the NORDRAD network, we describe which methods can be used to improve the quality of an operational radar network.

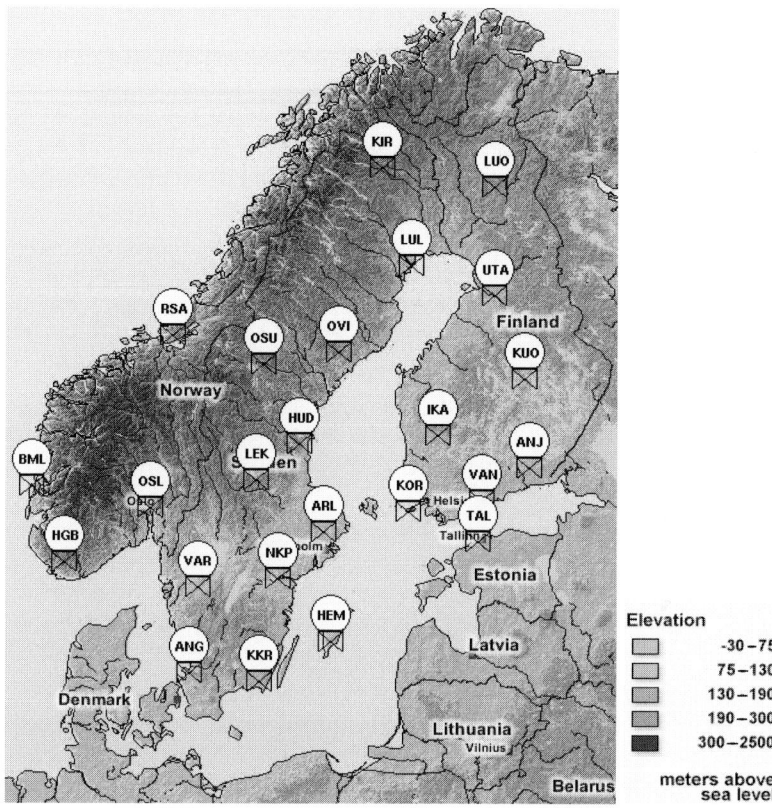

Fig. 1.15. The location of NORDRAD radars in relation to the topography of the Nordic region which they cover. Background map courtesy of UNEP GRID-Arendal

Field Calibration

After several years of operational use in the mid-1990s, it became clear that in the NORDRAD network-individual radars did not perform at a common calibration level. The magnitude of the reflectivity level differences, particularly those between neighbouring Swedish and Finnish radars, could even be of the order of 10 dBZ. A three-year joint project among the three NORDRAD countries was established to determine the main causes of the permanent and substantial observed differences (Koistinen et al., 1999).

Two identical sets of calibration experiments were performed at two overlapping radars: Ikaalinen in Finland and Hudiksvall in Sweden. The methods included standard gain horn or reference feed-horn tests with an independent signal generator and sphere calibrations. These special calibrations were preceded by ordinary receiver calibrations using the output of a signal generator injected into the waveguide. All loss figures were measured or calculated, but there was no tool to measure the loss in a dry radome. We had to rely on the manufacturers' figures which can be too small, especially in aged radomes.

The aim of a feed-horn calibration is to calibrate the whole receiving chain from the radome to the A/D converter. During all calibrations, the frequency of the emission from the signal generator was tuned to be equal to the radar receiver frequency. The horn had been calibrated together with the injection cable in a so-called anechoic room at the Technical University of Helsinki. The range to the feed horn was measured with a differential GPS system and from topographical maps. The measurement range was calibrated using fixed TV mast targets at known ranges. The received microwave power P_r at the radar (in dBm units) was measured. It was also calculated from the equation given by Smith (1968). The repeatability of the results was found to be accurate and stable within 0.2–0.3 dB in the upper part of the receiver calibration curve. The differences between the measured and calculated power at Ikaalinen and at Hudiksvall were 1.7 and 0.6 dB, respectively. These small values revealed that there were no major problems in the receiving part of the Ericsson and Gematronik radars.

The aim of a sphere calibration is to absolutely calibrate the whole transmitter–receiver chain. Such measurements were found to require extreme care and accuracy concerning the measurement geometry and methods, the materials, weather, and communication between the groups at the radar and at the test site. An aluminium sphere (diameter 20.2 cm, weight 90 g) or a metal net sphere (diameter 60 cm) filled with a balloon, was raised using balloons, a kytoon, or a kite depending on the prevailing winds. During the test phase, preceding the actual radar pair calibration, it was found that winter is the best season for sphere calibrations for several reasons. First, clutter in the boundary layer due to dust, birds, and insects is much less (5–25 dBZ less at the same locations) than during summer. Secondly, ground clutter intensity is lowest (often less than −20 dBZ) over frozen lakes located behind hills when viewed from the radar. Thirdly, it was found that typical signal variations are of the order of 8–15 dB during a summer day, whereas they may be only 1–3 dB in stable winter conditions from sample to sample.

The sphere calibration method is quite sensitive to the assumption that the sphere is actually at the midpoint of the pulse volume located at the measured range bin. The results were calculated using the point target radar equation. As with the feed-horn measurements, the radar performance is best described by the difference between the measured and calculated received power. The difference varied between −1.5 and 3.8 dB. These figures show that at present the method did not achieve a better accuracy level than 2-4 dB. However, the results suggest that no major calibration problems exist in the Ikaalinen and Hudiksvall radar systems.

In relation to the calibration field tests, signal processing in both countries was analyzed carefully to reveal any possible system errors. In the Ericsson systems, the sign of the standard correction of −2.5 dB, done in log channel signal averaging was found to be incorrect. Repairing this removed a systematic overestimation of 5.0 dB of all measured dBZ values. In northern Swedish radars, an erroneous antenna gain value (by 1.8 dB) was found and corrected. In Finland, it was found that the two way gaseous attenuation correction factor of

0.016 dB/km, was actually ten times too low in the RVP-6 signal processors in all Finnish Doppler radars. As a consequence, all dBZ values were 0–4.0 dB too low. Thus, the largest part of the permanent differences of 6–13 dB between the Ericsson and Gematronik systems was due to errors in signal processing.

These experiences clearly revealed that a careful system analysis of signal processing and calibration routines is very important in any radar network. A feed-horn calibration is recommended to calibrate the receiving part of the system. A sphere calibration is ideal, but quite a laborious and inaccurate tool so far in practice. In spite of the good results in removing permanent biases, transient steps between the reflectivity levels of neighboring radars have appeared occasionally in the NORDRAD network after the calibration project. The main reasons for such inhomogeneities are assumed to be

- changes in the actual elevation angles
- electrical malfunctions in the system components
- wetting or icing of the radome.

In order to monitor quantitatively transient, long-term variations in radar systems, a new NORDRAD Quality Assurance Project was established in 1999.

Sun Calibration

One of the difficulties in radar calibration and the monitoring of calibration stability is finding a stable reference target with well-known characteristics. A tower or building located at close range (a few km) is much easier to manage compared to using balloons or kites, yet such structures should be avoided, if possible, when siting a radar due to the obvious risk that they will corrupt operational data. The sun is an interesting potential reference target since it emits energy over the whole radio spectrum, it moves slowly and is easy to track, it is available to many radars simultaneously, and solar flux measurements at various frequencies are readily available (Tapping, 2001). Experiments have been conducted on the ability to utilize operational polar volumes of reflectivity along with solar flux measurements at C-band (Learmonth Solar Observatory, Australia) and S-band (Dominion Radio Astrophysical Observatory, Canada) both to monitor radar signal calibration and to monitor the antenna's pointing accuracy (Andersson, 2000).

Atmospheric refraction of the radar beam at elevation angles above 5° (Bean and Dutton, 1968) was taken into account when colocating radar data with the sun's position. The sun was sampled through the use of a scan strategy with several overlapping scans. This provided the ability to measure the sun with a resolution of ±0.1° in elevation and ±0.43° in azimuth. With the solar flux measurement available, this allowed the estimation of the equivalent radar measurement and, thus, the ability to derive the error in antenna positioning.

A comparison between Australian solar fluxes and averaged radar data from four Swedish radars during a rare solar burst (Fig. 1.16) shows that the burst is clearly evident in the radar data. A comparison of estimated and achieved azimuthal (Fig. 1.17a) and elevation (Fig. 1.17b) angles, performed with the

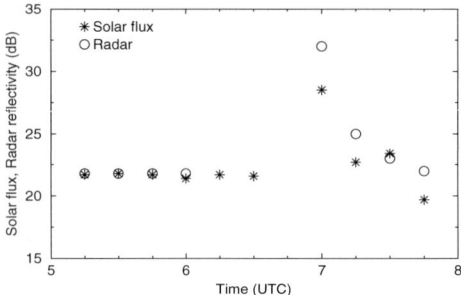

Fig. 1.16. A strong solar burst on September 23, 1998, as measured at Learmonth, Australia, and radars Norrköping, Göteborg, Hemse, and Leksand

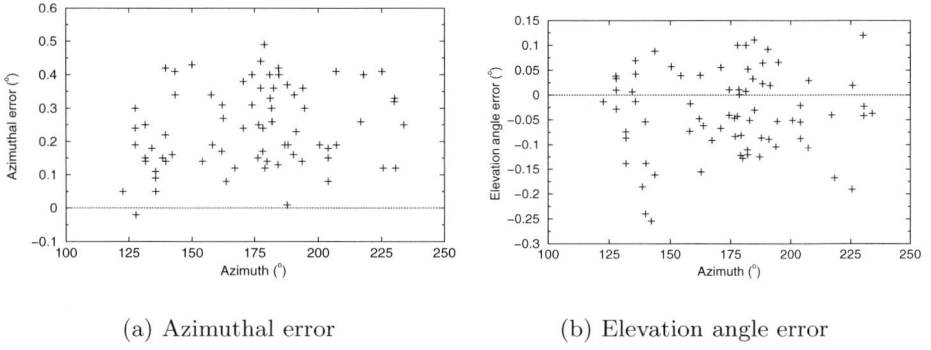

(a) Azimuthal error   (b) Elevation angle error

Fig. 1.17. Azimuthal (a) and elevation angle (b) errors in radar data from Norrköping as a function of azimuth angle, based on data from December 1998 to March 1999

Norrköping radar between December 1998 and March 1999, shows that the mean errors are 0.24° in azimuth and −0.02° in elevation, with standard deviations of 0.12° and 0.09° respectively. No periodicity in the data is apparent which would indicate a tilted, worn or warped turntable.

Deficiencies in the Swedish study were that the sun's signal strength was close to the minimum detectable signal of the radars, which implies lack of data in some cases where only part of the solar disc was measured. Improvements should be possible with increased radar sensitivity. Also, instead of tailoring the operational scan strategy for sun measurements, the use of a dedicated sun sampling strategy, where the radar antenna is nodded in elevation and wagged in azimuth, appears more suitable for use with operational systems (Tapping, 2001).

Gauge–Radar Comparisons
The concept of using an external reference target for calibration and monitoring of calibration stability becomes more attractive if the same target may be shared by several radars, hence the interest in using the sun as such a target. Another

form of external reference target may be in the form of conventional precipitation measurements by gauges. The principle is that overlapping radars have common coverage areas with common gauges and that comparing gauge and radar data may reveal systematic differences between different radars' calibration levels. This knowledge could then be used to derive a means of normalizing radar data to a common level defined by the gauge measurements.

This type of gauge–radar comparison was conducted using radar data from Norway, Sweden, Finland, and Danish data from Copenhagen, along with daily measurements from around 1600 gauges in the climate station networks in Norway, Sweden, and Finland (Michelson, 2001). In order for this kind of comparison to be meaningful, long integration periods must be used. Doing so has the effect of smoothing out much of the noise in both data sources. If there are no known climatologically forced precipitation gradients in a given radar's coverage area, then a long radar data integration in an ideal case will be isotropic about the radar itself while the gauge data integration will be more or less uniform in space. The basis for the comparison becomes the gauge-to-radar ratio (F), according to (1.18), as a function of distance from the radar.

$$F(\mathrm{dB}) = 10\log(G/R). \tag{1.18}$$

In terms of radar calibration, the most valuable piece of information in such a derived relation becomes the Y-axis intercept, which can in such a context be referred to as the system bias. This value gives the difference in calibration level between the gauges and the radar at the radar's location. Deriving the system bias for each radar provides the most elementary basis for normalizing radar data. The shape of the curve illustrates the bias as a function of distance from the radar. If the relations from several radars are reliable enough, then the combined use of the system and distance biases may be incorporated into the normalization process.

This strategy is illustrated in Fig. 1.18 using three-month winter gauge and radar integrations for the Kuopio area in Finland. The radar accumulation is virtually isotropic, and the derived relation with distance is strong. The system bias is 3.1 dB, and the distance bias increases to 21.5 dB at maximum distance. Such biases were derived for NORDRAD radars and Danish radar Copenhagen for the three-month winter period (Fig. 1.18) and a three-month summer period (June–August 2000) in order to determine the maximal seasonal difference in the distance biases (Michelson, 2001). A by-product of this procedure is that it allows the detection of systematic differences in biases among data from different radar manufacturers.

The ability of this strategy to normalize data to a common level was evaluated as part of the evaluation of the method for gauge adjustment of radar data presented in Chap. 3. A normalization procedure incorporating both system and distance biases for each radar was applied prior to gauge adjustment using on-line SYNOP observations. The evaluation method involved calculating the same variable F using independent gauge data. This measure is intuitive and convenient from a radar perspective, since any mean value of F is the bias expressed

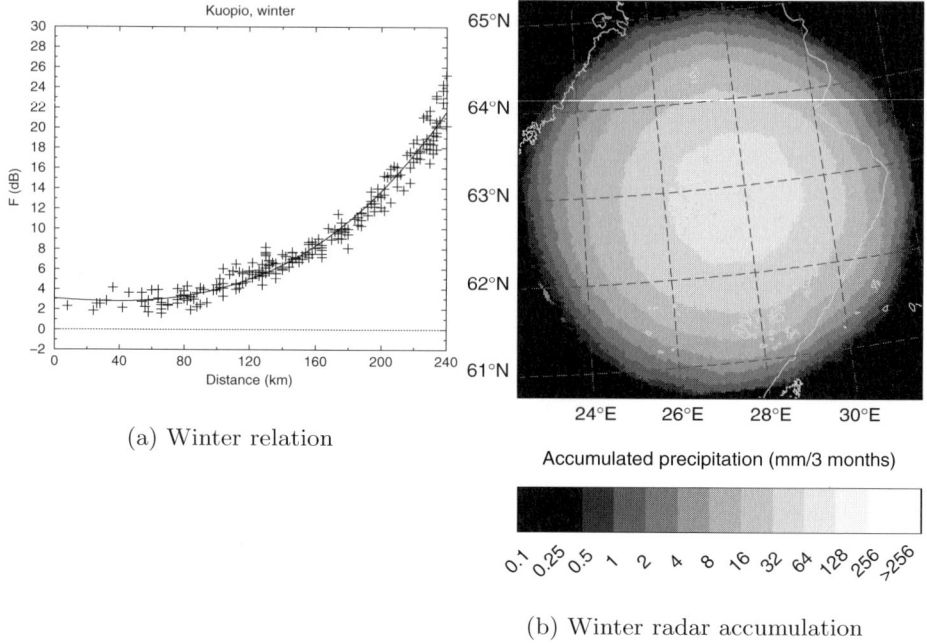

Fig. 1.18. Statistical relation between F and distance (a) and three-month integrated precipitation (b) for December 1999 – February 2000 for Kuopio radar

in dB. These results were compared with those derived from unnormalized and unadjusted radar accumulations. The results, also summarized and presented in Chap. 3, show not just that the biases are significantly reduced for all distances, but that the standard deviations are reduced as well, thus indicating that using the normalization procedure has led to a reduction in the variability in the comparison against gauges.

A prerequisite for conducting this kind of analysis is the availability of high quality gauge measurements at all ranges covered by radar. Unless this is achievable, any derived relation may be statistically significant yet physically unrealistic and therefore meaningless (Michelson, 2001; Michelson and Koistinen, 2000).

Poor Siting

A poor radar site will have a devastating effect on data quality in general and on the ability to perform activities such as those outlined above. In Chap. 2, problems associated with making radar measurements in Alpine environments are presented, and we share many of them with those radars located near or in the Scandinavian mountains (Fig. 1.15). However, common to all radar sites is the necessity of avoiding obstacles proximate to the radar, such as buildings, towers, trees, and higher terrain, since these will cause partial or total beam blockage, thus corrupting entire sectors of radar coverage. Paradoxically, data from a radar with a completely free horizon may also be contaminated by echoes generated from side lobes. An ideal site would place the radar antenna slightly higher than surrounding vegetation which serves to absorb side-lobe radiation.

Exemplifying the effects of unfortunate siting are the results of the analysis described above for radars Norrköping and Hemse (Fig. 1.19). Data from

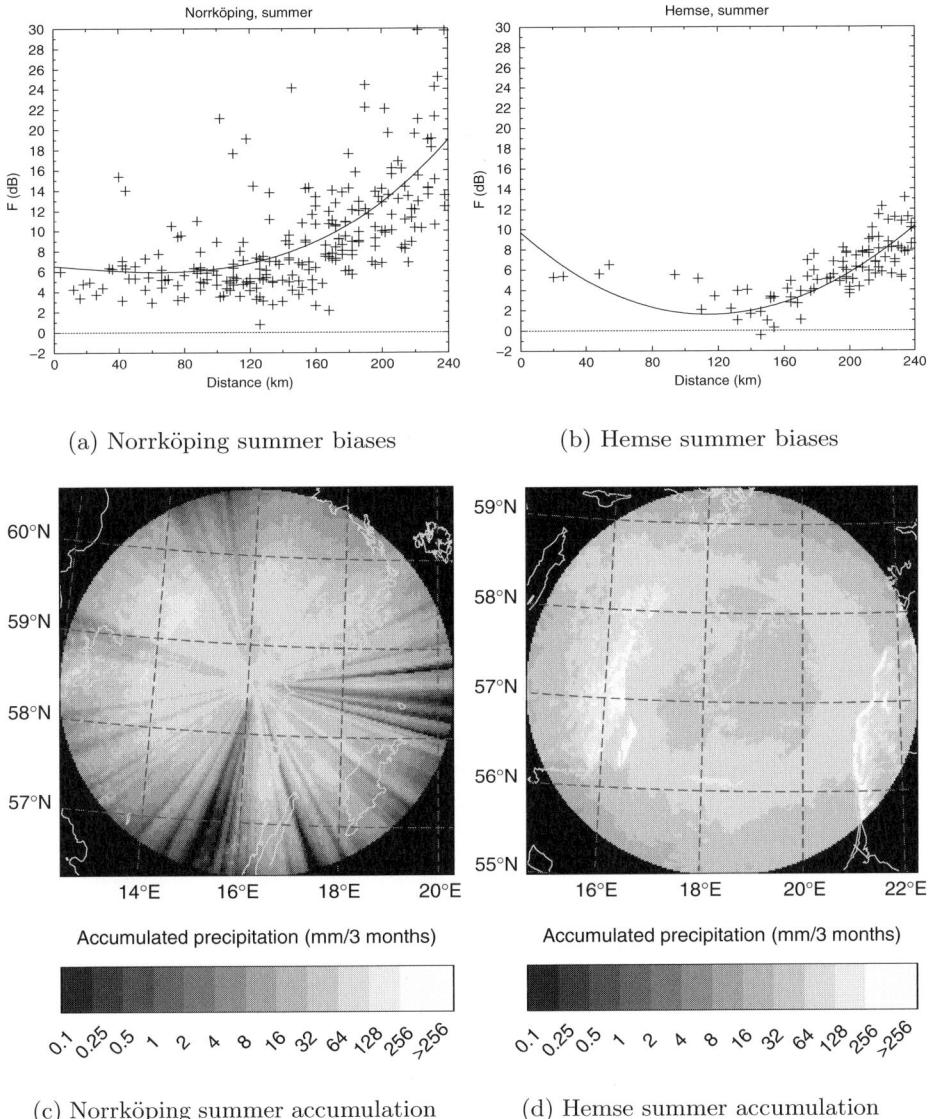

(a) Norrköping summer biases

(b) Hemse summer biases

(c) Norrköping summer accumulation

(d) Hemse summer accumulation

Fig. 1.19. Poor siting as reflected through three-month accumulated precipitation and comparison with gauges. Radars Norrköping in eastern Sweden and Hemse on the island of Gotland

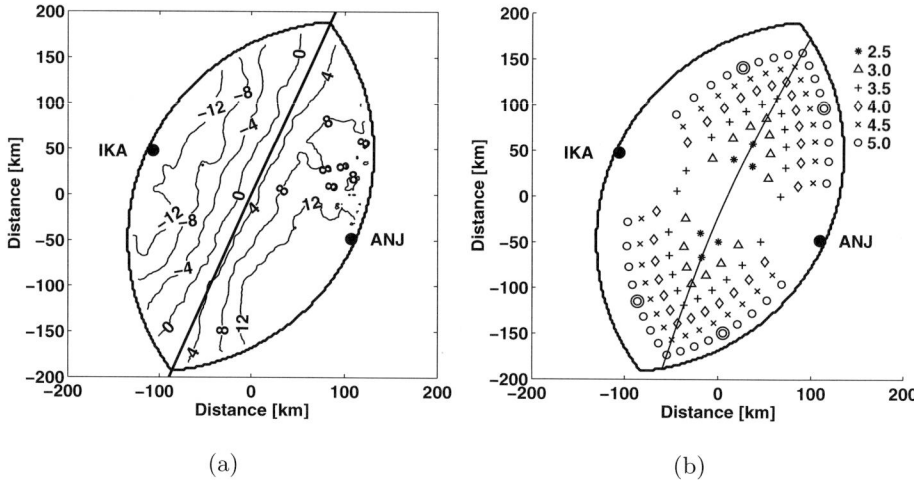

Fig. 1.20. Difference in reflectivity between the Anjalankoski (ANJ, 60°54' N, 27°6' E) and Ikaalinen (IKA, 61°34' N, 23°4' E) radars. Contours of constant difference are given in steps of 4 dB. The radar locations are given as *black dots*, and the *thick line* connects points equally far from both radars (a). A selection of points for numerical analysis when the IKA radar measures at an elevation angle of 0.5° and the ANJ radar at 0.7° (b). The *solid line* connects points equally far from both radars. Other symbols are explained in the text

Norrköping is contaminated in multiple sectors by blockages from all the obstacles mentioned in the previous paragraph. The relation with gauges suffers from scatter as a result which weakens our trust in it. Such contamination has also led to systematically lower sun measurements than those expected (Andersson, 2000). Such poor siting has been experienced with several Swedish radars and has resulted in decisions to move the radar previously located in Göteborg to a new site called Vara, and the radar in Norrköping to a new site called Vilebo. The Hemse data in Fig. 1.19b and 1.19d "suffer" from the radar having the combination of a completely free horizon and no Doppler capability between the 120 and 240 km range. Strong anomalous propagation echoes from eastern Sweden combined with few gauge observations at near and intermediate distances has caused physically unreasonable, bias-statistics. However, as we will see in Chap. 3 (Sect. 3.2), this radar's siting makes it ideal for studying anomalous propagation and clutter from both land and sea.

Monitoring the Calibration Level and Elevation Angles Using Reflectivity Data

When the main causes for the permanent discrepancy between Swedish and Finnish radars had been removed, a follow-up project was then set up with the aim of improving the intensity level harmonization of the NORDRAD network to within ±2 dBZ and establishing workable and efficient quality assurance and maintenance practices.

The main task was the creation of a numerical analysis program by which estimates of the calibration difference of the radars and difference of the lowest elevation angles used by the radars are obtained. Altogether, 15 radars from the NORDRAD network were chosen for the study. The input data in the paired-radar analysis are the pseudo-CAPPI reflectivity data projected onto a polar stereographic grid. Products at 15-minute intervals were processed to create estimates of the average reflectivity difference on the common field of view of the radars. An example of the paired-radar data is shown in Fig. 1.20a.

In the following, we briefly explain the numerical method by which the elevation angle and calibration differences are estimated based on these data (Huuskonen, 2001). The key issue in understanding the procedure is illustrated in Fig. 1.20b, where several sets of points are given. Those indicated with open circles, for instance, are points where the height of the beam from the distant radar is at the altitude of 5.0 km. Moving away from the solid line, where the altitudes from both radars are equal, the beam height from the closer radar decreases by 0.5 km for each successive point. The plot would be symmetrical if the collection angles were identical. As the collection angles are 0.5 and 0.7°, the points have moved considerably toward the Anjalankoski radar which is the radar with the higher collection angle.

The points at which the distant radar measures at 5.0 km and the closer radar at 3.0 km are denoted by doubled open circles. We note that there are four points altogether; one pair on either side of the borderline between the radars. For the data at these points, we get the following formulae:

$$m_l = dBZ(5) - dBZ(3) + \Delta(\text{ANJ}) - \Delta(\text{IKA}) \tag{1.19}$$
$$m_r = dBZ(3) - dBZ(5) + \Delta(\text{ANJ}) - \Delta(\text{IKA}), \tag{1.20}$$

where m_l and m_r refer to the measurement left and right of the dividing line, dBZ_r is the reflectivity at the altitude r, and Δ is the calibration error of the radar, which are assumed to be independent of the reflectivity. The ANJ signs are positive, because the IKA data have been subtracted from the ANJ data. Here we have made use of the assumption that the precipitation is uniform and the vertical reflectivity profile (VPR) the same at all locations. This assumption should be well satisfied as the method is applied operationally to 1–3 week long averaging periods. After addition and division by two, we get

$$\Delta(\text{ANJ}) - \Delta(\text{IKA}) = (m_l + m_r)/2, \tag{1.21}$$

which gives the calibration difference of the radars by taking the mean of the measurements m_l and m_r. This gives the reflectivity difference between the two radars. The determination of the collection angle is based on studying how much the calibration difference varies when we determine the difference for all possible altitude pairs. If the assumption of uniformity holds, each altitude pair should give the same result for the calibration difference, to within the error fluctuations. Thus, the collection angle, which produces the smallest variation of the

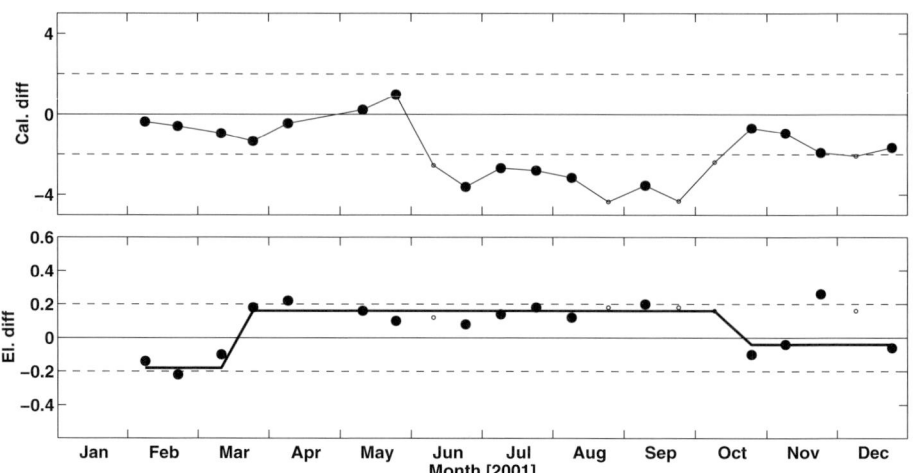

Fig. 1.21. Calibration differences (*upper panel*) and elevation angle differences (*lower panel*) for the Korpo-Vantaa pair during 2001. Good quality results are shown in large full circles

calibration difference around its mean, is the most probable angle. We search through all elevation angle differences with a small step and find the minimum of the standard deviation of the calibration differences. In a numerical implementation, one can process much more data and also make the grids denser than those in Fig. 1.20. Tests with data have shown that a grid having a step around 100–200 m is optimal.

Figure 1.21 shows results for the Korpo-Vantaa pair as an example. The upper panel shows the calibration differences. A positive value means that the first radar in the pair gives higher reflectivity values. We see that the calibration difference is within the ±2 dB limit except during some summer months. The error in the calibration difference is less than 1 dB, reaching 0.5 dB in the best cases.

The lower panel shows the elevation angle results. The thick line is based on the long-term best-fit value determined from all data from March to October, and on the two known adjustments of the Korpo antenna in March and October, which were 0.35° and −0.2°, respectively. These adjustments are also seen in the data. The scatter of points around the line gives us confidence that the method is able to give the elevation angle difference to better than 0.1°.

1.5.3 Methodical Errors

Quantitative precipitation measurement remains one of the most demanding applications of of weather radars around the world. Reflectivity based techniques as well as polarimetric techniques – but to lesser degree as will be shown – are prone to quite a number of errors like ground clutter, anomalous propagation, attenuation, enhanced reflectivity in the melting (bright band), and beam blockage especially in mountanious areas. Such errors need to be corrected for, if we are to have a chance of estimating rain with the high accuracy required for hydrology and for data assimilation in numerical weather prediction models.

This requirement of up to 10% accuracy goes beyond the limits of standard weather radars; it may only be reached if we use precisely calibrated and maintained radar systems, correct carefully for measurement errors and, more promisingly, use advanced techniques such as polarimetry. Methodical error sources will now be addressed briefly, together with basic strategies to correct for them. More detailed discussions are provided by Germann and Joss (2003), Koistinen et al. (2003), Illingworth (2003), Testud (2003), all this book.

Ground Clutter

When measuring weather signals, we are interested in that part of the transmitted radiation that is returned from the atmosphere and from hydrometeors such as cloud and precipitation particles. Returns coming from fixed targets like mountains, buildings or just ground – when hit by the main beam or by side lobes – may, however, contribute to the received signal. They are summarised as ground clutter and need to be removed or corrected for.

The fundamental physical difference between ground clutter targets and meteorological objects is that they do not move and that they exist independently of any weather events. These properties are used for clutter recognition and correction.

One approach is to use a **clutter map**. With the radar scan modus as used for precipitation measurements – usually a PPI scan at low elevations, say 1° – measurements are performed on clear days without any precipitation. The intensity and position of the reflectivity pattern is then removed from the reflectivity field when measuring precipitation. The shortcoming of this method is that ground target reflectivities change when the ground becomes wet. In any case, the clutter map used needs to be updated frequently; see Germann and Joss (2003), this book.

A **Doppler clutter filter** makes use of the fact that ground targets, in contrast to cloud and precipitation particles, do not move. The velocity power spectrum then will contain a zero velocity component above the background noise level, which can be filtered out as described above. It is, however, important to realise that there are other sources of zero velocity power, for example, from mixer offsets, COHO leakage, and A/D converter errors which need to be minimised. Further, when weather targets move perpendicularly to the radar beam, they will not give a Doppler velocity signal. These lines of zero Doppler velocity, however, are easy to recognise in the Doppler velocity displays.

A special case of ground clutter is caused by so-called anomalous propagation of a transmitted radar beam. Under "normal" atmospheric conditions, the refractive index of the atmosphere decreases with height above ground such that the radar beam is bent downward from the horizontal and follows more closely the earth's surface. When temperature increases with height, as in inversion situations or when cold precipitation reaches the ground beneath a thunderstorm, the radar ray may be bent downward more strongly than normal. Such superrefraction situations lead to the visibility of far away ground clutter, see Germann and Joss (2003) and Koistinen et al. (2003), this book.

Because the received signal time series from precipitation and ground clutter show different statistics, statistical filtering may be used to correct for clutter contamination; see Aoyagi et al. (1978); Coveri et al. (1993).

Attenuation
Microwave radiation suffers attenuation when propagating through the atmosphere from a transmitter to the target of interest and then back to the receiver. Attenuation depends strongly on the radar frequency used and on the gases and particles along the propagation path. At common radar frequencies, attenuation by the clear atmosphere is mainly caused by oxygen and water vapor. These contributions are known but small in relation to those of cloud and especially precipitation particles. A standard correction procedure to be updated daily, usually is provided with the radar system.

For cloud particles, attenuation additionally depends on their state. Attenuation caused by ice particles is generally two orders of magnitude below that of water droplets with the same size distribution. As a rule of thumb for most meteorological applications, attenuation by cloud particles can be neglected for radar wavelengths above 5 cm but might be important for wavelengths below 1 cm. For cloud liquid water contents of 1 g m^{-3} and a radar wavelength of 1 cm, attenuation is of the order of 1 dB/km.

Attenuation by rain, however, always needs caution when quantitative rain estimates are desired. A great number of attenuation estimates for rain of different intensities and drop size distributions have been published. For 3.2, 5.5, and 10 cm wavelengths, the attenuation is about 0.005, 0.003 and 0.0009 (dB/km)/(mm/h), respectively.

The correction for measurements in actual rain, known to have high variability in space and time, is, however, difficult because the actual rain properties should be known for each resolution volume along the path. Strategies of gate to gate correction using the estimated rain rate in one resolution volume to correct for the following one turn out to be stable only for short distances up to some 10 km (Bringi and Chandrasekar, 2001). Strategies developed for airborne and spaceborne precipitation radars with wavelengths of 1 cm or less, however, look promising and await applications for ground based Doppler and polarimetric weather radars, as shown by Testud (2003), this book. Finally, polarimetric weather radar data such as the differential reflectivity Z_{DR}, the depolarisation ratio LDR, and the differential propagation phase ϕ_{DP}, have outstanding potential for correcting propagation effects (Bringi and Chandrasekar, 2001), as discussed by Illingworth (2003), this book.

Bright Band

At mid and higher geographical latitudes, most of the precipitation formation occurs via the ice phase. Through a broad variety of collision and collection processes within the clouds aloft, pristine ice particles grow to more and more complicated ice particle aggregates, as can be recognised when observing the variety of snowflakes. If such particles fall below the zero degree isotherm, they melt and reach the ground as rain.

This melting process increases the scattering cross section of the snow flakes, and therewith the measured reflectivity, since when large snow aggregates become wet at the surface or edges, the radar wave is scattered as by a large, quasi-liquid particle instead of the ice particle. With further melting, the ice or snow particle collapses to a drop, small in size compared to the initial particle. Because reflectivity goes with the sixth power of the particle radius, the reflectivity strongly decreases below the melting layer. It further decreases because of the increased fall speeds of liquid drops compared to tumbling snow flakes, thus lowering the concentration of particles within the resolution volume.

With stratiform precipitation systems, the melting layer and the corresponding radar 'bright band', some 100 m in depth, are relatively easily recognised, especially when observed polarimetrically.

If the slanting radar ray some distance from the radar system crosses the melting layer, the reflectivity there is orders of magnitude higher than for the corresponding rain below. This would cause an overestimation of rain below, if not corrected for. Strategies for corrections are given by Germann and Joss (2003), Koistinen et al. (2003) and Illingworth (2003), all this book.

Vertical Profile

The precipitation formation process – mainly coalescence and coagulation of liquid and frozen cloud particles – and the fall of existing precipitation particles down through cloud regions cause a significant increase of precipitation intensity from aloft down to earth. Indeed, the measured radar reflectivity generally increases by some 10 to 20 dBZ from, say, 6000 m altitude down to the surface. Because, with increasing distance from the radar, the measurements are taken at increased altitudes, this may cause errors in precipitation estimates at ground that significantly exeed the instrumental measurement errors of a well-calibrated radar system. The same holds if measurements can only be taken at some height because of beam blockage by nearby mountains. Correction strategies either using actual reflectivity profiles or climatological profiles are presented by Germann and Joss (2003) and Koistinen et al. (2003), both this book.

1.6 Products of Modern Weather Radars and Networks

1.6.1 Standard Products

So far we have briefly described the state of measurements and data processing techniques with recent weather radar productions. The quantities provided

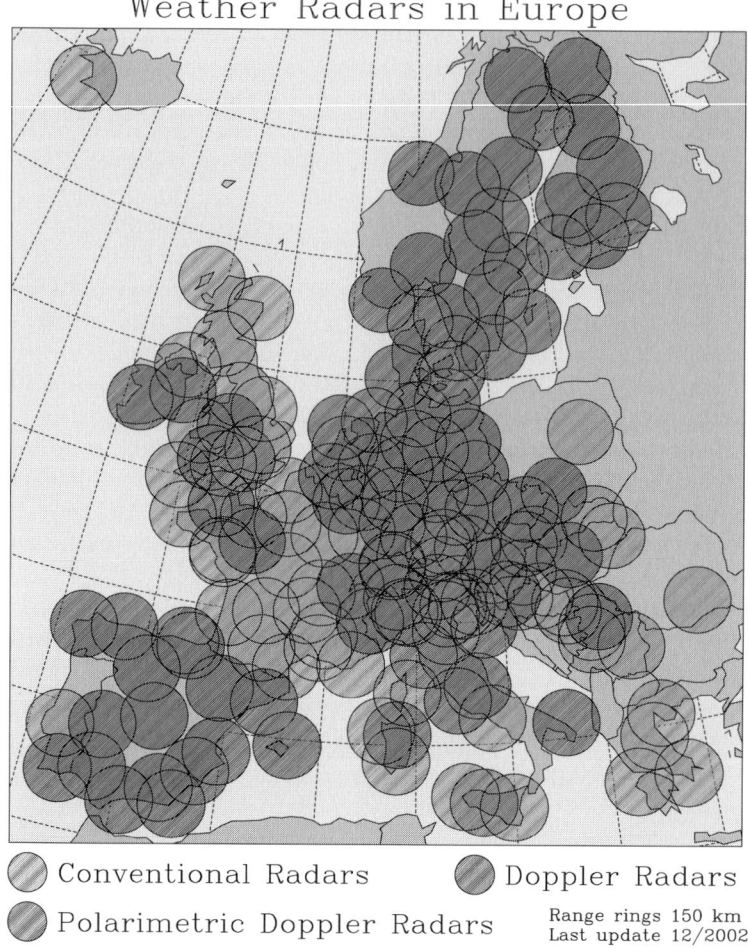

Fig. 1.22. Coverage of Europe with different types of weather radars. The circles indicate 150 km radius. See also the color Fig. 2 on page 314

are reflectivity, Doppler velocity and Doppler spectral width, and further polarimetric quantities. As indicated by Fig. 1.22, Europe is rather well covered with weather radars of a variety of standards and in different operational environments. Products are available for single sites as well as from the different networks. This radar coverage is continuously being upgraded in coverage and system performance, as can be followed at http://www.chmi.cz/OPERA/. The envisaged European standard is a C-band Doppler radar, although the Spanish network is composed of C- and S-band Doppler radars. Italy already operates several polarimetric C-band Doppler radars for local applications. The radars are networked for the Nordic countries, the UK and Ireland, central Europe and Spain, as shown above in Fig. 1.14. Southern Europe is planning to establish a further regional network.

From single radar sites, the commonly available and archived products are the measured and calibrated reflectivity pattern in CAPPI or PPI projections with 2×2 km resolution in space and 15-min resolution in time. These are the basic products required to follow precipitation structure and development. In some cases, volume scans of reflectivity and reflectivity profiles are provided, as well as measured maximal reflectivities and echo tops, which are the altitudes of a defined reflectivity value. More specialised products include different severe weather warnings and Doppler wind profiles. Intensity, resolution, and data quality standards, however, differ from country to country.

Twelve European countries estimate the precipitation intensity from reflectivity measurements at single sites as well as precipitation accumulations for a grid spacing of 2×2 km and time resolution from 5 min onward.

Composites of precipitation intensity or accumulations are available only for some countries and for the NORDRAD network.

A very few countries provide and archive more advanced products such as radial Doppler winds, Doppler spectral width, simply called 'turbulence,' and vertical wind profiles at single sites.

1.6.2 Wind Fields by Doppler Radar Measurements

Wind is an essential element of what we call weather, and wind fields describe the exchange of air masses horizontally and in the vertical. They need to be known to understand and follow atmospheric circulation and for weather forecasting on all scales. Doppler weather radars at single sites and composited into networks have great potential for estimating the wind field. The Doppler velocity is a measure of the mean velocity component of all hydrometeors within the resolution volume in the direction of the radar beam. By proper scan strategies together with a priori assumptions of the actual wind field and under the assumption that the particles move with the wind, wind field estimations are possible. We briefly discuss possible measurements with a single Doppler radar, with multiple Doppler radars, and bistatic Doppler radar configurations.

The Doppler velocity is estimated from the phase shift between the return signals of consecutive transmitted pulses. This gives the mean particle displacement during that time interval. Therefore, the Doppler velocity can be estimated unambiguously only within the so-called Nyquist interval given by the wavelength of the radar λ and the pulse repetition frequency (PRF):

$$-V_a \leq 0 \leq V_a \quad \text{with} \quad V_a = \lambda \cdot \text{PRF}/4. \tag{1.22}$$

The PRF further gives the maximum distance $r_{\max}$, up to which unambiguous velocity estimates are possible:

$$r_{\max} = c/2 \cdot \text{PRF} \tag{1.23}$$

with c the velocity of light.

For a given PRF, the echo from the previous transmit pulse, but coming from a more distant target, would also arrive at the receiver within the same time gate, see Fig. 1.4. The range for unambiguous measurements of the Doppler velocity may be extended significantly by use of two different pulse repetition frequencies (Doviak and Zrnić, 1984). If τ_1 and τ_2 are the two pulse intervals, the extended unambiguously estimated velocity is

$$v = \lambda/4 \cdot (\tau_2 - \tau_1). \tag{1.24}$$

If the actual wind velocity exceeds the Nyquist interval, unfolding prior to using the measurements for real wind field analysis is necessary. Techniques are given in Bergen and Albers (1988) and James and Houze (2001).

The most simple but effective way of estimating 2-D wind fields, for example, for nowcasting applications, is by visual inspection of a PPI or CAPPI presentation of the measured Doppler velocities. Some typical wind fields will be discussed.

Figure 1.23 shows a PPI for a rather homogeneous field of northwesterly winds of 10 m/s measured at Oberpfaffenhofen.

The wind direction is perpendicular to the line of zero velocity. Recall that the radar measures the radial component only, so if the beam is pointing to the southwest or northeast, as in this case, the measured value is zero. Because this structure remains similar with distance – which means with increasing height above ground – there is no change of wind direction with height. To the south, some inhomogeneities, caused by the ground clutter of the Alps and local convection, are also seen within the reflectivity pattern.

Within storms, wind fields show areas of convergence, divergence and rotation. Figure 1.24 gives a schematic presentation of the associated Doppler pattern, typical within the lower levels of a thunderstorm.

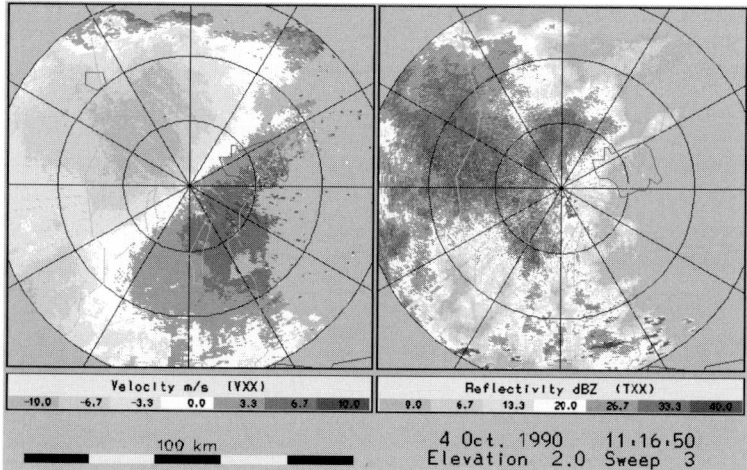

Fig. 1.23. PPI of Doppler velocity and reflectivity of a rather homogeneous wind situation in southern Germany. The color version, color Fig. 3 on page 315 shows the structures more clearly

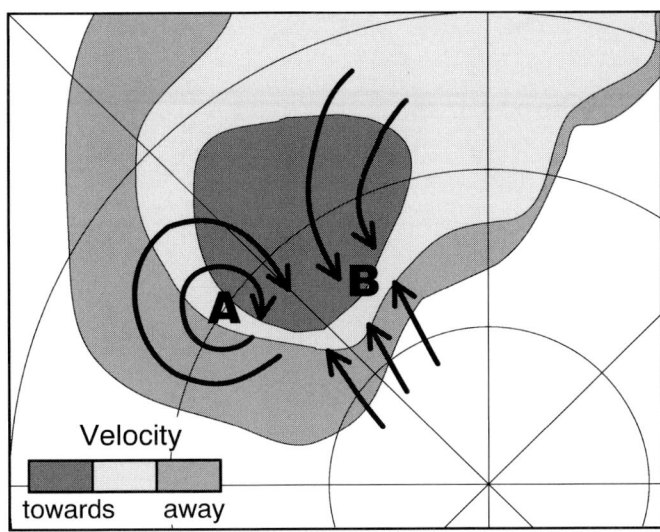

Fig. 1.24. A typical Doppler signature indicating areas of convergence and anticyclonic rotation in the lower part of a thunderstorm

If PPI scans around the radar site are performed with different elevation angles, wind profiles can be estimated for that area. For elevations between 10° and 30°, the measurement altitude increases by 0.17 and 0.51 km, respectively, for each kilometre of distance. This technique is known as VAD – Velocity Azimuth Display (Lhermite and Atlas, 1961; Browning and Wexler, 1968). Figure 1.25 schematically gives some Doppler velocity patterns for an elevation scan of, say, 10° that would be observed for typical wind profiles: (a) represents a homogeneous wind field, with no change in direction or velocity for the whole measured height; (b) is a wind field of constant direction but increasing velocity with height; (c) shows the case of velocity constant but direction backing by 180° with height; (d) shows a low level jet but with constant velocity above the jet; (e) presents a directional shear layer with 90° rotation but constant velocity; while (f) represents directional shear with a low level jet and increasing wind velocity above the shear layer. More complicated examples are given in Wood and Brown (1986).

In addition, vertical RHI Doppler velocity displays from measurements across a frontal system or a thunderstorm already give good indications of the 3-D wind fields within that system. Examples are given by Meischner et al. (2003), this book, for fronts and thunderstorms. However, visual inspection and interpretation needs experienced observers, and such forecasts are time-consuming. Operational applications call for automatic methods.

Analytical estimates of the wind field a few tens of kilometres around the radar can be performed by VAD and, in a more advanced way, with Velocity Volume Processing (VVP) (Waldteufel and Corbin, 1979; Koscielny et al., 1982). A special approach for operational applications is the Uniform Wind method

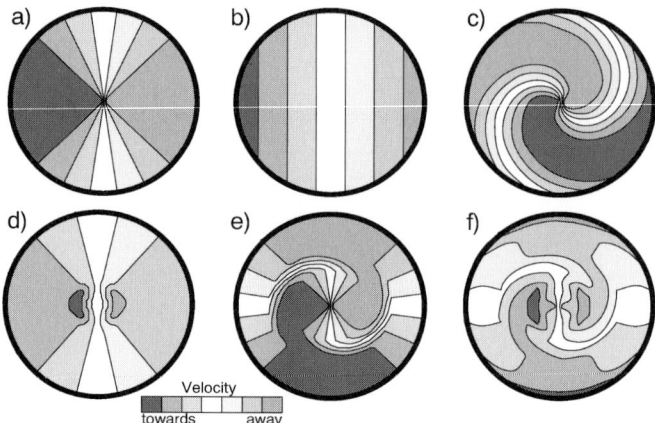

Fig. 1.25. PPI Doppler velocity pattern for typical wind fields; see text

(UW) (Persson and Anderson, 1987; Hagen, 1989). There, high quality measurements of the radial wind components are used for neighbouring sector elements of some $15° \times 10$ km along a circle around the radar. With the assumption that the wind field is constant within each sector element, the horizontal wind field for the covered area can be estimated with an accuracy of about 0.5 m/s. Figure 1.26 shows an example of such an estimated horizontal wind field, and Fig. 1.27 displays an operational product, a time–height cross section from Finland.

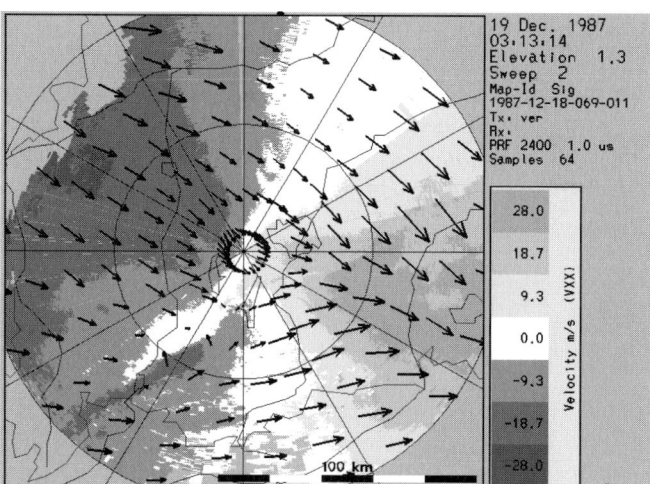

Fig. 1.26. Horizontal wind field around Oberpfaffenhofen, as analysed with the Uniform Wind method and overlaid on the Doppler measurements used. See color Fig. 4 on page 315

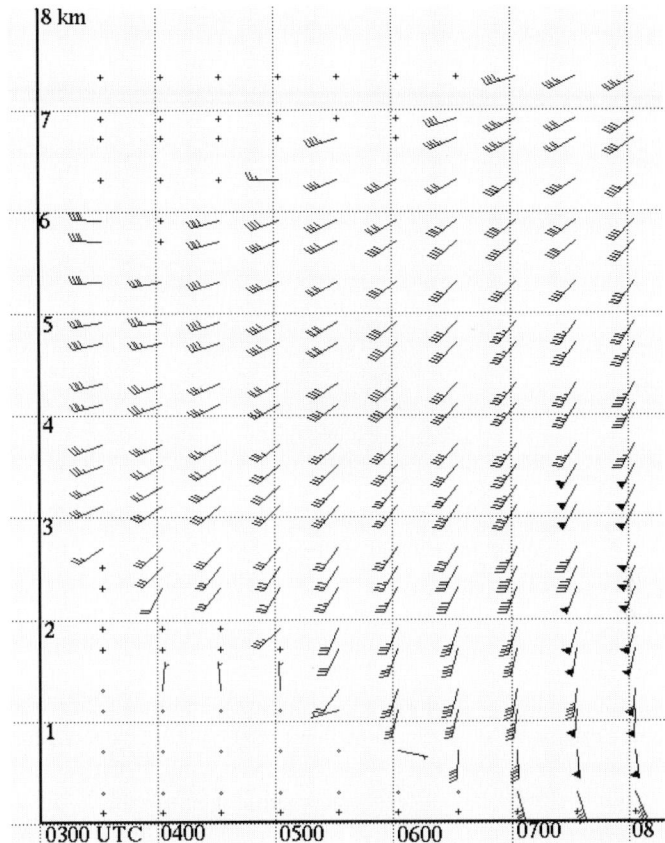

Fig. 1.27. An example of operational time–height cross section of VVP winds from the Finnish Korpo Doppler radar on March 3, 2000 at 03 - 08 UTC. Wind speed and direction are expressed as traditional meteorological wind barbs, for example, at 08 UTC at the height of 1.9 km, the direction is 190° and speed 55 knots. Wind field is obtained from an approaching occluded front containing strong warm advection. Plus signs (+) denote regions where the measured signal is either too weak (clear air) or does not pass the quality thresholds. Note that a few erroneous wind barbs at the lower edge of the frontal precipitation have passed the selected quality thresholds

If more than one Doppler radar is covering the area of interest, multiple Doppler methods may be applied. A convenient distance between the two radar systems is 30 to 40 km. Wind field estimation by multiple Doppler radar measurements has been described in much detail following the basic methods of Ray et al. (1980) for the retrieval of 3-D wind fields. Laroche and Zawadzki (1994), Shapiro and Mewes (1999), Chong et al. (1983), Chong and Testud (1983), and Gal-Chen (1978) additionally show how to retrieve thermodynamic fields. See also Carbone et al. (1980). Figure 1.28 gives a midlevel horizontal wind field of a thunderstorm, estimated by dual Doppler measurements.

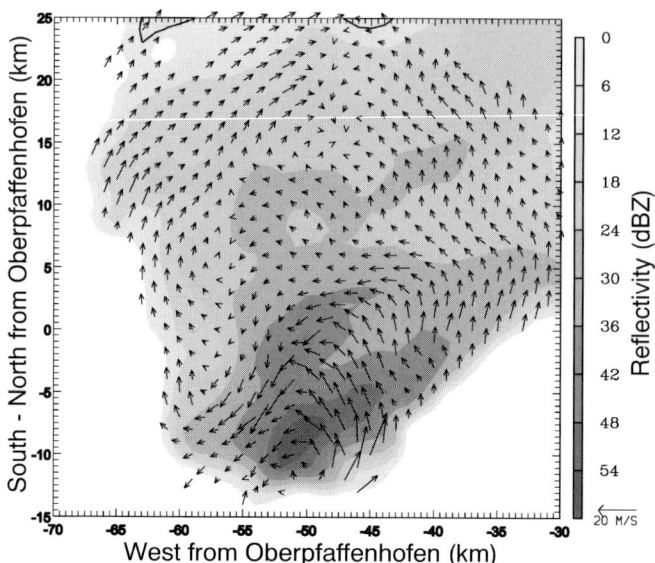

Fig. 1.28. Dual Doppler wind field estimate at 5500 m agl altitude of a thunderstorm by measurement with the DWD radar Hohenpeissenberg and the DLR radar POLDIRAD

Recent technological developments enable more efficient measurements of different wind components if a single Doppler radar system is complemented with one or more separate remote and coherently receiving bistatic antennas. Such additional receiving sites are relatively economical and easy to handle, and the overall effort is small compared to the operation of multiple complete Doppler radars. **Bistatic multiple Doppler radar networks** have been described by Wurman et al. (1993) and Wurman et al. (1994) and developed further since. McGill University, Montreal, Canada, NCAR, Boulder, USA, as well as the DLR Oberpfaffenhofen Germany, have already been operating such systems for research for several years.

The performance of such systems has been improved, the potential for operational applications demonstrated, and the error structure of 3-D wind field estimates analysed and documented in detail (Protat and Zawadzki, 1999; de Elia and Zawadzki, 2000; Friedrich and Hagen, 2001; Friedrich, 2002; Takaya and Nakazato, 2002). The decisive advantages of a bistatic Doppler network compared to multiple Doppler systems are the comparatively low cost of installation, the absolutely simultaneous measurement of different wind components of the same resolution volume, and the flexibility in locating and setting up the relatively simple bistatic antennas in order to optimise the coverage and data quality for an area of special interest. Figure 1.29 shows the principle.

The location of the observed volume is defined by the pointing and the range gating of the active radar. Sideward scattered radiation is received coherently by the bistatic receivers, which consist of an antenna and a Doppler processing unit, giving estimates of reflectivity and corresponding Doppler velocity. For

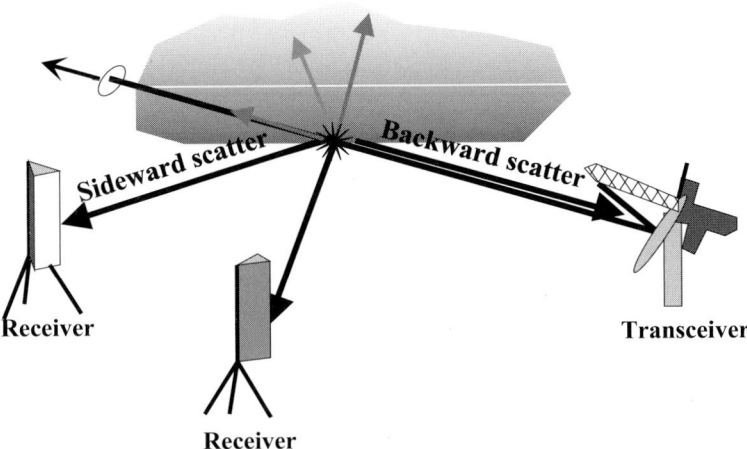

Fig. 1.29. Prinziple of a bistatic Doppler radar network consisting of one transmitting radar and two separate passive receivers. Thus, different Doppler velocity components can be measured simultaneously for the same resolution volume

coherent processing, the phase of the transmitted pulse is either known, if a klystron transmitter is used, or monitored by a receiver listening to the active radar when a magnetron transmitter is used. This phase information, together with the timing and range information, goes to each bistatic receiver. At these remote receivers, data are processed and sent back to the central computing unit, the bistatic hub, which, together with the Doppler velocities from the remote receivers, estimates the wind field. Frequency and time synchronisation are achieved by using stable local oscillators at each site controlled by GPS.

For the estimation of the Doppler velocity components by the remote sites, some special characteristics of a bistatic in contrast to a monostatic configuration need to be known (Friedrich, 2002). For the passive remote sites, the surfaces of constant delay time form ellipsoids with foci at the transmitting site and at the passive remote sites, respectively. This means that with a bistatic system, only velocity components perpendicular to that ellipsoidal surface are measured, whereas in the monostatic case, only radial components perpendicular to a sphere around the radar are measured. Each remote receiver together with the transmitting radar defines such an ellipsoid.

Unlike the monostatic case, where only radiation backscattered to the radar is measured, observations are made at different scattering angles, depending on the location of the bistatic receiver in relation to the transmitting radar and the target location. For Rayleigh scattering, the intensity depends on the scattering angle as well as on the polarisation of the transmitted wave. This allows one to optimise the system for different applications, including low or high elevation measurements (Wurman et al., 1994; Friedrich, 2002).

The sample size and volume, and hence the received power, also depend on the scattering angle. And most importantly, the Nyquist interval depends on the

Table 1.1. Radar parameters for bistatic compared to monostatic Doppler configuration. γ is the scattering angle between the incident beam and the direction to the bistatic receiver in the plane given by the transmitting radar, the scatterer, and the bistatic receiver. τ is the pulse length and λ the wavelength used. $T_s = 1/\text{PRF}$ is the sampling time

	Monostatic	Bistatic
Scattering angle	$\gamma = 0°$	$0° \leq \gamma \leq 180°$
Surface of constant delay	Sphere	Ellipsoid
Sample volume length	$a = \frac{c\tau}{2}$	$a = \frac{c\tau}{2 \cdot \cos^2(\gamma/2)}$
Velocity component	$v_r \perp$ Sphere	$v_e \perp$ Ellipsoid
Nyquist interval	$v_{nt} = \pm\frac{\lambda}{4\,T_S}$	$v_{ne} = \pm\frac{\lambda}{8 \cdot T_S \cdot \cos(\gamma/2)}$

scattering angle, too. Table 1 characterises differences between monostatic and bistatic Doppler measurements.

The accuracy of wind field measurements can be increased by using several bistatic receivers overlapping the observation area (Friedrich, 2002). Due to the wide opening angle of the receiving antennas, high side lobes of the transmitting antenna may need attention if areas of high reflectivity gradients are observed. This requires knowledge of the antenna patterns for the transmitting as well as for the receiving systems (de Elia and Zawadzki, 2000, 2001; Friedrich, 2002).

Due to the limited vertical antenna aperture of the bistatic receiving antennas at ground, the measured wind field is dominated by the horizontal components. The vertical component can be retrieved by means of a variational analysis method. (Protat and Zawadzki, 1999, 2000).

Figure 1.30 shows a wind field estimated at 4.35 km msl for a convective cloud. It displays the horizontal wind vector underlaid by the vertical velocity. A distinct downdraft with two separated maxima at 25 km distance with an azimuth of 250° can be recognised. The eastern maxima, a precipitation shaft, forces the flow from northwest around this obstacle toward easterly directions.

The recent experience and advances gained with experimental bistatic Doppler radar networks show the challenging potential for local 3D wind field estimation. Some further improvements in data quality need to be obtained in the near future. Envisaged applications include assimilation in local weather forecast models, e.g. for airport operations or local events, as discussed by Macpherson (2003), this book. The established weather radar networks of the national weather services might await such bistatic complementations.

1.6.3 Turbulence by Doppler Spectral Width Measurements

Monitoring turbulence levels in cloud systems is of interest for different applications. The dynamic state of cloud systems, or parts thereof, may be described and their further development forecasted, as suggested by Hardaker and Collier

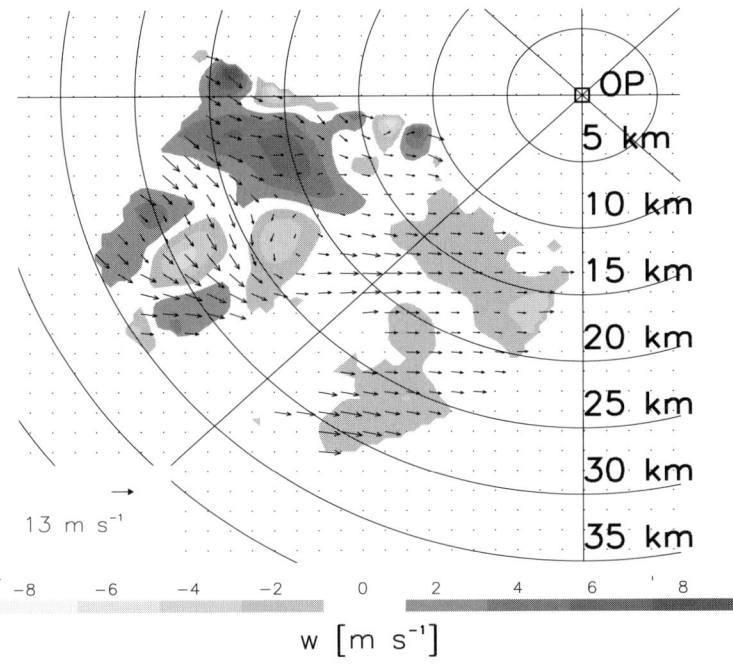

Fig. 1.30. Horizontal wind vector underlaid by vertical motion at 4.35 km msl within a moderately active convective cloud estimated by a bistatic Doppler system (Friedrich, 2002). See also color Fig. 5 on page 316

(1995) and Bohne et al. (1997). Diffusion and transport of trace gases, for example, the nitrogen oxide produced by lightning within storms, may be estimated and modelled more precisely when the actual and local dynamical structure is known (Höller et al., 1999). Small scale turbulence may force cloud droplet growth, accelerating the precipitation formation process (Pinsky and Khain, 1997; Shaw et al., 1998). This process thereby increases the probability of forming large supercooled drops, known to have severe potential for aircraft icing (Pobanz et al., 1994, Hauf and Schröder, 1998). Identification of such cloud areas near airports and especially within holdings would help for warnings to be issued. At areas of aircraft approach, avoiding turbulent zones would help to avoid aviation hazards and improve passengers' comfort.

Doppler weather radars provide Doppler velocity and Doppler spectral width as operational products and thus have the potential to detect turbulence in precipitation filled air and to estimate the energy dissipation rate ϵ, as was shown by Frisch and Strauch (1976), Brewster and Zrnić (1986), and Istok and Doviak (1986).

The accuracy and reliability of such radar detected information recently has been demonstrated by comparision with aircraft in situ measurements (Meischner et al., 2001).

According to Doviak and Zrnić (1984), the contributions to spectral broadening σ_M, as actually measured by the radar, are turbulence σ_T, antenna motion σ_A, different vertical speeds of falling hydrometeors σ_D, and shear σ_S. Because they are independent of one another, the variances contribute to the measured value σ_M as

$$\sigma_M^2 = \sigma_T^2 + \sigma_A^2 + \sigma_D^2 + \sigma_S^2. \tag{1.25}$$

It can be estimated that errors inherent in the radar system and its operation together with uncertainties from cloud inherent processes may sum up to a total error for σ_M of about 1.5 m s^{-1}. The contribution of shear however, remains to be estimated.

In order to relate the estimated Doppler spectral width caused by turbulence, σ_T, for the resolution volumes along the beam to the eddy dissipation rate ϵ, some assumptions need to be fulfilled. A first fundamental assumption is that the turbulence is homogeneous and isotropic, because with the radar, we only sense the radial velocities and their fluctuations from which we want to estimate the 3-D turbulence strength. The second important assumption is that within the resolution volume, we cover only eddy sizes within the inertial subrange; in other words, we assume a Kolmogorov spectrum throughout that volume. It has been estimated by Istok and Doviak (1986) that the outer scale of the inertial subrange must be about four to five times larger than the radar resolution volume. If this is not fulfilled, the measured Doppler spectral width, σ_M, will contain contributions from turbulence of scales within the input energy containing range and from shear of the ordered flow. Then σ_M cannot be related accurately to ϵ because eddies within the input energy containing range are not isotropic. For thunderstorms, the ordered updraft and downdraft regions typically show scales of some kilometres, as verified by numerous Doppler measurements, whereas the radar resolution volume at a distance of 100 km from the radar has a typical size of 600 m in depth and 1800 m in diameter.

According to Istok and Doviak (1986), for every resolution volume, the turbulent kinetic energy dissipation rate ϵ can be estimated as

$$\epsilon = \frac{2.4\sigma_T^3}{R\theta A^{3/2}}, \tag{1.26}$$

with R the range from the radar in metres; θ the beam width in radians, σ_T the Doppler spectral width in metres per second caused by turbulence only, and A the Kolmogoroff constant = 1.6.

The contribution of wind shear σ_S to the actually measured Doppler spectral width, σ_M, now needs to be estimated and subtracted:

$$\sigma_T^2 = \sigma_M^2 - \sigma_S^2, \tag{1.27}$$

where the index M indicates the measurement and S the contribution of shear. This shear contribution, according to Doviak and Zrnić (1984), is given by

$$\sigma_S^2 = \frac{2\theta^2}{16 \ln 2} R^2 K^2 \tag{1.28}$$

with θ the beamwidth, R the range, and K the shear in inverse seconds.

This contribution of shear to the measured spectral width for one resolution volume is estimated from the Doppler velocity measurements neighbouring that volume; see Mayer and Jank (1989).

Radar measurements have been compared with well-coordinated, high resolution aircraft in situ measurements by the FALCON research aircraft (Bögel and Baumann, 1991). An example is given in Fig. 1.31.

The measurements show that Doppler weather radars are well suited to estimate energy dissipation rates above a certain level of turbulence. For operational C-band systems, this will be around 10^{-3} m^2 s^{-3} for ϵ. Because from single Doppler radar measurements, no straightforward distinction between energy production areas and energy dissipation areas will be possible, as well as no information on isotropy, care is necessary for further quantitative use of estimated dissipation rate ϵ. For general warning applications, however, the available information will be of high value. This is particularly underlined by the general agreement of radar estimated wind variances σ_R^2 with the in situ estimates σ_{AC}^2 (Fig. 1.31). The Doppler spectral width, as measured, includes turbulence and shear for outer scales, too. Both affect aircraft handling and reduce aircraft passenger comfort. So this operationally available parameter already is of great value for warning applications.

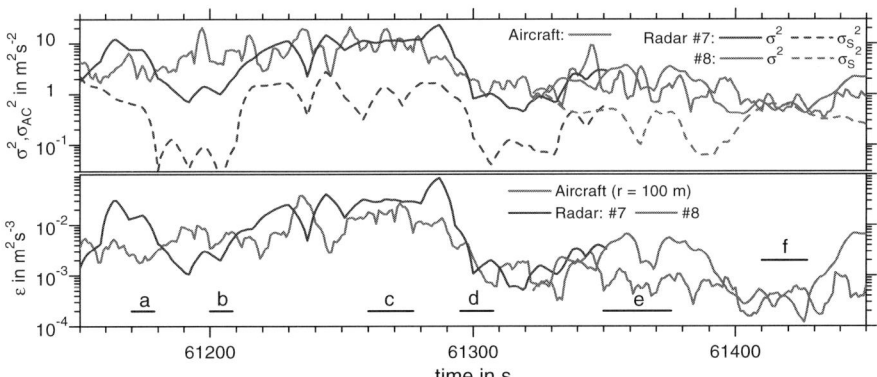

Fig. 1.31. Wind field variances – *upper panel* – and energy dissipation rate ϵ – *lower panel* – estimated from Doppler spectral width measurements in a thundercloud anvil compared with high resolution and high precision aircraft in situ measurements. a ... f indicate areas for which power spectra were analysed to check for an inertial subrange and homogeneity, see Meischner et al. (2001)

Detailed comparisons between high precision and high resolution aircraft measurements and carefully interpolated radar measurements, however, indicate some overestimation of both eddy dissipation rate and variances by radar (Meischner et al. 2001).

References

1. Andersson, T., 2000: Using the sun to check some weather radar parameters. RMK 93, SMHI, SE-601 76 Norrköping, Sweden.
2. Andersson, T., 1992: Image mosaics from Swedish weather radars. In Newsome, D. H., editor, *COST 73 International Weather Radar Networking. Final Seminar of the COST Project 73*, 139–142. Kluwer, Dordrecht.
3. Aoyagi, J., 1978: Ground clutter rejection by MTI weather radar, *Proc. 18th Conf. Radar Meteorol.*, AMS.
4. Bean, B.R. and E.J. Dutton, 1968: *Radio Meteorology*. Dover Publications, New York.
5. Bergen, W.R. and S.C. Albers, 1988: Two- and three-dimensional de-aliasing of Doppler radar velocities. J. Atmos. Oceanic Technol., **5**, 305–319.
6. Blake, L.V., 1990: System noise temperature, in M. Skolnik: *Radar Handbook* 2nd Ed., McGraw-Hill, New York 1990, Chapter 2: Prediction of Radar Range.
7. Bögel, W., and R. Baumann, 1991: Test and calibration of the DLR Falcon wind measuring system by maneuvers. J. Atmos. Oceanic Technol., **8**, 5–18.
8. Bohne, A.R., F.I. Harris, S.L. Tung and D.J. Smalley, 1997: Radar studies of aviation hazards. Part 4: Doppler spectrum width analysis. Final Report 26 April 1993 – 28 May 1997. Phillips Laboratory, Air Force Material Command, Hanscom MA.
9. Brewster and D.S. Zrnić: 1986: Comparison of eddy dissipation rates from spatial spectra of Doppler velocities and Doppler spectrum width. J. Atmos. Oceanic Technol., **3**, 440–452.
10. Bringi, V.N. and V. Chandrasekar, 2001: *Polarimetric Doppler Weather Radar*. Cambridge University Press, Cambridge.
11. Browning K.A. and R. Wexler, 1968: The determination of kinematic properties of a wind field using Doppler radar. J. Appl. Meteorol., **8**, 105–113.
12. Carbone R.E., R.I. Harris, P.H. Hildebrand, R.A. Kropfli, L.J. Miller, W. Moninger, R.G. Strauch, R.J. Doviak, K.W. Johnson, S.P. Nelson, P.S. Ray and M. Gilet, 1980: The multiple Doppler radar workshop, November 1979. Bull. Am. Meteorol. Soc., **61**, 1169–1203.
13. Carlsson, I., 1995: NORDRAD – weather radar network. In Collier, C.G., editor, *COST 75 Weather Radar Systems*, pp. 45–52. European Commission, Brussels.
14. Chong, M., J. Testud, and F. Roux, 1983: Three-dimensional wind field analysis from dual-Doppler radar data. Part II: Minimizing the error due to temporal variation. J. Climatology Appl. Meteorol., **22**, 1216–1226.
15. Chong, M., and J. Testud, 1983: Three-dimensional wind field analysis from dual-Doppler radar data. Part III: The boundary condition: An optimum determination based on a variational concept. J. Climatology Appl. Meteorol., **22**, 1227–1241.
16. Collier, C.G., 2001: Advanced weather radar systems, *COST Action 75*, Final report, EUR 19546.
17. Coveri, C., P. Caldini, G.F. Vezzani and R. Lee, 1993: Statistical ground clutter filter, *Proc. 26th Conf. Radar Meteorol.*, AMS.

18. de Elia R. and I. Zawadzki, 2001: Optimal layout of a bistatic radar network. J. Atmos. Oceanic Technol., **18**, 1184–1194.
19. de Elia R. and I. Zawadzki, 2000: Sidelobe contamination in bistatic radars. J. Atmos. Oceanic Technol., **17**, 1313–1329.
20. Doviak R.J. and D.S. Zrnić, 1984: Doppler radar and weather observations. Academic Press.
21. Friedrich, K., 2002: Determination of three-dimensional wind-vector fields using a bistatic Doppler radar network. Ph.D. thesis, Fakultät für Physik, Ludwig-Maximilians-Universität München.
22. Friedrich, K. and M. Hagen, 2001: Wind vector field determination with bistatic multiple-Doppler radar network. *Proc. 30th Conf. Radar*, AMS, pp. 133–135.
23. Frisch and Strauch, 1976: Doppler radar measurements of turbulent kinetic energy dissipation rates in a northeastern Colorado convective storm. J. Appl. Meteorol., **15**, 1012–1017.
24. Gal-Chen, T., 1978: A method for the initialization of the anelastic equations: implications for matching models with observations. Mon. Weather Rev., **106**, 587–606.
25. Hagen, M., 1989: Ableitung von Windfeldern aus Dopplermessungen eines Radars und Anwendung auf eine Kaltfront mit schmalem Regenband. DLR Forschungsbericht FB 8961.
26. Hardaker, P. and C. Collier, 1995: Extracting weather forecast information from polarisation and Doppler radar data. *COST 75 Weather Radar Systems*, C.G. Collier, Ed., European Commission, pp. 440–462.
27. Hauf, T. and F. Schröder, 1998: Supercooled large drops and aircraft icing. Results from Research Flights in Germany, March 1997. Proc. IWAIS-98, Reykjavik, 8.–11.6.98.
28. Höller, H., U. Finke, H. Huntrieser, M. Hagen and C. Feigl, 1999: Lightning produced NOx (LINOX)-Experimental design and case study results. J. Geophys. Res., **104**, D11, 13911–13922.
29. Huuskonen, A., 2001: A method for monitoring the calibration and pointing accuracy of a radar network. *Proc. 30th Conf. Radar Meteorol.*, AMS, pp. 29–31.
30. Istok, J.M. and R.J. Doviak, 1986: Analysis of the relation between Doppler spectral width and thunderstorm turbulence. J. Atmos. Sci., **43**, 2199–2214.
31. James, C.N. and R.A. Houze, Jr., 2001: A real-time four-dimensional Doppler dealiasing scheme. J. Atmos. Oceanic Technol., **18**, 1674–1683.
32. Keeler, R.J. and R. E. Passarelli, 1990: Signal processing for atmospheric radars, in D. Atlas: *Radar in Meteorology*, AMS, Boston.
33. Koistinen, J., R. King, E. Saltikoff and A. Harju, 1999: Monitoring and assessment of systematic measurement errors in the NORDRAD network. *Proc. 29th Conf. Radar Meteorol.*, AMS, pp. 765–768.
34. Koscielny, A.J., R.J. Doviak and R. Rabin, 1982: Statistical considerations in the estimation of divergence from single-Doppler radar and applications to prestorm boundary-layer observations. J. Appl. Meteorol. **21**, 197–210.
35. Laroche, S., and I. Zawadzki, 1994: A variational analysis method for retrieval of three-dimensional wind field from single-Doppler radar data. J. Atmos. Sci., **51**, 2664–2682.
36. Lhermite, R.M. and D.A. Atlas, 1961: Precipitation motion by pulse Doppler radar. *Proc. 9th Weather Radar Conf.*, Boston, Am. Meteorol. Soc., Boston, pp. 498–503.

37. Manz, A., T. Monk and J. Sangiolo, 1998: Radome effects on weather radar systems, *Int. COST 75 Seminar "Advanced Weather Radar Systems,"* Locarno, 23–27 March, 1998.
38. Mayer, W., and T. Jank, 1989: Dopplerspektren aus inkohärenter Rückstreuung, Prinzip und Anwendungsbeispiele. DLR-FB 89-48. English translation: ESA-TT-1197, 1990.
39. Meischner, P.F., R. Baumann, H. Höller and T. Jank, 2001: Eddy dissipation rates in thunderstorms estimated by Doppler radar in relation to aircraft in situ measurements. J. Atmos. Oceanic Technol., 1609–1627.
40. Michelson, D.B., 2001: Normalizing a heterogeneous radar network for BALTEX. AMS Radar Calibration Workshop: RADCAL 2001 Presentations CD-ROM. AMS, Boston. version 1.0 (http://cdserver.ametsoc.org/cd/010430_1/ RADCAL_main.html).
41. Michelson, D.B. and J. Koistinen, 2000: Gauge-radar network adjustment for the Baltic Sea experiment. Phys. Chem. Earth (B), **25(10–12)**: 915–920.
42. J. Paul and P.L. Smith, 2001: Summary of the radar calibration workshop. *Proc. 30th Conf. Radar Meteorol.*, AMS, pp. 174–176.
43. Persson P.O.G. and T. Anderson, 1987: A real-time system for automatic single-Doppler wind field analysis. *Proc. Symp. Mesoscale Analysis Forecasting*, Vancouver, ESA Publication SP-282, pp. 61–66.
44. Pinsky M. and A. Khain, 1997: Formation of inhomogeneity in drop concentration induced by the inertia of drops falling in a turbulent flow, and the influence of the inhomogeneity on the drop-spectrum broadening. Q.J.R. Meteorol. Soc., **123**, 165–186.
45. Pobanz, B. M., J. D. Marwitz and M. K. Politovich, 1994: Conditions associated with large-drop regions. J. Appl. Meteorol., **33**, 1366–1372.
46. Protat A. and I. Zawadzki, 2000: Optimization of dynamic retrievals from a multiple-Doppler radar network. J. Atmos. Oceanic Technol., **17**, 753–760.
47. Protat A. and I. Zawadzki, 1999: A variational method for real-time retrieval of three-dimensional wind field from multiple-Doppler bistatic radar network data. J. Atmos. Oceanic Technol., **16**, 432–449.
48. Ray S.P., C.L. Ziegler, W. Baumgarner and R.J. Serafin, 1980: Single- and multiple-Doppler radar observations of tornadic storms. Mon. Weather Rev., **108**, 1607–1625.
49. Shapiro, A., and J.J. Mewes, 1999: New formulations of dual-Doppler wind analysis. J. Atmos. Oceanic Technol., **16**, 782–792.
50. Shaw, R.A., W.C. Reade, L.R. Collins and J. Verlinde, 1998: Preferential concentration of cloud droplets by turbulence: Effects on the early evolution of cumulus cloud droplet spectra. J. Atmos. Sci., **55**, 1965–1976.
51. Siggia, A.D., 1993: Random phase codes for Doppler weather radars. Thesis, MIT 1983.
52. SIGMET Inc., 2000: *RVP7 User's Manual*, Westford, Oct. 2000.
53. Smith, P.L., 1968: Calibration of weather radars. *Proc. 13th Conf. Radar Meteorol.*, AMS, pp. 60–65.
54. Takaya, Y. and M. Nakazato, 2002: Error estimation of the synthesized two-dimensional horizontal velocity in a bistatic Doppler radar system. J. Atmos. Oceanic Technol., **19**, 74–79.
55. Tapping, K., 2001: Antenna calibration using the 10.7 cm solar flux. AMS Radar Calibration Workshop: RADCAL 2001 Presentations CD-ROM, 29 pp. AMS, Boston. version 1.0 (http://cdserver.ametsoc.org/cd/010430_1/RADCAL_main.html).

56. Waldteufel P. and H. Corbin, 1979: On the analysis of single-Doppler radar data. J. Appl. Meteor., **18**, 532–558.
57. Weipert, A. M. and Pierce, C., 2000: Multi-sensor supported flood detection and monitoring system for Poland, *2nd Eur. Conf. Radar Meteorol.*, Delft, 18.–22. Nov. 2002.
58. Wood V.T. and R.A. Brown, 1986: Single Doppler velocity signature interpretation of nondivergent environmental winds. J. Atmos. Oceanic Techol., **3**, 114–128.
59. Wurman J., M. Randall, C.L. Frush, E. Loew and C.L. Holloway, 1994: Design of a bistatic dual-Doppler radar passive receiver. In *Proc. of the IEE* – Special issue on *Remote Sensing Instruments for Environmental Research*, **82**, 1861–1872.
60. Wurman J., S. Heckman and D. Boccipio, 1993: A bistatic multiple-Doppler radar network. J. Appl. Meteorol., **32**, 1802–1814.

2 Operational Measurement of Precipitation in Mountainous Terrain

Urs Germann[1,2] and Jürg Joss[2]

[1] McGill University, Montréal, Canada
[2] MeteoSwiss, Locarno-Monti, Switzerland

2.1 Introduction with a Focus on the Variability of Precipitation in the Alps

To put a weather radar in a mountainous region is like pitching a tent in a snowstorm: The practical use is obvious and large – but so are the problems.

The orography interferes both with what we want to observe and with the how we can observe it. First, by lifting, channelling, blocking, supplying moisture, and heating on sun-exposed slopes, the mountains influence the flow and the stability of air masses from the synoptic down to the microscale. The orography thus plays a key role in precipitation mechanisms. The result is a complex picture of precipitation regimes and high variability on many scales. The vast influence of an orographic barrier on the distribution of precipitation is glaring when looking at climatological maps (Frei and Schär, 1998). Second, the orography complicates precipitation measurements by radar because of beam occultation, overshooting, severe ground clutter, and partial shielding, as well as difficult operating conditions on mountain sites. The result is a complex error structure. So, on the one hand, a weather radar with its high resolution in time and space provides the meteorologist and the hydrologist of a mountainous region with unique observations of a highly variable quantity, but, on the other hand, poses particular difficulties to the radar scientists and engineers.

The meteorologist wants to have observations from the first signs of precipitation of 0.1 mm/h to heavy hailstorms with over 100 mm/h, within the full range of radar coverage, for example, between 1 and 230 km as for the Swiss radars. Including the range of signal fluctuations this corresponds to 20 orders of magnitude in terms of signal power between the transmitted and the weakest signal to be received and analysed. This is more than we can cope with using today's receiver technology. The hydrologist, on the other hand, may need high precision because 10% more rain may break the dam! A really challenging engineering application.

The dominant sources of errors involved in quantitative estimates of precipitation by radar are ground clutter including anaprop, shielding combined with the vertical profile of reflectivity (Joss and Waldvogel, 1990), partial shielding, overshooting, beam-broadening and partial beam filling, variations in the relation between radar reflectivity and rainfall rate (Z–R relation), hardware faults, as well as attenuation in heavy rain, in the melting layer and in the water cover on the radome (Germann, 1999). Variations in the Z–R relation are attributed

Fig. 2.1. The real challenge in the Alps: What does the radar see? Is it the tail of a dragon, or is it just the tail of a mouse?

to changes in the phase and size distribution of hydrometeors. Although all these errors occur both with and without orography, some of them are much more severe in a mountainous context. There, the most challenging problem is the combination of shielding, partial shielding and ground echoes, which often inhibit a direct view of precipitation close to the ground (Figs. 2.1 and 2.2, Gabella and Perona, 1998; Germann and Joss, 2002; Pellarin et al., 2000). If we determine, for each pixel of the radar volume, to what extent precipitation at that location can be observed from the radar without being disturbed by the horizon or clutter, we obtain the radar visibility map. In a mountainous region, this map is rather complex and shows pronounced spatial discontinuities, particularly, close to the ground. To fill the holes in badly visible regions, measurements from several kilometres above the ground or from neighbouring regions must be extrapolated. Yet, this implies a high resolution in azimuth, in range and in elevation. In order to limit the size of holes produced by the elimination of ground echoes, for instance, in the Swiss network, clutter removal is done at the highest possible radial resolution of 83 m. Any type of extrapolation procedure used to fill holes requires a careful analysis of the four-dimensional map of visibility and considerably reduces the accuracy of ground level precipita-

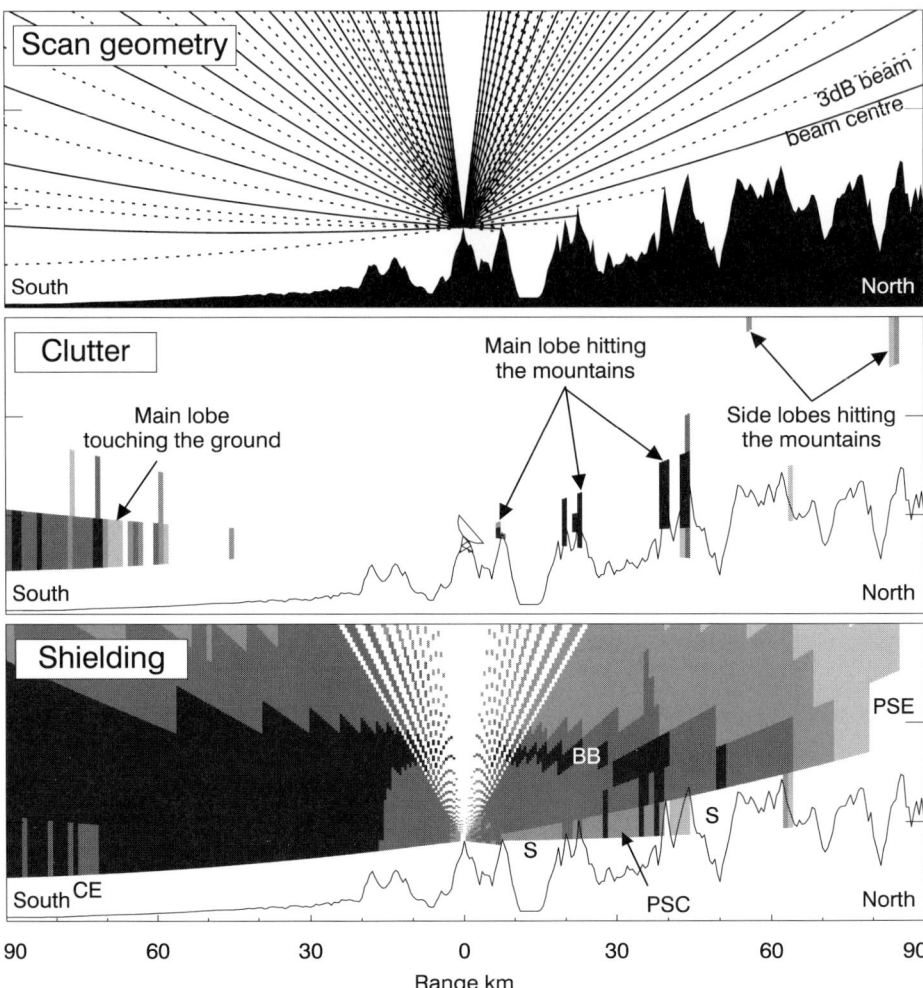

Fig. 2.2. The challenge in the Alps: combination of shielding, partial shielding and ground echoes inhibits a direct view of precipitation close to the ground. The three panels show vertical cross sections from 0 to 6 km above sea level of (*top*) scan geometry of Monte Lema radar, (*centre*) intensity of ground echoes during fine weather with the clutter elimination algorithm turned off, and (*bottom*) 24-hour accumulation of stratiform precipitation on 3 August 1998 illustrating the problem of shielding. To the north, the lowest two elevations are blocked at a range of 8 and 44 km, respectively. Everything behind is shielded (S). Between 8 and 44 km, the second-lowest elevation is partially shielded and occasionally contaminated by ground clutter not eliminated (PSC). The notch of the third lowest elevation at 80 km north (PSE) is caused by partial shielding, by evaporation, or by a combination of both. To the south beyond 70 km, the lowest elevation shows gaps that result from eliminating frequent ground clutter (CE). The melting layer is at a height of around 3.3 km (BB)

tion estimates. The holes caused by clutter elimination and shielding also limit the use of echo tracking and advection nowcasting (Mecklenburg et al., 2000; Germann and Zawadzki, 2002).

For the old second-generation radars of Meteo-Swiss, the overall average error of a radar point measurement for a daily rainfall amount was found to be 3 dB(R). The 3 dB(R), which corresponds to a factor of two, is the standard deviation of the logarithm of the ratio between daily accumulations of gauge and radar measurements. It is based on 6 years of data of 60 automatic gauge stations distributed all over Switzerland and two radars, one close to Geneva and the other one close to Zürich. For the three new third-generation radars the 3 dB(R) overestimates the error because much effort has been spent to improve the clutter elimination (Sect. 2.2) and to introduce correction procedures for visibility (Sect. 2.3) and profile effects (Sect. 2.4).

2.1.1 Variability of Precipitation in the Alps

Before discussing, in the following sections, specific problems and solutions of radar operation in the Alps, let us first have a look at the spatial continuity of orographic precipitation and its important implications for the design of observing systems. We use the term *continuity* rather than *variation* to place emphasis on the fact that close measurements are similar. Owing to the continuity, we can compare measurements in different locations, extrapolate point observations to neighbouring areas, for instance, when producing rainfall maps from gauge network data, or estimate vertical profiles of reflectivity in shielded regions.

A study of the spatial continuity of Alpine precipitation based on radar data has been presented in Germann and Joss (2001). To quantify the spatial continuity, the authors chose the variogram, defined as the variance of the difference between two reflectivity measurements as a function of the separation distance. The variogram indicates how much the precipitation rate varies in space.

Why are we particularly interested in the spatial continuity of precipitation? First, holes in the radar image caused by ground echoes and shielding must be filled by interpolating and extrapolating nearby pixels, both horizontally and vertically. The uncertainty introduced by this procedure depends on the spatial continuity. Second, an observing system should provide high spatial resolution in regions where the observed quantity exhibits high spatial variation. Therefore, a climatology of precipitation variograms is a prerequisite to designing a gauge network. And it may also be helpful when choosing the site of a radar. Third, the representativeness of a point observation of an area average is solely a function of the spatial continuity. We can determine from the variogram within what accuracy a gauge measurement is representative of a basin average. High variation means low representativeness. Finally, extrapolating the variogram to the zero-lag (nugget variance) is an elegant way to obtain an estimate of the observation error, the sum of instrumental errors plus the sampling error.

It is for these reasons that we provide, in the remaining part of this section, a short summary of the results of the spatial continuity studies presented in Germann and Joss (2001).

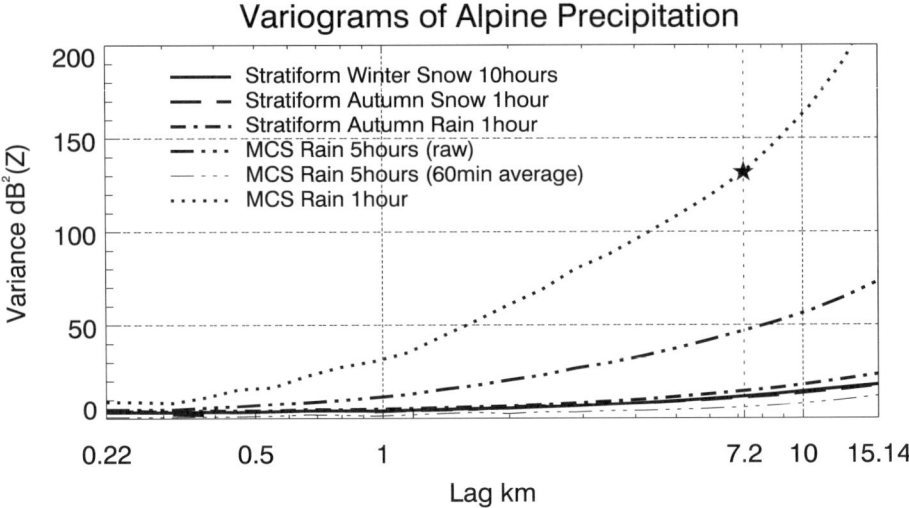

Fig. 2.3. Variograms of Alpine precipitation: stratiform autumn (30 September 1998), stratiform winter (2 December 1997), mesoscale convective system (MCS, 19 August 1998). The variograms are calculated from raw reflectivity except for the MCS 5-hour period, for which both raw and one hour average reflectivity are used. The two variograms in snow (*solid and dashed line*) are so closed that the dashed line can hardly be seen. From Germann and Joss (2001)

Figure 2.3 shows a set of variograms determined from radar observations of Alpine precipitation under different synoptic conditions. Two things are striking. The variograms are smooth compared to those commonly obtained from gauge data, and, comparing the different events, the variance varies by up to one order of magnitude. High variability is typically associated with convective activity, here orographically induced convection. The variances of the variograms of the mesoscale convective system (MCS) are up to ten times larger than those of stratiform variograms. At the moment of maximum convective activity in the MCS (labelled in the figure MCS Rain 1hour), we read for a lag of 7.2 km, an average variance of $132\,\mathrm{dB}^2(Z)$. By taking the square root, we obtain an average difference in logarithmic reflectivity of $11.5\,\mathrm{dB}(Z)$. Assuming in the Z–R relation an exponent of 1.5, this corresponds to a factor of 5.8 in terms of rainfall rates. In other words, rainfall rates measured at two points separated by 7.2 km, on average, vary by a factor of 5.8. For the same lag, we obtain $6.8\,\mathrm{dB}(Z)$, $3.8\,\mathrm{dB}(Z)$, and $3.3\,\mathrm{dB}(Z)$ in the MCS 5-hour period, in the stratiform rain, and in the stratiform snow, respectively.

These numbers confirm that spatial continuity of precipitation depends on the type of precipitation and thus on the synoptic conditions. Is there also a dependence on the location with respect to the orographic barrier? Such a dependence would reflect the influence of the mountains on precipitation physics. Comparing variograms along a cross section from the southern foothills to up-

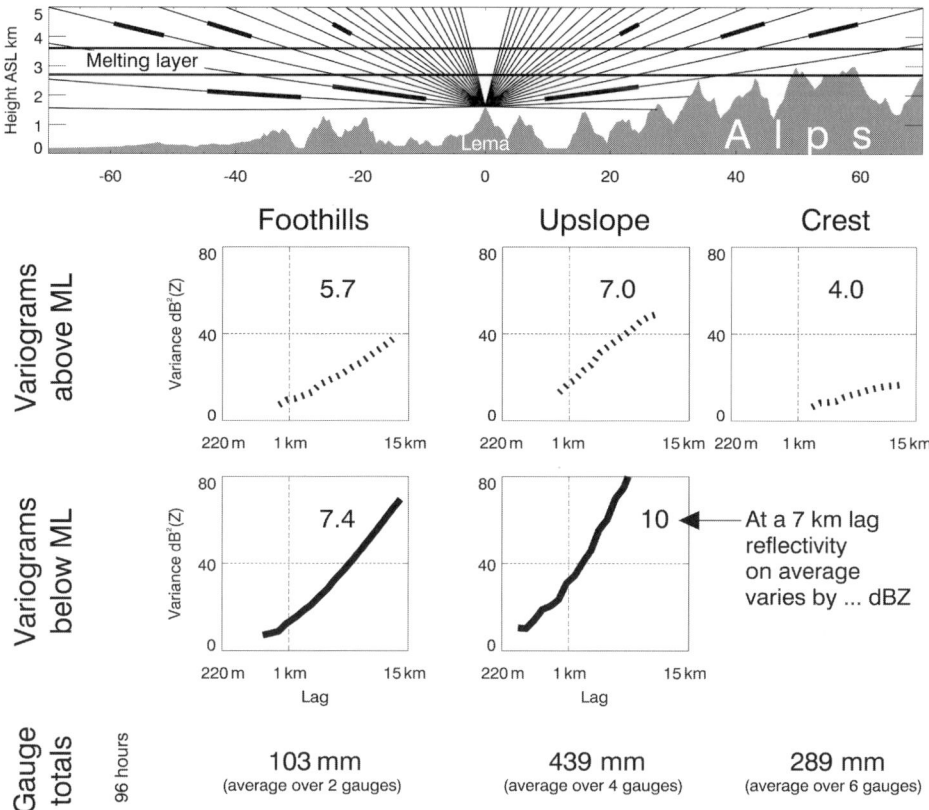

Fig. 2.4. Variograms below and above the melting layer (ML) along a cross section through the southern Alps. The variograms are derived from radar reflectivity of Monte Lema radar and represent average conditions over 96 hours of heavy rain in September 1999. For more details, see Germann and Joss (2001)

slope regions and up to the crest of the Alps during 96 hours of heavy and long-lived orographic precipitation revealed a surprisingly stationary pattern: moderate variation upstream and over the foothills, maximum variation upslope, and weak variation close to the crest of the mountain range (Fig. 2.4). A different ranking is obtained when looking at gauge totals of the same period and along the same cross section. In the foothills, the gauge totals are relatively small. Maximum totals coincide with maximum variation in upslope regions. Further north in the valleys close to the crest of the Alps, we found moderate to high totals, and this despite the weak variation indicating stratiform precipitation. It is not yet clear to what extent the persistent stratiform rainfall associated with moderate to high rain amounts close to the main divide is related to the frequent convection in upslope regions. Tall convection upslopes possibly feed adjacent regions downstream with ice crystals. While falling through rising saturated air

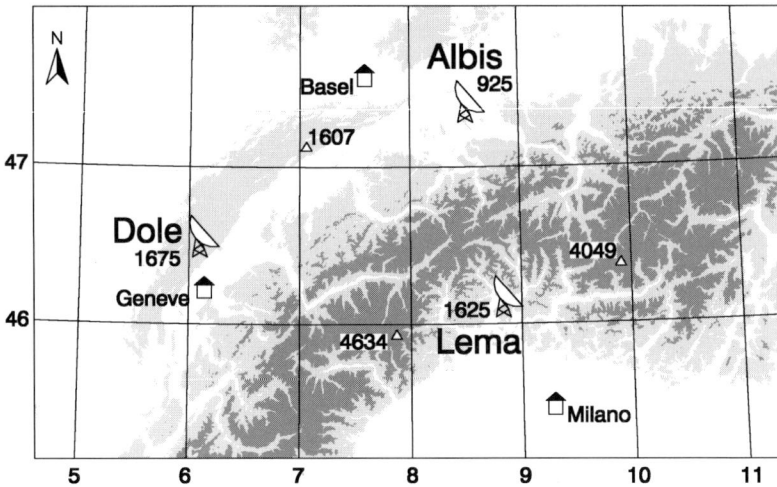

Fig. 2.5. Situation plan of Meteo-Swiss C-band radar network. Levels of shading correspond to terrain height below 800 m (*white*), between 800 and 2000 m (*light grey*), and above 2000 m (*dark grey*). All heights are in meters above sea level

masses, the crystals rapidly grow and may produce persistent stratiform precipitation with relatively high totals.

To sum up, spatial continuity of precipitation depends both on the synoptic conditions and on the position with respect to the Alpine barrier. As outlined above, this is crucial knowledge when designing a system to observe Alpine precipitation.

After this short excursus into spatial continuity of orographic precipitation and its implications for observing systems, we now move to the discussion of practical solutions in an operational context. The Meteo-Swiss C-band radar network is situated in an exemplary region to study the problems of radar meteorology in the mountains (Fig. 2.5). For a description of the Swiss network, the reader is referred to Joss et al. (1998). The current infrastructure is based on 40 years of work with radar operation in the Alps. Although the material of this chapter is illustrated mostly with examples from the Swiss network, we expect it to apply to other mountainous regions as well.

2.2 Clutter Elimination

Elimination of ground clutter is a prerequisite for the use of radar data, both in a quantitative and a qualitative way. In mountainous regions, where the radar sites are often on the tops of mountains and the lowest elevation angle is close to or even below the horizon, ground clutter poses a serious problem (Fig. 2.2, centre).

2.2.1 Avoiding Clutter

In regions with frequent clutter, the sensitivity for detecting precipitation is reduced (discussed later on in Sect. 2.3.2), regardless of whether the clutter is eliminated or not. Therefore, avoiding clutter is better than the best elimination technique. For example, using short pulses may often allow us to see between clutter echoes and interpolate from the in-between spaces. This must not be forgotten when discussing the specifications of a new radar system. Also, regarding contamination by clutter, C-band wavelengths are to be preferred to S-band for two reasons: Assuming a constant beam width, the ratio between the echo of hydrometeors in the Rayleigh region and the echo of a ground target is proportional to the fourth power of the transmitter frequency. So, a C-band radar has a much higher chance of seeing the weather rather than the ground target in regions where both coexist. The second argument assumes a fixed antenna diameter, and thus fixed antenna costs, a reasonable constraint. For a fixed antenna diameter, the beam width is proportional to the wavelength, and the pulse volume proportional to the square of the wavelength. A small pulse volume helps to limit the radius of contamination by clutter around a ground target. The same is true for high resolution in range and low side-lobe power. A thorough study of these relationships is extremely important in a mountainous region in order to avoid clutter as much as possible.

2.2.2 Eliminating Clutter

Clutter varies from site to site and in time and space, which makes its elimination a challenging task. A straightforward solution for clutter elimination is a static clutter map determined from a series of radar images in clear-sky conditions. The drawback of this solution is residual clutter, or the loss of valid precipitation measurements in regions with weak or variable clutter. Examples for variable clutter are ground echoes from anomalous propagation (anaprop) or side-lobe clutter. More sophisticated approaches combine all available information and use a dynamic clutter map. An example is the decision tree algorithm proposed by Lee et al. (1995) and implemented at the three radar sites of Meteo-Swiss Monte Lema, La Dôle, and Albis (Joss et al., 1998). The algorithm takes, for each 83 m raw gate, a clutter/nonclutter decision using the radial velocity, the spectrum width, the minimum detectable signal (MDS), one-lag and two-lag signal fluctuations (noncoherent statistical tests), the vertical gradient of reflectivity, as well as a continuously updated clutter map. For the use of polarimetric measurements to detect clutter and anomalous propagation, the reader is referred to Illingworth (2003), this book.

2.2.3 Improving Clutter Elimination in the Alps

In 1998 and 1999, several experiments were performed to further improve the operational clutter elimination algorithm of the Meteo-Swiss radar network. A modified version was implemented on all three radar sites in summer 1999. The unmodified and the modified decision tree are hereafter referred to as the old and the new elimination algorithm, respectively. In the remaining part of this section, we briefly present and explain the modifications made in the operational algorithm. Details of the old algorithm can be found in Joss et al. (1998), Joss and Lee (1995) and Lee et al. (1995). The modified, currently implemented decision tree is depicted in Fig. 2.6.

The goal of the experiments was twofold: to reduce the amount of residual clutter, and to reduce at the same time the number of precipitation signals erroneously eliminated by the old algorithm. For lack of a better way to sufficiently reduce clutter, in the old algorithm, the threshold T5 of the first statistical filter was set to a somewhat high value of 4 dB. With this setting, the first statistical filter was rejecting a considerable number of valid precipitation signals. Especially at high elevations where clutter is rare and weak, there is no need for setting T5 to a high value. An alternative way to reduce the level of residual clutter allowing for a lower value for T5 on all elevation angles has to be found.

Based on the observations described afterward, we conclude that clutter remaining after the old elimination has mainly nonzero Doppler velocity and preferably occurs in the vicinity of highly coherent zero-velocity clutter.

To eliminate this type of clutter, a neighbour test is added in the Doppler test of the new clutter decision tree (Fig. 2.6). Gates with nonzero velocity only pass the decision tree if the adjacent gates have low clutter map entries. Adding the neighbour test allows us to make the thresholds of the statistical tests less rigorous, thus eliminating fewer valid precipitation signals. The neighbour test is active at the lowest two elevation angles only.

Here is an illustration of how we explain the success of the new clutter tree. In Switzerland, the lowest elevation scan at $-0.3°$ contains 23% cluttered pixels (Table 2.1: $10927/46800 = 23\%$), contributing about 70% of the total clutter of the volume scan (10927/15503). A volume scan consists of 20 elevation angles. The lowest two elevation angles, $-0.3°$ and $0.5°$, together, contribute more than 90%. These estimates are obtained on La Dôle during fine weather by switching off the clutter elimination, then counting all pixels ($1\,\text{km} \times 1°$) with reflectivity larger than 13 dBZ. The numbers are averaged over half-hour periods, and, to some extent, may depend on the weather situation (refraction). The experiments have been repeated using the old and the new clutter elimination algorithm.

Table 2.1 lists the total counts. Except for the group 'Whole volume,' data are considered up to the maximum range of coherent analyses of 130 km from the radar site. N_{all} is the number of all cluttered pixels, whereas N_{moving} only counts pixels with nonzero velocity, that is velocities larger than 0.125 times the Nyquist velocity. The total number of pixels considered is indicated in parenthesis. For more details, see Sect. 1.3.1 of Germann (2000). Similar results have been obtained for the Monte Lema and Albis radar.

With the old elimination algorithm, the residual clutter has mainly nonzero velocity and preferably occurs in the vicinity of highly coherent zero velocity clutter. This hypothesis is based on the following two observations: First, 96% of the residual clutter has nonzero velocity (946/990). Second, N_{moving} increases

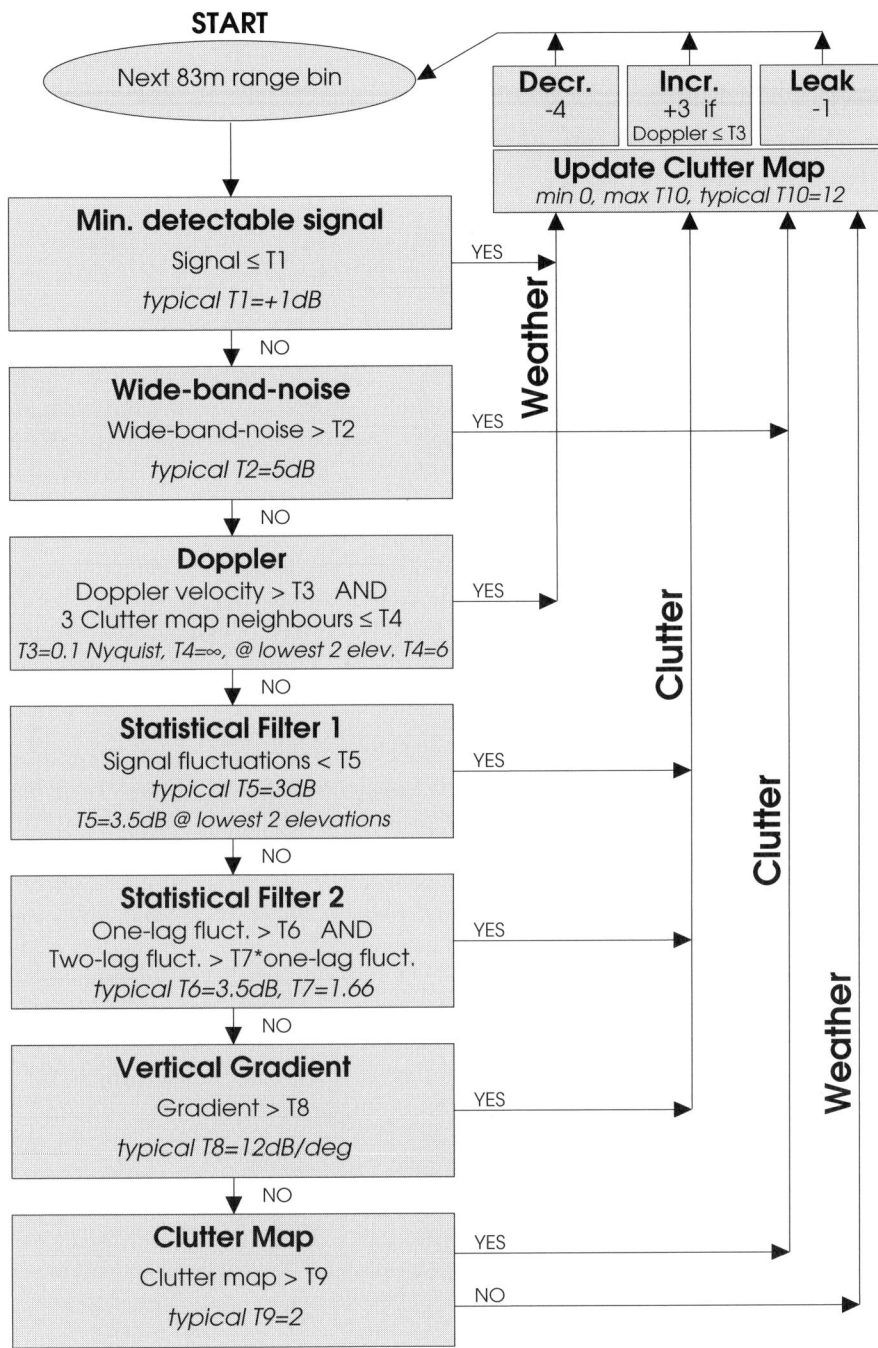

Fig. 2.6. Decision tree for clutter elimination at Meteo-Swiss radar sites

Table 2.1. Frequency of clutter of La Dôle radar using no, old, and new elimination

	Volume of Coherent Analyses				Whole Volume
	$-0.3°$ elev. (46800)		All 20 elev. (549720)		All 20 elev. (697320)
Algorithm	N_{all}	N_{moving}	N_{all}	N_{moving}	N_{all}
No	**10927**	114	**15503**	342	**19476**
Old	**613**	584	**990**	946	**1395**
New	**40**	16	**402**	279	**458**

when using the old instead of no elimination. How can this be? For each 1 km gate, we have data of twelve 83 m raw gates. For the Doppler product, which is used to determine N_{moving}, only the velocity of the 83 m raw gate with the maximum reflectivity is taken. The Doppler product shows that 15161 of the 15503 clutter-contaminated raw pixels originate from zero-velocity clutter. The statistical tests of the old algorithm, based on the signal fluctuations, easily detect and eliminate this type of clutter. That is why zero-velocity clutter mainly disappears when activating the old elimination. But from the fact that at the same time N_{moving} increases, we conclude that zero-velocity clutter is then in part replaced by neighbouring nonzero velocity clutter. Clutter with nonzero velocity may be explained by antenna movement or simply by chance: since the number of range gates is huge, even with a small probability of occurrence, some of the clutter signals have nonzero velocity.

When using a *dynamic* clutter map, particular attention must be paid to dead-ends and positive feedbacks. In our case, a continuous clutter map leak was needed to avoid a no-go situation for gates with reflectivity above the MDS and zero Doppler velocity: in the old elimination algorithm, once the gate was flagged as clutter, there was no way to decrease the clutter map entry (Lee et al., 1995). This means a no-go, as long as reflectivity is above the MDS and the Doppler velocity is zero. Note that zero-velocity areas can be large and stationary. Examples are weather situations with weak winds, weak winds in the planetary boundary layer, or regions where the wind blows perpendicularly to the radar beam. Unless a gate is clearly identified as clutter, the clutter map leak (Fig. 2.6, top-right edge) continuously subtracts a small value from the map.

2.2.4 Closing Remarks

The last column of Table 2.1 summarises the effectiveness of clutter elimination. Without elimination, the radar volume contains 3% clutter (19476/697320), with the old elimination 0.2%, and with the new one 0.07%. Considering the lowest elevation only, we obtain 23%, 1.3%, and 0.09%, respectively. On the lowest elevation, the new algorithm is reducing the number of residual clutter by 573 from 613 to 40. When considering all elevations, it may be a surprise that the corresponding difference is only slightly larger, namely, 588. This is caused by the superposition of two opposite effects. The new neighbour test, which is active at the lowest two elevation angles, reduces the amount of residual clutter. But the threshold T5 of the first statistical filter has been lowered, in particular at higher elevations, in order to avoid erroneously eliminating precipitation signals. Previously, it was set to 4.0 dB. Now T5 is 3.5 dB at the lowest two elevation angles, and 3.0 dB elsewhere (Fig. 2.6). The disadvantage of a lower T5 is an increase of residual clutter, which compensates in part for the improvement achieved by the new neighbour test.

In addition to the neighbour test and the clutter map leak, a speckle filter has been added. It eliminates isolated echoes, which rarely originate from precipitation. This filter is not part of the decision tree. It is implemented at a later stage of data processing, and works at the 1 km $\times$ 1° $\times$ 1° resolution.

2.3 Correction for Visibility

A map that indicates, for each pixel of the radar volume, to what extent precipitation is visible from the radar is called a radar visibility map. It can be a simple binary map, indicating 'visible' or 'nonvisible,' or it can be a more precise estimate indicating the visibility as a percentage ranging from 0% visible up to 100% visible. Factors that reduce the visibility are shielding, partial shielding, and clutter (Fig. 2.2).

For two practical reasons, we are interested in the visibility map:

1. For further data processing, it is crucial to know whether a pixel is (a) perfectly visible, (b) completely shielded and thus missing, or (c) frequently contaminated by clutter or partially shielded and thus less accurate and less reliable. Pixels of different visibility have to be treated in different ways, for instance, when generating ground level precipitation maps or when calculating the vertical profile of reflectivity used for profile correction (next section).
2. The bias in partially visible regions can be corrected if there is a good estimate of the visibility. For example, for our best radar estimate of precipitation, the value of a pixel that is only 50% visible is multiplied by a factor of 2.

2.3.1 Simulated Visibility

A first guess of the visibility map can be obtained by a geometrical simulation. The visibility is calculated ray by ray by combining the scan geometry of the radar, the antenna gain pattern, a digital terrain model, the earth's curvature, and a model of how electromagnetic waves propagate through the lowest 10 km of the atmosphere. Refraction is simulated by using a fictive earth radius 4/3 times the correct radius. Figure 2.7 depicts visibility maps of Monte Lema radar obtained from a geometrical simulation for the lowest four elevation angles -0.3, 0.5, 1.5, and 2.5°. The result does not perfectly correspond to the real visibility map because of small errors in the pointing angle, uncertainties in the simulation

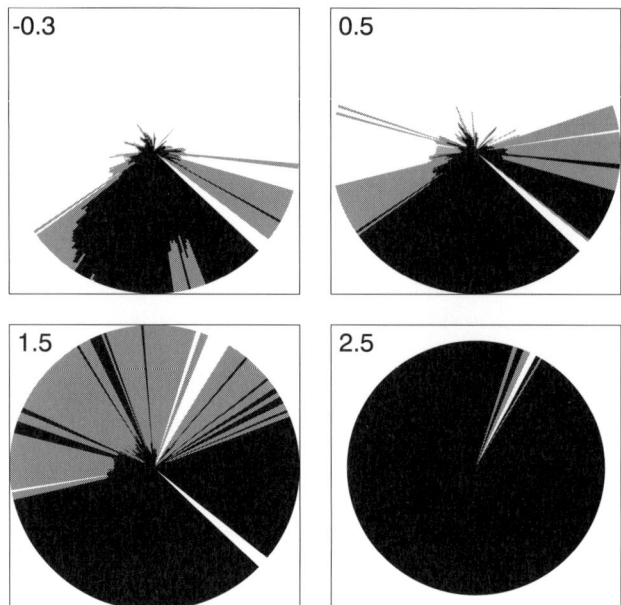

Fig. 2.7. Maps of the radar visibility obtained from a geometrical simulation for the lowest four elevations of the Monte Lema radar (−0.3, 0.5, 1.5, and 2.5°). Maximum range is 230 km. Levels of shading correspond to less than 10% visible (*white*), between 10 and 90% visible (*grey*), and more than 90% visible (*black*). See also Fig. 2.2, bottom

of the refraction of the radar beam, and insufficient resolution of the digital terrain model, in particular at close ranges. Buildings, trees, and other obstacles close to the antenna limit the accuracy of a geometrical simulation, if their position and shape are not precisely known.

2.3.2 Clutter Hiding Precipitation

Also, clutter may hide precipitation, regardless whether detected or not. The reduction of visibility caused by clutter can be estimated from clutter maps, either determined during fine weather with the clutter elimination turned off, or from output statistics of the elimination algorithm itself. It is important to consider that only weather signals that are weaker than the clutter signal are actually affected. Thus, to consider that all the cluttered pixels during fine weather are invisible during precipitation is too pessimistic and overrates the role of clutter.

2.3.3 Real Radar Visibility

An alternative approach to estimate the radar visibility is to examine long-term accumulations of radar images during precipitation. In perfectly visible regions, the accumulation tends to climatology, in completely shielded regions, it is zero;

and in partially visible regions, the value is somewhere between. Hence, by dividing long-term accumulations by climatological maps, we obtain an estimate of the real radar visibility. Here, the problem is not the uncertainty of a geometric simulation. It is the lack of knowing for each pixel of the radar volume, its climatological value of precipitation. There are systematic changes, both horizontally and vertically, that have to be accounted for when following this approach. In the vertical direction, we have the profile of reflectivity, reflecting the growth, the change of phase and the fall speed of hydrometeors. In the horizontal direction, we expect a strong dependence on the position with respect to the Alpine barrier; see, for instance, the gauge totals in Fig. 2.4. So, the spatial distribution of a long-term accumulation of radar images is a superposition of two factors: systematic changes attributed to the climatology of precipitation and the real radar visibility. To determine the latter contribution, we have to estimate and subtract the former one. This is not an easy task.

Also, beam-broadening, partial beam filling, and attenuation affect accumulations of precipitation. The result is a monotonic decrease of the echo with range at constant height. This effect is, according to the definition from above, part of the real radar visibility. If it is included in the visibility map, the average bias caused by beam-broadening, partial beam filling, and attenuation is corrected by the two steps of visibility correction described at the beginning of this section.

Figure 2.8 compares for the 0.5° elevation of Monte Lema radar, the geometrically simulated visibility with the visibility, calculated from accumulated precipitation.

The period used for the accumulation is short, and climatological variation including vertical growth has not been corrected yet. In spite of the simplistic and preliminary character of this analysis, the comparison reveals interesting facts. The simulated map is a good first guess, providing a global picture. To a first approximation, the two panels of Fig. 2.8 reveal the same line of demarcation between the visible and the nonvisible parts of the elevation scan. The largest difference between the two range–azimuth displays is at far ranges. It is caused by the vertical profile of reflectivity and by beam-broadening provoking partial beam filling. The simulated map seems to overestimate the real visibility in the first few range gates next to the radar and close to mountain peaks where weather echoes are hidden by frequent strong ground clutter. The bottom part of Fig. 2.8, on the other hand, seems to overestimate the real visibility in the first tens of kilometres behind obstacles. There, accumulated side-lobe clutter may look like accumulated precipitation. When doing the accumulation, we must take care not to accumulate clutter and clear-air echoes. They will spoil the results. Further analyses with longer periods and correction for climatological variation are needed to get a quantitative picture of the true visibility.

2.3.4 Synthesis

Geometrical simulations and long-term accumulations of precipitation may be used to estimate the radar visibility map. We propose to combine both approaches in order to get a better result, that is, to make a first guess from a

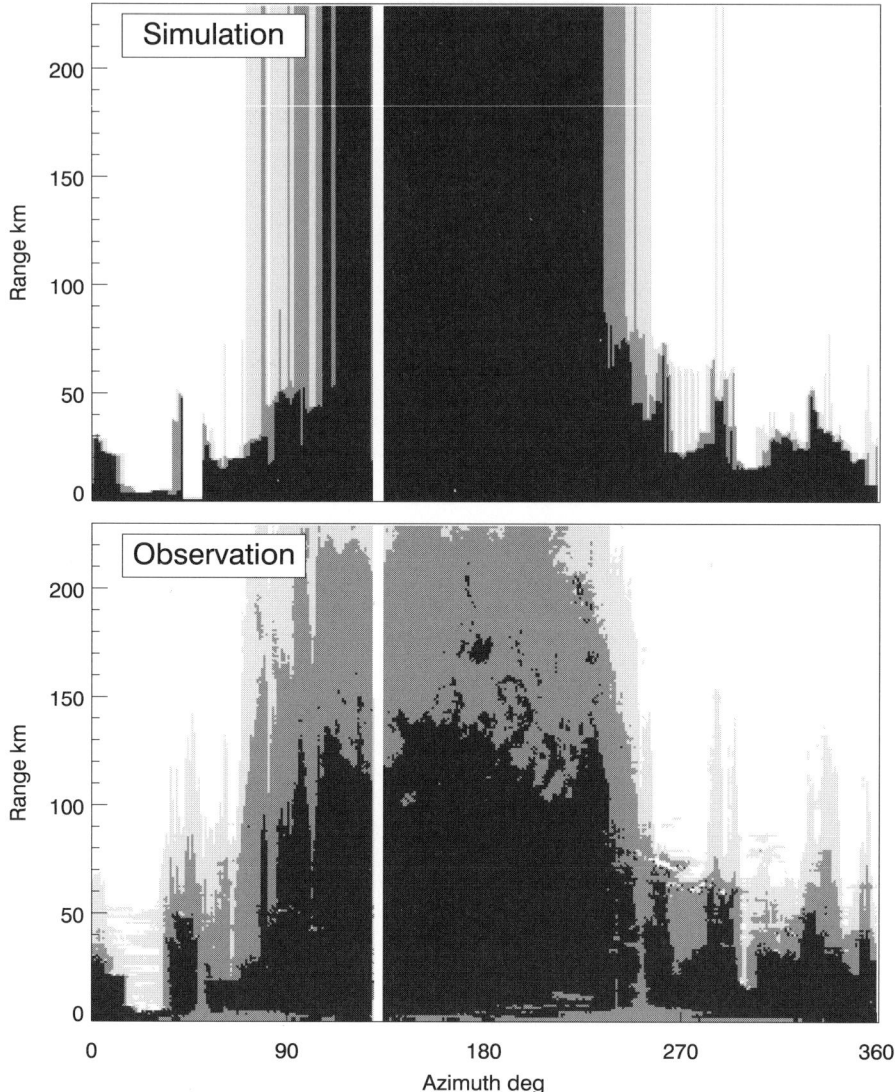

Fig. 2.8. Simulated versus observed visibility. (*upper*): Simulated visibility at 0.5° elevation of Monte Lema radar obtained from geometrical simulation. Levels of shading correspond to less than 10% visible (*white*), between 10 and 50% visible, between 50 and 100% visible, and 100% visible, respectively. (*lower*): 22-day accumulation of warm-season precipitation at the same elevation angle. Grey shades change at 0.6, 6, and 60 mm. Because of an important telecommunication link the radar power is switched off between 129 and 134° azimuth. Note that this type of plot showing range versus azimuth does not preserve area. The short ranges are overrepresented compared to far ranges

geometrical simulation, to add the information from clutter maps, and to check against the results obtained from long-term accumulations.

For the approach using long-term accumulations, estimating climatology over complex orography is a difficult step. A reasonable first guess can be obtained by combining operational radar data with data sets from field experiments such as the Mesoscale Alpine Programme (Bougeault et al., 2001), during which several mobile radars were operated at different locations within the southern Alps. The results from long-term accumulations are complementary to those from geometrical simulations and, in spite of the uncertainty, are expected to improve our estimate of the radar visibility map.

To determine, for instance, all pixels that are, at a high level of confidence, well visible for precipitation rates above 1 mm/h, we propose the following criteria: Select those pixels for which (a) the simulated geometrical visibility is at least 90%, (b) the long-term accumulation lies within the range of climatological variability, and, (c) clutter is expected to be weaker than 1 mm/h.

Geometrical simulations of the radar visibility can be determined for any fictitious site without having to install a radar. It is an important step when choosing the site of a new weather radar and provides, as shown in Fig. 2.8, a good first guess of its performance.

2.4 Profile Correction

In the previous section, we discussed how a variable horizon, earth curvature, and ground clutter reduce the portion of the radar volume that is visible from the antenna site. These factors inhibit a direct view of precipitation close to the ground and thus leave holes in the radar image. As long as the holes are small, they can be filled by interpolating nearby pixels at the same height. But above the Alps, the holes are often so large that there is no way of horizontal interpolation (Fig. 2.7). The same is true for flat countries at far ranges. Then, the only thing we can do is to extrapolate measurements from above made at higher altitudes, down to the ground level. A measurement from aloft is better than no measurement at all, and for qualitative use often sufficient, provided that precipitation reaches the height of the lowest radar measurement. For quantitative use, however, measurements from aloft must be corrected for the vertical change of the radar echo, caused by the growth and the change of phase and fall speed of hydrometeors. The radar echo as a function of height is called *profile*, and, the vertical extrapolation of measurements from above to the ground level is called *profile correction*.

2.4.1 What Can Be Achieved

Instead of beginning with a detailed technical discussion of profile correction schemes, we start with looking at the results of the operational algorithm implemented in the Swiss network. The objective is to get a map of the best possible estimate of ground level precipitation rates. Figure 2.9 compares the results of

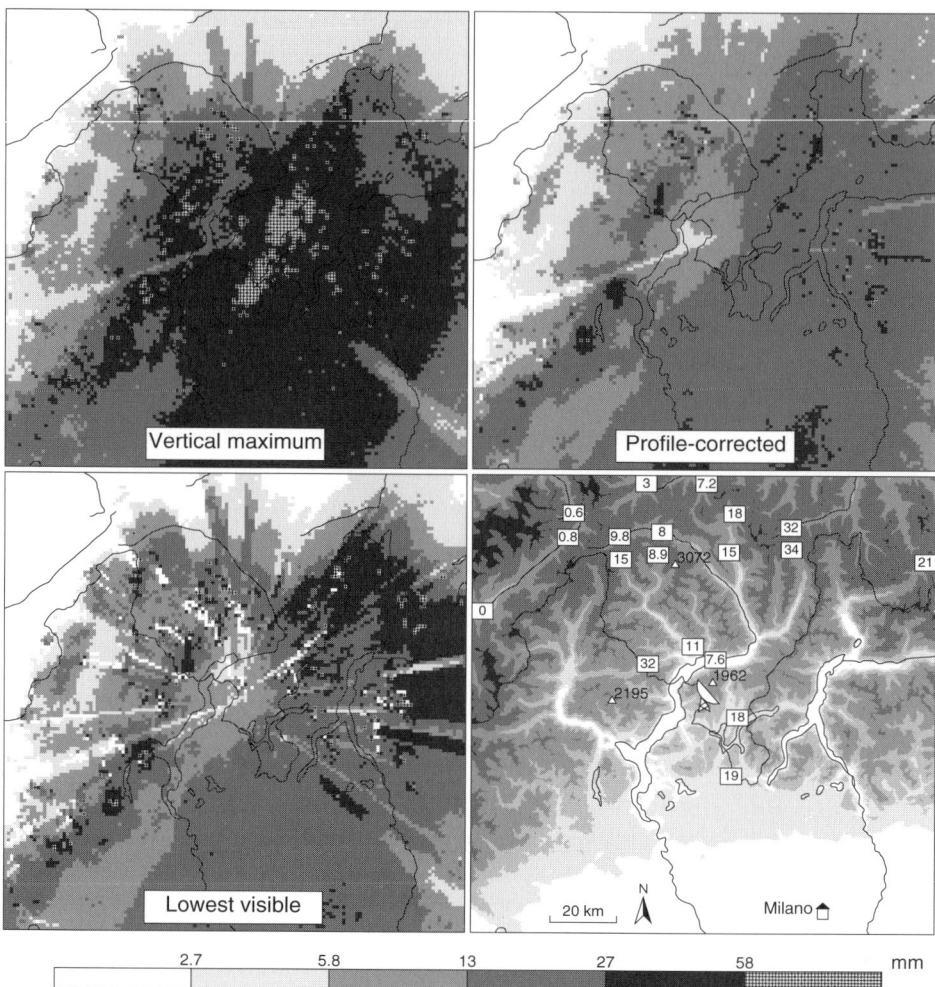

Fig. 2.9. Comparison of three methods to estimate rainfall amounts by radar in a mountainous region. The three estimates are based on the vertical maximum echo (*upper left*), the lowest visible echo (*lower left*), and estimates corrected using the mesobeta profile (*upper right*). Partial shielding has been corrected for the two products 'lowest visible' and 'profile-corrected,' but not for the 'vertical maximum.' See text. Rainfall amounts are daily accumulations of 3 August 1998. Topography and gauge totals of the same period in millimeters are depicted (*lower right*). Levels of shading correspond to terrain height between 0-250 m (*white*), 250–500 m, 500–1000 m, 1000–2000 m, 2000–3000 m, and above 3000 m, respectively. The frame is a 140 km × 140 km area centred over the radar site Monte Lema (1625 m). The heights of three mountains are also indicated: Pizzo Campo Tencia (3072 m), Cima della Laurasca (2195 m), and Monte Tamaro (1962 m). This figure also appears as color Fig. 6 on page 317.

profile correction with two conventional products, the vertical maximum echo and the lowest visible echo. The two images 'lowest visible' and 'profile-corrected' are based on radar data that have been corrected for partial shielding. Both products require an estimate of the radar visibility map: first, to correct for partial shielding and second, to consider for each location only those measurements that are visible. The 'vertical maximum' is based on uncorrected measurements, and, the only information needed is the coordinates of the radar measurements. It is thus straightforward to implement. The differences among the three products become most apparent when looking at accumulations. Thus, Fig. 2.9 depicts daily accumulations of rainfall. The lower right of Fig. 2.9 shows the accumulated amounts estimated by gauges.

Vertical maximum echo: A comparison with the lower right plot of Fig. 2.9 shows that the 'vertical maximum echo' overestimates the amounts reported by the gauges in large areas. The reason for this is contamination by the bright band (most of the area with totals above 27 mm). In some places, the overestimations are also caused by residual, not eliminated clutter (small crosses corresponding to totals above 58 mm). In shielded regions, on the other hand, the amounts are too small because the measurements are made in the snow above the melting layer, where echoes are weak. The superposition of these errors leads to an error picture, which is typical for the 'vertical maximum' product: overestimation in the centre and underestimation in the periphery.

Lowest visible echo: This estimate suffers from several factors: (a) from the complex visibility map in a mountainous region; (b) from bright-band contamination leading to overestimation, when the lowest visible echo happens to be in the melting layer; (c) from underestimations in shielded regions; and (d) also from some residual ground clutter. Strong discontinuities perpendicular to the radar beam occur where either the lowest visible pixel rapidly jumps from low to high elevation angles because of sharp obstacles or where the estimated visibility maps do not correspond to the real radar visibility (white areas). The area in dark grey in the upper right part of the image is caused by the bright band. The few small crosses again mark the places of residual ground clutter.

Profile-corrected image: This product obviously shows the best agreement with the amounts reported by the gauges. It is the only product out of the three that is almost free of bright-band contamination and residual clutter. Underestimations in shielded regions have been reduced. The obvious artefacts of the other two products are almost completely eliminated.

2.4.2 How to Do It Operationally

Now, that we have seen the possible gain from profile correction, let us have a closer look at how it works. If we knew the local vertical profile of radar echo and the visibility at any place in the whole volume, profile correction would be straightforward. It is the high variability of the profile in space and time and the difficulty of estimating it in shielded regions that renders profile correction a difficult task.

The need for profile corrections has been stated in Joss and Waldvogel (1990). Since then, a variety of correction schemes has been proposed in the literature, but only few of them have actually been applied under operating conditions, and little attention has been paid to the particular situation in a highly mountainous region.

The reflectivity profile is strongly variable in time and space. Therefore, the decisive point of profile correction schemes is the estimation of a representative profile. Such an estimation procedure may use profiles measured in well-visible regions (Andrieu et al., 1995; Vignal et al., 1999), for example, in the vicinity of the radar (Germann and Joss, 2002; Joss and Lee, 1995), profile climatology (Joss and Pittini, 1991), or, in more general terms, the four-dimensional information of reflectivity aloft plus knowledge on its correlation with the profile (Smyth and Illingworth, 1998; Germann and Joss, 1999), as well as other data sources such as gauges (Collier et al., 1983) or models (Mittermaier and Illingworth, 2002; Kitchen et al., 1994). Koistinen (1991), Joss and Pittini (1991), Joss and Lee (1995), and Germann and Joss (2002) determine the profile from the mean of measured radar echoes as a function of height. Such mean profiles are, because of beam convolution, smoother than the true profile. Andrieu et al. (1995) and Vignal et al. (1999) propose an inversion technique to retrieve a profile that is not affected by beam convolution. Because of high processing costs and small improvements compared to the mean profile approach, the inversion technique is not suitable for operation in a mountainous region (Vignal et al., 2000; Germann and Joss, 2002).

Profile variation can be found down to the space–time scale of single instantaneous radar measurements. Such microscale profiles exhibit variations caused by single convective cells up- and downdrafts, drop sorting, fallstreaks, slanted cells and remaining signal fluctuations, etc. They look to some extent random, and their interpretation is difficult, unless detailed in situ measurements are available. Principally, profile corrections would achieve the best results when using profiles at the radar pixel scale. First attempts to take into account the spatial variability of the profile have been presented in Kitchen et al. (1994), Germann and Joss (1999), Vignal et al. (1999), and Sánchez-Diezma et al. (2001). The information available from volume scanning radars is so coarse in highly mountainous regions that the optimum *realistic* scale is much larger.

In an operational context, the following points are important: (a) The technique must continuously yield a well-defined profile. (b) Only small changes can occur between the profiles used to correct successive scans. This avoids discontinuities in the corrected images when animated. (c) The estimated profile must be representative of the true profile present in the region where it is applied. We consider the profile to be representative, if the error caused by the difference between the estimated and the true profile is small compared to the gain achieved by profile corrections.

These points were used when the profile correction algorithm was implemented at the Meteo-Swiss radar sites. The aim is to correct for the large errors related to the profile in an Alpine environment, that is, underestimation in shielded regions, and overestimation at close ranges in the bright band. The operational profile is determined from volumetric radar data integrated over a few hours in time and within a 70 km range of the radar antenna (meso-beta scale). In space, only the well visible part of the 70 km cylinder is used. Integration in time is done with a precipitation-volume-weighted exponentially decaying function. The weight of a volume scan continuously decreases with time. The time 'constant' of the weighting procedure is variable and depends on the amount of new information available. It is small, that is, fast adapting to the new profile, when there is precipitation in the whole region. It is infinite, and thus the profile is kept constant in the absence of precipitation. The same meso-beta profile is applied to the whole radar image. The profile-corrected image in Fig. 2.9 illustrates the improvement obtained by this correction procedure. More details can be found in Germann and Joss (2002).

2.4.3 Beam-Broadening and the Profile

The volume of a radar sample increases with the square of the range, often referred to as beam-broadening. It has no effect on the measurement, as long as the hydrometeors are uniformly distributed over the pulse volume, but this is rarely the case. Its effect becomes serious at far ranges, or in case of beam occultation, especially in the mountains, when the upper part of the beam is in the region of weak snow echoes or even above the echo top. Beam-broadening combined with the vertical decrease of the radar echo may lead to serious underestimations. This can be corrected if we know the profile and the radar visibility at the point of interest. Future work has to show to what extent this type of error can be reduced in an operational and mountainous context.

2.5 Adjustments by Gauges

Within the correction schemes discussed in the previous sections, the role of gauge adjustment is to make sure that radar estimates in well-visible regions are, on a long-term basis, unbiased against gauge measurements. Thus, gauge adjustment is the last step in a logic chain of orthogonal correction schemes, including hardware calibration using microwave equipment, clutter elimination, correction for visibility and profile errors, and, if relevant and feasible, also correction for attenuation.

2.5.1 Calibration, Then Adjustment

Calibration of the hardware (automatic?) as discussed in some detail by Gekat et al. (2003), this book, has to compensate for short-term variations of the radar equipment, that is, to provide stable conditions for precipitation measurements. It may also be needed after the replacement of components to ensure that no relative change in sensitivity has occurred. In other words, high *reproducibility* is the aim. To strive for high absolute accuracy and thus for absolute calibration is neither the required goal nor may it be possible, at least not at a good cost-benefit ratio. Absolute calibration would mean knowing the actual transmitted power in watts, the detailed antenna pattern, the noise power of the receiver in electrical units, and the true pointing angles. Relative accuracy defines our ability to reproduce in the future the values we measure today. If the hardware remains stable, we can adjust our equipment with experience gained in the domain of our variable of interest. This is the role of long-term gauge adjustments. In short, hardware calibration guarantees reproducibility, and gauge adjustments make the radar observations bias-free. Of course gauge measurements also have errors, which sets an upper limit to what can be achieved by adjustment with gauges.

Real-time adjustment using short periods, for instance, hourly amounts of gauges, is not recommended because of the poor representativeness of the point observations. We may make things worse. In the introduction of this chapter, we mentioned the 3 dB(R) standard deviation between daily accumulations of radar and gauge measurements. The value can be reduced by integrating over longer periods and by combining several gauges to calculate one adjustment factor. When combining, for example, 25 gauges to determine one adjustment factor, the value becomes up to 5 times smaller (square root of 25). But, on the other hand, the 3 dB(R) becomes larger when using time periods shorter than 24 hours. For hourly amounts, for instance, the value is up to 5 times larger (square root of 24). To get consistent improvement by adjustment, the standard deviation from the uncertainty discussed above must be smaller than the errors we want to correct for. This is certainly not the case when using hourly amounts and only a few gauges.

The lack of representativeness for operational short-term adjustment, however, does not exclude the use of a number of close-by gauges for correcting radar data in a case study. But the products operationally distributed on the network to the many users are not adjusted from day to day. Operational adjustment is done on a long-term basis. The resulting continuity creates a stable environment, which is known to all users. Operational users are mainly people from meteorological, hydrological, and geological offices. It is easier to relate on a stable environment, than on one which is constantly being adapted to measurements of questionable short-term representativeness.

2.5.2 Weighted Multiple Regression

A more sophisticated way to combine radar and gauge data is the weighted multiple regression analysis (WMR) developed in cooperation with people from Italy and the Czech and Slovak Republics; see, for instance, Gabella et al. (2001), Kracmar et al. (1999), and Boscacci (1999). It can be used either in addition or as an alternative to the correction procedures discussed in the previous sections. When used in addition to other corrections, the objective is to correct for residual errors and to check whether there remain significant influences after correction. Note that a perfect correction procedure will lead to a lack of correlation between the dependent and the independent variables of the WMR. Replacing the chain of correction procedures by the WMR may be a cheap way to correct for several types of error in one step, especially, in case no volume data are available, for instance, because the antenna scans only one elevation.

A successful WMR was found by taking, as the dependent variable, the logarithm of the correction factor of each gauge. The correction factor is the ratio between gauge and radar totals accumulated over a given period. This period must be long enough to eliminate variations on small time-scales not explained by the regression equation. The dependent and the independent variables are related by power laws and exponential functions. Therefore, the logarithm creates the basis for a linear regression. For the three independent variables, we used the logarithm of the distance between radar and gauge, the height above sea level of the gauge, and the height above sea level of the lowest visible radar measurement over the gauge.

The idea is to separate the influence of calibration, beam-broadening, visibility and orography by calculating the four regression coefficients that lead to best agreement between a radar image, such as the vertical maximum echo, and gauge observations. The first regression coefficient is a constant offset in decibels taking together calibration errors and other overall multiplicative biases. The remaining three coefficients are for the three independent variables, respectively. Distance between radar and gauge is transformed to a logarithm, whereas the other two independent variables are not. This is done because range effects are expected to follow a power law, and the decrease of radar echoes with height is approximately exponential.

Weights are introduced in the regression analysis to make the results unbiased in terms of linear rainfall amounts.

Once the regression coefficients have been determined by using a set of gauges, we can calculate the complete map of correction factors and apply it to the whole radar image. Of course, the map of correction factors depends on the weather situation, in particular, on the profile. And the correction factors must be recalculated for best results, if the weather changes.

A thorough study of the performance of the WMR in heavy precipitation in the southern Alps is presented in Gabella et al. (2000) and Gabella et al. (2001). A flood event in Piedmont (northern Italy) in November 1994 is selected, and measurements of two C-band radars 140 km apart and about 60 gauges are examined. Both validations based on radar–gauge and radar–radar comparisons show a significant reduction of errors inherent in radar measurements in complex terrain. For the radar–gauge analysis, calibration of the regression coefficients and validation is done with two independent subsets of data. The amount measured by the radar turned out to be a good criterion for the weights in the regression analysis. In addition to reducing errors, the map of correction factors obtained from WMR also provides a measure of the quality of radar measurements; pixels in shielded regions or at far ranges are associated with large correction factors (Gabella et al., 2001).

2.6 What Next?

Before ending this chapter, we would like to emphasize that a weather radar is a unique solution to provide a qualitative overview of precipitation at a high resolution in time (5 min), in space (1 km within 150 km), and in intensity (a factor of two). Furthermore, the information is obtained in three dimensions from a single instrument. This opens the way for a variety of nowcasting applications and the initialisation and verification of numerical weather prediction models.

We end this chapter with an outline of three future tracks that may be followed to further improve operational measurements of precipitation by radar. The three tracks are (a) make use of the idea of the minimum variance estimator, (b) toward a set of radar products providing a more complete description of precipitation, and (c) determine the error structure of radar precipitation maps.

Minimum variance estimator: The minimum variance estimator is a common statistical technique for making the best possible estimate of a quantity by combining different data sources by means of a weighted linear average. The weights are chosen on the basis of the covariance matrices in a way that minimises the final error variance. Variational data assimilation of numerical models is based on this principle. There, the background field obtained from a previous model run is pushed in the direction of the observations until the analysis error variance reaches (under certain conditions) a minimum. The minimum variance estimator may be a solution to further improve the radar estimates of precipitation rates in poorly visible regions such as over the mountains or at far ranges. Possible information sources for a poorly visible location are (a) nearest neighbours at the same height (horizontal extrapolation); (b) measurements from aloft (vertical extrapolation); (c) Lagrangian advection of observations made in well-visible regions in the past few tens of minutes (Germann and Zawadzki, 2002; Mecklenburg et al., 2000; and Li et al., 1995); (d) nearby gauge measurements, and finally, if all else fails, also (e) climatology. It depends on the particular situation which of these data sources best correlate with the unknown rate at the poorly visible location. The minimum variance estimator yields the best estimate of the unknown rate by combining all those information sources that show significant correlation.

Toward a set of radar products providing a more complete description of precipitation: The mountains, earth curvature, and ground echoes inhibit a direct view of precipitation close to the ground. Then, what we usually do is to extrapolate measurements from neighbouring regions. For decades, we were looking for the extrapolation procedure that yields the best possible estimate of the *expected value* of the ground level precipitation rate. The expected value of the precipitation rate at the ground is undoubtedly an important quantity. But it is not the only one. And, for some applications, it is not at all the most important one. To estimate the risk of a landslide, for instance, we need to know the probability of encountering, within a given region and time, a precipitation amount larger than a certain threshold, that is, we need probabilistic information. Or to estimate the spatial representativeness of a point observation, we would need a parameter that describes the spatial continuity of precipitation, that is, the variogram (Germann and Joss, 2001). In both examples the expected value is of little use. So, the idea is to define new products providing information that is complementary to the map of the expected ground level precipitation rate. Here, the radar with its high resolution in space and time is a unique source of information, which has not yet been exhausted. In severely shielded regions, where only measurements from aloft are available, it may actually be easier to provide probabilistic information than an estimate of the mean.

Error structure: Although a lot of attempts have been made to quantify the errors involved in precipitation measurements by radar, we are still far away from having a complete error map providing, for each pixel in time and space and for each type of product, the corresponding uncertainty. For many practical applications, a measurement without an estimate of its uncertainty is of little use. Think of a hydrologist who is responsible for regulating a reservoir high up in the mountains in the case of a flood alarm. Is it a disastrous precipitation that is going to cause serious damage if the dam is not immediately opened, or is it just a strong event and the dam need not be opened? To make such a decision, probabilistic information is needed together with a safe estimate of the uncertainties of the observations.

References

1. Andrieu, H., G. Delrieu, and J. D. Creutin, 1995: Identification of vertical profiles of radar reflectivity for hydrological applications using an inverse method. Part II: Sensitivity Analysis and Case Study. J. Appl. Meteorol., **34**, 240–259.
2. Boscacci, M., 1999: Quality checks for elaborate radar measures. *COST-75 Final International Seminar on Advanced Weather Radar Systems*, edited by C. G. Collier, EUR 18567 EN, Comm. of the European Commun., Locarno, Switzerland, pp. 280–287.
3. Bougeault, P., et al., 2001: The MAP Special Observing Period. Bull. Am. Meteorol. Soc., **82**, 433–462.
4. Collier, C. G., P. R. Larke, and B. R. May, 1983: A weather radar correction procedure for real-time estimation of surface rainfall. Q. J. R. Meteorol. Soc., **109**, 589–608.
5. Frei, C., and C. Schär, 1998: A precipitation climatology of the Alps from high-resolution rain-gauge observations. Int. J. Climatology, **18**, 873–900.
6. Gabella, M., and G. Perona, 1998: Simulation of the orographic influence on weather radar using a geometric-optics approach. J. Atmos. Oceanic Technol., **15**, 1486–1495.
7. Gabella, M., J. Joss, and G. Perona, 2000: Optimizing quantitative precipitation estimates using a noncoherent and a coherent radar operating on the same area. J. Geophys. Res., **105**-D2, 2237–2245.
8. Gabella, M., J. Joss, G. Perona, and G. Galli, 2001: Accuracy of rainfall estimates by two radars in the same Alpine environment using gage adjustment. J. Geophys. Res., **106**-D6, 5139–5150.
9. Germann, U., 1999: Radome attenuation – a serious limiting factor for quantitative radar measurements? Meteor. Z., **8**, 85–90.

10. Germann, U., 2000: *Spatial Continuity of Precipitation, Profiles of Radar Reflectivity and Precipitation Measurements in the Alps*, PhD Thesis, Swiss Federal Institute of Technology (ETH), online at http://e-collection.ethbib.ethz.ch, 120 pp.
11. Germann, U., and J. Joss, 2001: Variograms of radar reflectivity to describe the spatial continuity of Alpine precipitation. J. Appl. Meteorol., **40**, 1042–1059.
12. Germann, U., and J. Joss, 2002: Mesobeta profiles to extrapolate radar precipitation measurements above the Alps to the ground level. J. Appl. Meteorol., **41**, 542–557.
13. Germann, U., and J. Joss, 1999: Meso-gamma reflectivity profiles correlate with horizontal features, a help to improve precipitation estimates? *Proc. 29th Conf. Radar Meteorol.*, AMS, pp. 848–851.
14. Germann, U., and I. Zawadzki, 2002: Scale-dependence of the predictability of precipitation from continental radar images. Part I: Description of the methodology. Mon. Weather Rev., **130**, 2859–2873.
15. Joss, J., et al., 1998: *Operational Use of Radar for Precipitation Measurements in Switzerland*, vdf Hochschulverlag AG, ETH Zürich, online at http://www.meteoswiss.ch, 108 pp.
16. Joss, J., and R. Lee, 1995: The application of radar-gauge comparisons to operational precipitation profile corrections. J. Appl. Meteorol., **34**, 2612–2630.
17. Joss, J., and A. Pittini, 1991: Real-time estimation of the vertical profile of radar reflectivity to improve the measurement of precipitation in an Alpine region. Meteorol. Atmos. Phys., **47**, 61–72.
18. Joss, J., and A. Waldvogel, 1990: Precipitation measurements and hydrology. *Radar in Meteorology*, edited by D. Atlas, Am. Meteorol. Soc., pp. 577–597.
19. Kitchen, M., R. Brown, and A. G. Davies, 1994: Real-time correction of weather radar data for the effects of bright band, range and orographic growth in widespread precipitation. Q. J. R. Meteorol. Soc., **120**, 1231–1254.
20. Koistinen, J., 1991: Operational correction of radar rainfall errors due to the vertical reflectivity profile. *Proc. 25th Conf. Radar Meteorol.*, AMS, pp. 91–94.
21. Kracmar, J., J. Joss, P. Novak, P. Havranek, and M. Salek, 1999: First steps towards quantitative usage of data from Czech weather radar network. *COST-75 Final International Seminar on Advanced Weather Radar Systems*, edited by C. G. Collier, EUR 18567 EN, Comm. of the European Commun., Locarno, Switzerland, pp. 91–101.
22. Lee, R., G. Della Bruna, and J. Joss, 1995: Intensity of ground clutter and of echoes of anomalous propagation and its elimination. *Proc. 27th Conf. Radar Meteorol.*, AMS, pp. 651–652.
23. Li, L., W. Schmid, and J. Joss, 1995: Nowcasting of motion and growth of precipitation with radar over a complex orography. J. Appl. Meteorol., **34**, 1286–1300.
24. Mecklenburg, S., J. Joss, and W. Schmid, 2000: Improving the nowcasting of precipitation in an Alpine region with an enhanced radar echo tracking algorithm. J. Hydrol., **239**, 46–68.
25. Mittermaier, M. P., and A. J. Illingworth, 2002: Comparison of model-derived and radar-observed freezing level heights: Implications for vertical reflectivity profile correction schemes. Q. J. R. Meteorol. Soc., in press.
26. Pellarin, T., G. Delrieu, J. D. Creutin, and H. Andrieu, 2000: Hydrologic visibility of weather radars operating in high-mountainous regions: A case study for the Toce Catchment (Italy) during the Mesoscale Alpine Programme. Phys. Chem. Earth (B), **25**, 953–957.

27. Sánchez-Diezma, R., D. Sempere-Torres, J. D. Creutin, I. Zawadzki, and G. Delrieu, 2001: Factors affecting the precision of radar measurement of rain. An assessment from an hydrological perspective. Proc. 30th Conf. Radar Meteorol., AMS, pp. 573–575.
28. Smyth, T. J., and A. J. Illingworth, 1998: Radar estimates of rainfall rates at the ground in bright band and non-bright band events. Q. J. R. Meteorol. Soc., **124**, 2417–2434.
29. Vignal, B., H. Andrieu, and J. D. Creutin, 1999: Identification of vertical profiles of reflectivity from volume scan radar data. J. Appl. Meteorol., **38**, 1214–1228.
30. Vignal, B., G. Galli, J. Joss, and U. Germann, 2000: Three methods to determine profiles of reflectivity from volumetric radar data to correct precipitation estimates. J. Appl. Meteorol., **39**, 1715–1726.

3 Operational Measurement of Precipitation in Cold Climates

Jarmo Koistinen,[1] Daniel B. Michelson,[2] Harri Hohti,[1] and Markus Peura[1]

[1] Finnish Meteorological Insitute (FMI), Helsinki, Finland
[2] Swedish Meteorological and Hydrological Institute (SMHI), Norrköping, Sweden

3.1 Introduction

Over the last 50 years, severe weather such as thunderstorms, squall lines, tornadoes, hurricanes, and extreme precipitation events with ensuing flash floods have dominated both scientific and operational radar meteorology (Atlas, 1990, Collier, 2001). This is natural, as one of the main benefits of weather radars is very dense sampling of precipitating systems in time and space, facilitating real-time warning and nowcasting of mesoscale severe weather which may have huge socioecomonic impacts. Also, an implication has been that some commercial radar manufacturers have paid little attention to the sensitivity of weather radar. Maximal sensitivity of a radar system is required for making snowfall measurements. Even greater sensitivity of operational radars is needed for the production of Doppler winds from clear air. The important topic of wind field estimation from Doppler measurements relates to Gekat et al. (2003), Meischner et al. (2003) and Macpherson (2003), all this book. The major proportion of the frequent boundary layer echoes found in summer, even in the north of Europe, originates from insects. For example, classification of 240 000 vertical profiles of reflectivity (VPR) from a one-year-long period in Finland revealed that 40% of all VPRs originated from clear air echoes reaching the ground, 20% from overhanging precipitation, i.e. ice crystal clouds or snowfall layers aloft, and only 40% involved precipitation reaching the ground level (Pohjola and Koistinen, 2002). The reflectivity of both nonprecipitating echo classes was typically between -20 and 5 dBZ. Both Canadian and Nordic radar networks (NORDRAD) mostly operate sensitive C-band systems. The operational Nordic radars are "standard" systems with sensitivities around -110 dBm, beamwidths of around $0.9°$ and pulse lengths from 0.5–2 μs. In Canada, the sensitivity requirement to detect snowfall and clear air echoes has led to the specification of a multi-pulse length capability (0.8–5 μs) and of a narrow antenna beamwidth ($0.65°$) (Joe and Lapczak, 2002).

Since the operational launching of the international NORDRAD network (Gekat et al., 2003, this book) in 1993, the heterogeneity of the composite products has deserved special attention. The initiating reason was the observation of large reflectivity differences between radars from different manufacturers. The Nordic experience proves that the real system calibration and signal processing in a network, even when the radars involved are "similar," can significantly affect the accuracy of network products. Continuous system monitoring and evaluation as described in detail by Gekat et al. (2003), this book, is required. This is probably an underestimated issue in many operational radar networks. Although it is important in any network, power level or elevation angle calibration errors are more clearly discerned in cold climate products than in products generated from severe convection in a warm climate. The reason for this is that cold precipitation often has strong vertical reflectivity gradients at lower heights and that reflectivity is often relatively homogeneous in large horizontal areas. Such precipitation structures easily introduce a visible reflectivity step where data from two neighbouring radars meet in a composite, for example, when the actual elevation angles differ by only 0.1° and power calibrations differ by 1 dB.

A problem in weather radar measurements is the occurrence of nonmeteorological echoes. It is often thought that proper Doppler filtering to full measurement range will remove all clutter. This is mostly valid for ground clutter but not for sea clutter, the sun, external emitters and military jamming, ship clutter and birds. Although clutter is a global problem in radar meteorology, large cold lakes and sea areas are favourable for the occurence of anomalous propagation conditions, which enhance clutter and which are a typical cold climate feature. Anomalous propagation and methods to eliminate nonmeteorological echoes are presented in Sect. 3.2.

The measurement of precipitation in a cold climate is not theoretically different compared to making measurements in warmer conditions. However, the operational aspect involves making measurements at long ranges, typically up to 240–250 km. Much confusion regarding the accuracy of precipitation measurements originates from the fact that range has been neglected as an influencing factor. Range increases the probability of severe attenuation. Nevertheless, the main issue is not an increased error in the reflectivity measured at distant ranges but rather the sampling difference between the actual precipitation at the surface and the radar estimate aloft. This fact was not widely accepted during the first 40 years of radar meteorology; for example, Wilson and Brandes (1979) concluded that the main reason for inaccuracies in radar-based precipitation measurements was the use of nonrepresentative Z–R relations. Back then, "Z–R factories" produced a plethora of "local optimal relations" (Battan, 1973), which couldn't solve the radar–gauge bias at longer ranges. It is not surprising that serious consideration of the underestimating sampling bias was established in environments which enhance the sampling difference, i.e. the relatively cold climates of Canada (Zawadzki, 1984), Finland (Koistinen and Puhakka, 1986) and in the temperate climate of Switzerland with enhanced shielding due to mountains (Joss and Waldvogel, 1990). Section 3.4 presents fresh Nordic experience in dealing with the sampling problem, and Sect. 3.5 introduces a working nowcasting tool for precipitation, applied operationally also in snowfall.

The "Z–R factories" have also created useful relations for snowfall measurements. The standard operational practice in Europe is for the radar system to measure the water equivalent reflectivity factor (Z_e) which is converted to rainfall intensity by applying a Z–R relation for rain and the radar equation for water throughout the year. Surprisingly often, if the hydrometeor water phase is considered, the snowfall intensity (S) is then calculated applying a Z–S relation (Sekhon and Srivastava, 1970) without adjusting Z_e to Z. This inaccuracy in estimating solid precipitation is further amplified by the fact that very little attention is usually paid to the actual three-dimensional (or even to the two-dimensional) distribution of the water phase of hydrometeors in the volumetric radar data. In Finland, for example, the major part of the scanned polar volume is always, even in July, above the bright band. The Nordic experience in radar based snowfall measurements is further discussed in Sect. 3.3.

3.2 Problems Due to Ducting over Cold Surfaces

Propagation of electromagnetic waves was comprehensively addressed several decades ago (Kerr, 1951) and has also been presented in a weather radar context (Watson, 1996; Alberoni et al., 2001; Steiner and Smith, 2002). Figure 3.1 illustrates normal propagation and ducts giving rise to echoes from the earth's surface. This entrapment and propagation of a small amount of radiation, offset from the main beam axis and at small grazing angles in such ducts, and even in less severe superrefractive conditions, is referred to as anomalous (AP or anaprop), and the echoes are referred to as AP echoes. Such echoes are strong and highly variable on small spatial scales over land (Alberoni et al., 2001). Over the sea, where they are referred to as sea clutter (Collier, 1998), they are more homogeneous and generally weaker in strength (Alberoni et al., 2001). Other types of nonprecipitation echoes originate from insects, birds, aircraft, chaff, military jamming, and the sun. All types of nonprecipitation echoes are referred to as spurious.

In the NORDRAD network, the radar systems with good coherency and effective Doppler filtering to full measurement range do not suffer from ground clutter, even in common cases when unfiltered reflectivity is severely contaminated by clutter. However, systems with limited Doppler range suffer from ground clutter, and even the coastal Doppler systems suffer from sea clutter contamination. A large statistical analysis of the occurrence of clutter has not been performed in the Nordic countries. Nevertheless, the vast operational experience gained by human eyes viewing image products confirms that ducting over cold and moist surfaces is the main reason for outstanding clutter episodes. In the NORDRAD radars, the lowest elevation angle in most systems is 0.4–0.5°. Thus, the major part of the measured clutter is introduced by ducting of a small fraction of the lower beam. One might expect that the duct is generated in common temperature inversions above snow and ice covered terrain in winter. The magnitude of a surface inversion in Lapland is often 10–20°C/300 m and in extreme cases 30°C/300 m. Despite such inversions, strong winter ground clutter due to AP is relatively rare in radar measurements. The first reason is that radar antennas are often located at least 30–50 m above the average ground level and thus above the strongest lower part of the surface inversion. The second reason is that strong temperature inversions occur in very cold air masses with surface temperatures below −20°C. Obviously the temperature gradient alone is not

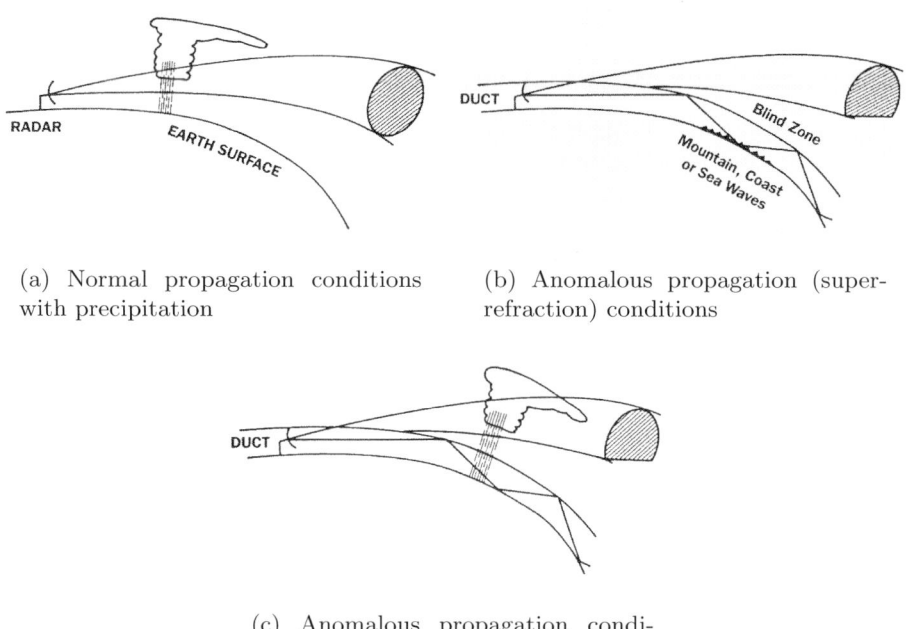

(a) Normal propagation conditions with precipitation

(b) Anomalous propagation (super-refraction) conditions

(c) Anomalous propagation conditions with precipitation

Fig. 3.1. Normal propagation conditions with precipitation and anomalous (super-refraction) propagation conditions giving rise to radar echoes from the earth's surface with and without precipitation. From Alberoni et al. (2001)

sufficient for generating AP. The absolute moisture content is very low in the inversion layer due to the low temperature, and the vertical moisture gradient is small since the air above the inversion is dryer in relative humidity but not much different in absolute terms.

The operational experience in the Nordic countries shows that the major part of clutter generated by anomalous propagation in ducting conditions occurs in spring and summer (April – August). The dominating favourable factor is a warm air mass above the cold Baltic Sea, generating a cool, moist and shallow ducting layer of air at the surface. During such periods, nocturnal moist layers over land areas also introduce AP relatively often, although the temperature inversion is much weaker than in winter. During favourable periods, very strong clutter is typical (in unfiltered reflectivity) along the coasts on the opposite side of the sea seen from the antenna during both day and night. In operational use, Doppler filtering mitigates almost all AP ground clutter. The main remaining problem in Doppler radars is sea clutter and ships (Figs. 3.2–3.7). They can't be filtered with the existing signal processors since reflectivity, Doppler spectrum width and average Doppler velocity are similar to those from precipitating echoes (Fig. 3.2a–c). One might consider ships as an irrelevant source of error in operational radar-based services. Unfortunately however, reflectivities from huge

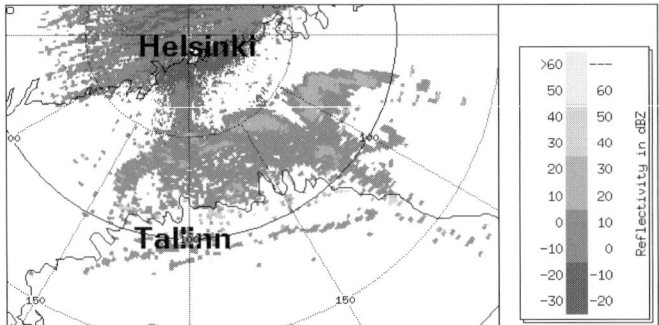

(a) Radar reflectivity factor in the 500 m pseudo-CAPPI product. Sea clutter is located in the Gulf of Finland, weak echoes in continental Finland are from insects and a few stronger spots in Estonia represent remaining coastal clutter, which has passed the Doppler filtering and quality thresholding in signal processing

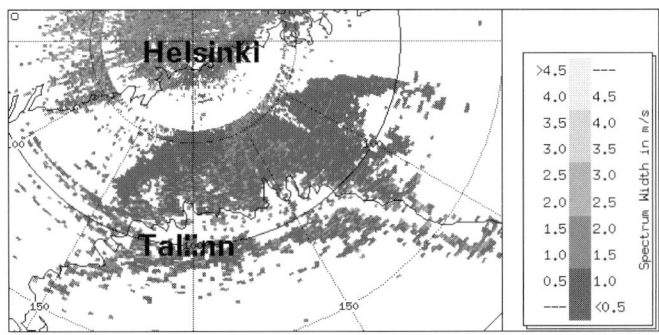

(b) Doppler spectrum width (m/s)

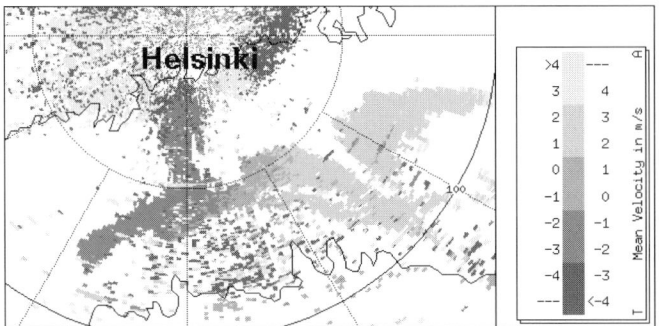

(c) Doppler velocity (m/s), PRF=570 Hz, at the lowest elevation PPI (0.4°). Dots off the Estonian coast represent ships and large flocks of migrating birds

Fig. 3.2. Examples of sea and ship clutter. From Vantaa on May 27, 2000 at 13:15 UTC. See also color Fig. 7 on page 318

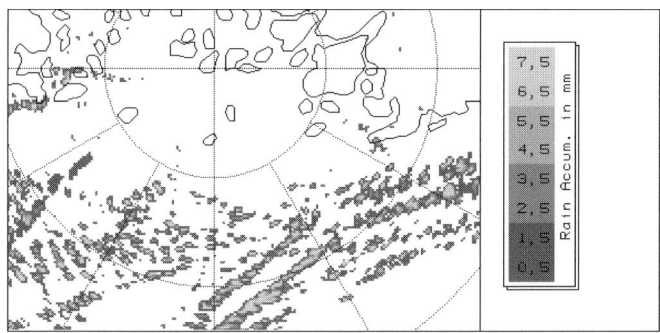

Fig. 3.3. Accumulated 12 hourly "precipitation" (mm) from ships, as seen from Korpo on June 21, 1997 at 06 UTC. Range ring interval is 50 km. See also color Fig. 8 on page 319

and frequently sailing ferries in the Baltic Sea easily reach 50–60 dBZ. As a consequence, image products and especially automatic shower warnings introduce misleading information to meteorologically inexperienced customers (Fig. 3.3). As long as operational Doppler signal processing or polarimetric measurements are incapable of diagnosing sea clutter and other nonmeteorological echoes, we have to use postprocessing methods on radar data together with multisource meteorological data for echo type classification. Such tools are presented in the following Sects. 3.2.1–3.2.3.

3.2.1 Image Analysis Methods

Many different methods have been formulated over the years to identify and remove spurious echoes. Radar-based methods using noncoherent radars have been developed based on combined signal processing of raw pulse data and image processing techniques which appear to perform well. One example is reported from the Netherlands (Wessels and Beekhuis, 1992). Customized signal processing is becoming easier with the introduction of new digital technology, and with it the ability to analyse the statistical properties of samples of pulses. This ability has been successfully utilized in both France and the UK (Sugier et al., 2002), using the signal decorrelation time as a means of distinguishing ground clutter and AP from rain. Image processing based quality control methods, many examples of which have been recently compiled (Steiner and Smith, 2002), and elaborate quality control systems such as those developed in Switzerland (Joss and Lee, 1995), the UK (Pamment and Conway, 1998; Harrison et al., 2000), and the USA (Fulton et al., 1998) all demonstrate the difficulties in identifying and suppressing spurious echoes while retaining echoes from true precipitation, especially where AP echoes are embedded in precipitation areas. Where Doppler signal processing is available, it has been demonstrated as being an effective means of removing clutter and AP echoes which are static in space, ie. from land (Koistinen, 1997). Sea clutter is much more difficult to remove using traditional Doppler techniques

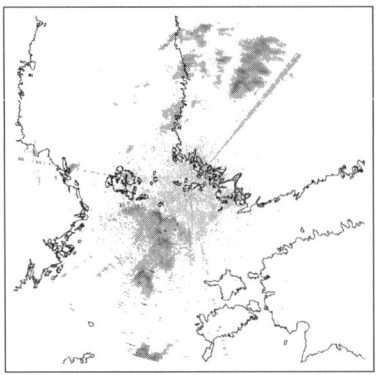 

(a) Cartesian image; inverted dBZ grey-scale

(b) Original data in polar coordinate system plotted as a b-scan

Fig. 3.4. A lowest elevation PPI image contaminated by AP: sea clutter (cloud-like speck in the southwest), emitter lines (south-southeast and west-northwest), ships (distinct specks of high intensity). The only actual precipitation is in the north-northeast. The image also contains low-intensity speckle noise, sun (the continuous line in the northeast) and insects (near the radar). Radar Korpo at 01:30 UTC on July 9, 2002

because the echoes are generated from sea waves which have true velocities and are therefore indistinguishable from precipitation (Fig. 3.2).

Other types of nonprecipitating echoes such as insects, birds, aircraft, chaff, military jamming, and the sun (Fig. 3.4) cannot be easily detected by standard radar parameters either. Instead, one may try image analysis methods concentrating on purely visual features such as speck size, elongation, orientation and grouping.

A recently developed scheme applies the set of detection algorithms listed in Table 3.1. Each detector concentrates on specific visual features in the image. Processing is carried out in polar space which is an image array of 500 bins by 360 rays in the case of the Finnish radars (Fig. 3.4b). In the scheme, anomaly detection and filtering tasks are carried out separately. In the detection stage, each detector outputs a response image which presents the spatial probabilities

Table 3.1. Detectors and their primary targets

Detector	Target
BIOMET	birds and insects near the radar
SPECK	noise, distinct specks
EMITTER	line segments
SUN	long line segments
SHIP	ships (and aircraft)
DOPPLER	non-continuous Doppler data

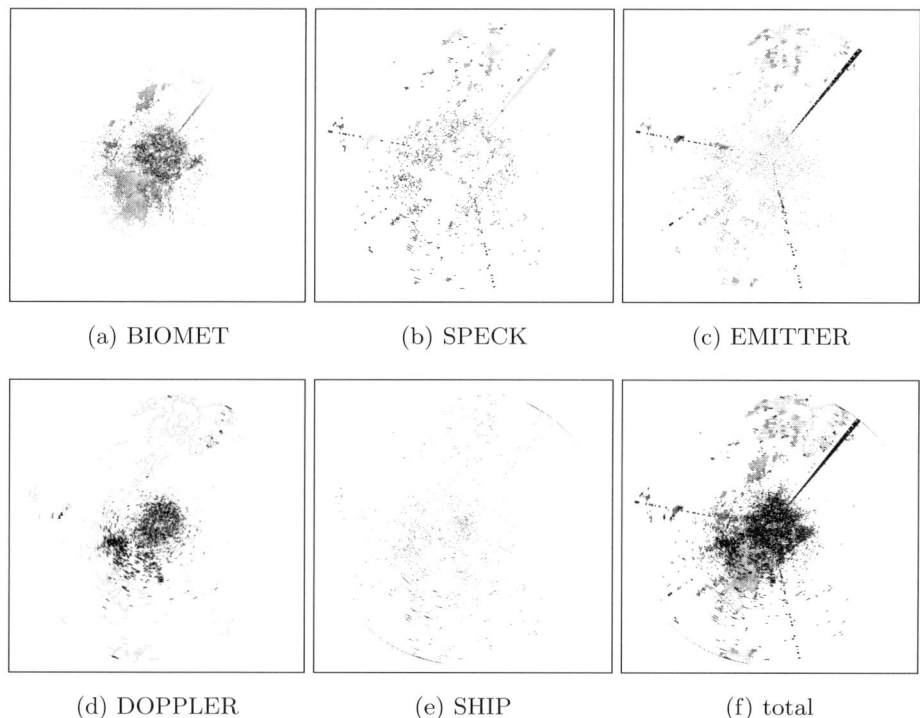

Fig. 3.5. Response images of anomaly detectors in Fig. 3.4. Dark areas indicate high probability of the given anomaly's occurence

of an anomaly class. Here, "probability" means a degree of confidence computed from detection sensitivity parameters set by an expert. More specifically, each such parameter is essentially a 50–50% threshold of anomaly versus precipitation; each detector typically needs a couple of such thresholds for reflectivities or visual features. Continuity is achieved by using fuzzy logic, i.e. continuous probability functions for classification instead of strict reflectivity thresholds (Sonka et al., 1993).

Response images of the detectors for the case of Fig. 3.4 are shown in Fig. 3.5a–e. Finally, the total probability of spurious echo occurence is obtained by taking the maxima over the specific response images (Fig. 3.5f).

The continuous-valued response image, having equal dimensions with the original data, allows cutting off anomalies with a tunable threshold. This is useful because the filtering always involves the compromise of removing anomalies and true precipitation, and applications have varying aspects for this compromise. For example, generation of warnings often requires conservative anomaly removal, whereas strongly filtered (yet sparse) precipitation data are preferred when computing motion vectors. The effect of changing the threshold is illustrated in Fig. 3.6.

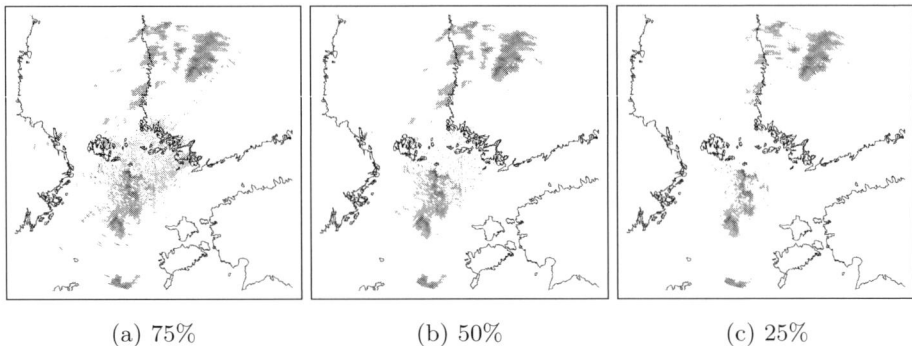

(a) 75%   (b) 50%   (c) 25%

Fig. 3.6. Filtering anomalies with BIOMET, SPECK, EMITTER, SHIP and DOPPLER detection algorithms using different thresholds of confidence in Fig. 3.4. Sea clutter in the southwest remains

A detailed discussion of the following detection algorithms is given in Peura (2002). The BIOMET algorithm detects birds and insects due to their proximity to the radar and their low reflectivities; the response image is essentially a fuzzy multiplication of the inverse distance and inverse reflectivity of each bin. The SPECK detector simply computes the area of each speck. In the response image, each speck (connected pixel segment) is marked with a value proportional to its area. EMITTER is designed for recognising beam-parallel lines of anomalous echoes originating from other nearby devices operating on the same frequency. The algorithm searches first for short, thin line segments and uses them as evidence in deducing the most contaminated directions. Then, the processing is repeated in those directions with increased level of sensitivity. SUN works similarly, but the contaminated direction is unique and can be precomputed. The SHIP detector is more complicated. Basically, it searches for small but intense peaks. Occasional side lobes reflected off ships, which are seen as associated echoes aligned perpendicularly to the ray, are detected as well. A DOPPLER detector observes discontinuities of the Doppler radial velocity field applying a small (3×3 bin) kernel. Such discontinuities are typically caused by birds. The discontinuities caused by transitions from data to areas void of data, or crossing the unambiguous velocity region, are not taken into account. The former is arranged by discarding the "no data" bins in the kernel. The latter effect is mitigated by computing the variance of the velocity values after transforming them on a unit sphere; hence the samples become treated as a set of points in two-dimensional space.

It should be noted that the detection stage is computationally much more demanding than the final filtering (thresholding) stage; one may consider designing end products that can be filtered online, interactively, applying a precomputed anomaly probability image. One may also design the detection scheme such that the contributions of each detector can be tuned invidually. In addition, processing detector responses separately allows for anomaly detection products which are of interest to some users (such as detecting birds for aviation).

3.2.2 Three-Dimensional Correction Methods

In recent years, with improved computer performance, filtering methods employing original polar volume data have been tested both in Europe (Alberoni et al., 2001) and in the USA (Steiner and Smith, 2002), which point toward limitations similar to those experienced with two-dimensional Cartesian data, i.e. that image/volume analysis techniques are often sensitive to individual cases and must be fine-tuned in order to gain maximum performance.

The European study (Alberoni et al., 2001) presented two methods, one applied to Italian and the other applied to Swedish data. Both methods are based on analysis of the vertical variability of radar echoes above a given point. The Italian method utilizes reflectivities from two scans V_1 and V_2 which are made at 0.5° and 1.4° scans, respectively. Using these, a rudimentary vertical reflectivity profile is characterized through the variable RD $= V_1 - V_2$ (dBZ). AP is diagnosed according to

$$\text{AP} = \begin{cases} \text{RD} > T_1 \\ \text{RD} > 0 \text{ and } V_2 < T_2 \end{cases} \tag{3.1}$$

where T_1 and T_2 are empirical thresholds. Although this robust technique is demonstrated as performing well in case studies, the frequency distributions for rain and for AP echoes overlap, which raises doubts about the method's applicability in an automatic, real-time environment.

The Swedish method utilizes reflectivities from many overlapping scans and introduces the mean square deviation (MSD) variable, calculated from low to high scans as

$$\text{MSD} = \frac{1}{n} \sum d_i^2, \tag{3.2}$$

where

$$d_i = V_i - V_{i+1} \tag{3.3}$$

and V_i is the reflectivity value in the scan with index i. Studies of the statistical properties of MSD in various conditions revealed that precipitation generally is characterized by smooth and unbroken reflectivity profiles and MSD ranging from 0 to 16 dBZ^2. AP echoes over land generally contain uneven and/or broken profiles and MSD either above 16 or below 0 dBZ^2. Sea clutter is characterized by broken profiles and MSD values under 0 dBZ^2. In order to distinguish among these characteristics, two other measures were formulated: the Frequency of unbroken profiles,

$$\text{FUP} = 100 \frac{\sum \text{up}}{\sum (\text{up} + \text{bp})} \tag{3.4}$$

and the frequency of accepted MSD,

$$\text{FAMSD} = 100 \frac{\sum \text{gp}}{\sum (\text{up} + \text{bp})} \tag{3.5}$$

both of which are calculated using a polar kernel ($n \times n$ bins) surrounding each polar bin. In Eqs. (3.4)–(3.5), up is the number of unbroken profiles, bp the number of broken profiles, and gp the number of profiles with an MSD value within a given range, all of which are calculated within the polar kernel. The Swedish method is more complex than the Italian one, and it contains a relatively larger number of potential thresholds for its parameters. These thresholds were found to require adjustment, not just for different cases, but even when analyzing a time series of polar volumes from a single case. The case shown in Fig. 3.7 shows that the method succeeds in identifying and removing virtually all AP echoes but that it also impacts on real rain. In comparison with the other cases presented (Alberoni et al., 2001), this case shows the method at its highest performance. Less AP echo removal and/or more real precipitation removal are severe limitations if the objective is to have a method which can be used automatically and in realtime.

The results from such studies show that AP echoes can be identified and removed using the three-dimensional information available in polar volume data but that such methods are subject to the same types of difficulties as with other methods, i.e. tuning them to gain optimal effect.

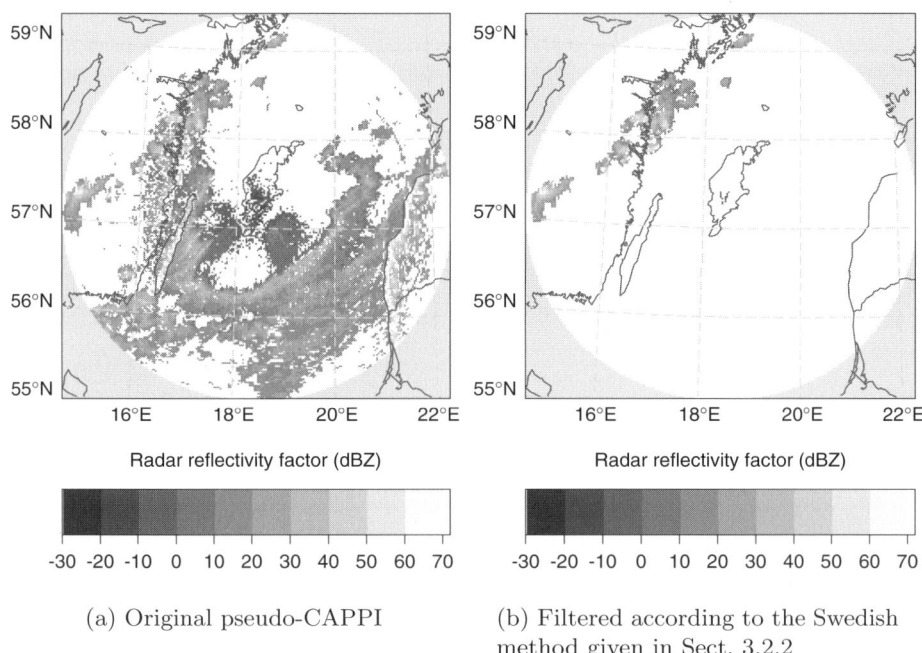

(a) Original pseudo-CAPPI

(b) Filtered according to the Swedish method given in Sect. 3.2.2

Fig. 3.7. Severe AP conditions as seen from Hemse on August 25, 1997 at 14:09 UTC. A line of convective cells stretches from west to north over the Swedish mainland. Strong AP echoes from land are seen covering the Swedish island of Öland and the Latvian and Lithuanian coasts. Severe sea clutter is organized as more or less concentric halfrings at different ranges, with the closest ring being the most discernible. See also color Fig. 9 on page 319

3.2.3 Multisource Methods

An interesting alternative to applying 2-D and 3-D image analysis techniques to radar data alone is to try to find data from other sources which may help to indirectly identify conditions where AP, and hence ground and sea clutter, may exist. An operational meteorological service commonly has a multitude of observed, modelled and analyzed meteorological variables, and these have been integrated in ways designed to identify and remove nonprecipitation echoes in radar data, for example, in the UK (Pamment and Conway, 1998).

SMHI's Mesoscale Analysis (MESAN) system (Häggmark et al., 2000) is a system designed to collect information from multiple sources and operationally produce hourly gridded meteorological variables for nowcasting purposes. The method used to generate these analyses is statistical (or optimal) interpolation (Daley, 1991), where first-guess fields taken from forecasts produced by the High Resolution Limited Area Model (HIRLAM) numerical weather prediction (NWP) model (Källén, 1996) are constrained by observations from a multitude of observation systems. In doing so, climatologically tuned and spatially variable structure functions are used. MESAN output consists of near-surface temperature and humidity, precipitation accumulations, visibility, wind, and cloud variables. For the 2-m temperature analyses, the first guess consists of a postprocessed forecast field from HIRLAM or the previous analysis, should the forecast field be unavailable. A structure function which is based on the relationship between temperature, water content and altitude above sea level is used in the interpolation (Häggmark et al., 2000). MESAN fields are generated on a rotated Plate Carée grid with a resolution of 0.2°. Another source of temperature information is the geostationary satellite Meteosat's infrared channel. A simple method combining these two information sources has been implememented with the objective of deriving a mask denoting areas with potentially precipitating clouds (Michelson et al., 2000). The basis for this method is an Australian one developed for use with expected minimum surface temperatures, based on climatological data, together with Geostationary Meteorological Satellite (GMS) brightness temperature composites, to derive observations of "no rain" to improve daily rainfall analyses (Ebert and Weymouth, 1999). Both methods depend on the difference between (near) surface and cloud top temperatures to distinguish between areas with and without precipitation.

Our method defines potentially precipitating clouds in areas where the difference in temperature is greater than or equal to 20°, the result being a so-called ΔT mask under which all radar echoes are retained and all others rejected as spurious. Due to our cold climate and the oblique viewing angle of the Meteosat platform, it is necessary to complement this definition with two "fail-safe" thresholds: $-5°C$ for the 2-m temperature and $0°C$ for the satellite temperature, under which all areas are defined as containing potentially precipitating clouds. The reason for this is that cold cloud tops risk being impossible to discriminate from snow and ice surfaces during the colder seasons. Fortunately however, the AP climatology is such that the greatest problems with AP echo contamination occur during warm seasons (Alberoni et al., 2001) and this is when the method seems to be usable with acceptable confidence.

The method was evaluated using data from July 2000. A set of 243 composite images containing data from NORDRAD, Danish, German and Polish radars was analyzed by an operator, and AP echoes were identified and masked manually. These masked composites were then compared with composites which were filtered using the ΔT mask and unfiltered composites. Qualitative statistics were derived using standard contingency tables for five classes of echoes: weak (≤ 10 dBZ), strong (>10 dBZ), land, sea and all. The derived statistics were False Alarm Rate (FAR), Percent Correct (PC), and Hanssen–Kuipers Skill (HKS) (Wilson, 2001).

The results, summarized in Michelson et al. (2002), show that FAR decreases with the use of the ΔT mask and PC increases for all but one class. The HKS is calculated by subtracting the Probability of False Detection (POFD) from the Probability of Detection (POD). The slightly lower values of HKS result from the POD being lower with the filtered data, while lower values of POFD resulting from the filtering are not lowered an equivalent amount or more. This means that the use of the ΔT mask successfully removes a significant amount of nonprecipitation at the expense of a small amount of true precipitation. While Doppler capability out to full operational range will succeed in removing static echoes and thus reduce the need to use this kind of multisource quality control method, it will fail to remove sea clutter since the waves from which these echoes originate move. The use of ΔT is successful in identifying and removing sea clutter but with a higher penalty to true precipitation compared to its use over land. Higher quality radar signal processing, combined with enhancements to this simple multisource method, will hopefully lead to higher quality products in the future. Data from the Meteosat Second Generation (MSG), with increased temporal and spatial resolutions and several more spectral channels, are anticipated as valuable in this context.

3.3 Precipitation Phase

3.3.1 Diagnosis and Measurement of Rain, Sleet and Snow

Over the past 50 years, the use of weather radars has concentrated on the measurement of summer rain. All Nordic countries, especially Finland, have severe snowfall climates; the average number of snow cover days varies from 100 in the southwestern archipelago to 220 in Lapland. In Norway, the number of snow days can be fewer, but the amounts of accumulated snowfall are large which, combined with the topography, introduce frequent avalanches. Both instantaneous and accumulated products of snowfall are important as the socioeconomic impacts of winter storm precipitation can be severe. For example, the average winter season road snow clearance costs in Finland are 100 million Euros. A drastic example of impact on traffic is the effect of the season's first winter storm in November 1998: heavy snowfall caused 200 car accidents in the Helsinki region, and extra costs for the airline Finnair at Helsinki airport were 1.7 million

Euros. In such cases, weather radar-based nowcasting and warning systems help to optimise the timing of snow clearance and road salting operations leading to considerable savings. Finnish and Swedish Road Authorities are mass users of dedicated weather radar snowfall products.

In principle, modern C-band radars are sensitive enough to detect even weak snowfall at distant ranges. Backscattering mechanisms in snowfall seem to be well established, although some uncertainty still exists on how the shape and density properties of large snowflakes and the structure of the ice–water–air mixture in partially melted hydrometeors influence microwave scattering (Matrosov, 1992; Zawadzki; 1995, Liao and Meneghini, 1999). The rainfall intensity (R) and snowfall intensity (S) depend on the water phase distribution (i.e. degree of melting), hydrometeor size distribution and the resulting terminal velocity distribution. As a consequence, several Z–S equations have been derived for snow (Battan, 1973). No doubt, some of these relations contain error factors related to instrument errors and to sampling differences between the radar contribution volume and the reference value (e.g., a gauge) at ground level (Kitchen and Blackall, 1992; Seed et al., 1996; Campos and Zawadzki, 2000). One confusing factor here is that the radar equation programmed in the signal processor of a radar system normally assumes that scatterers are liquid. As has been shown, by Smith (1984), the measured value, equivalent radar reflectivity factor in dry snow is $Z_e = 0.224Z$ or in logarithmic form $\mathrm{dBZ}_e = \mathrm{dBZ} - 6.5$ (dB).

In spite of the known Z–R and Z_e–S relations, the methods to switch from Z–R to Z_e–S are crude. Most European weather services still use values for rain all year round. Others switch between two relations according to the calendar. It is possible to produce double sets of precipitation products applying a Z–R relation for one, and a Z–S relation for the other, as is done in the Canadian radar network (P. Joe, pers. comm.). Even this strategy introduces quantitative errors when precipitation phase changes rapidly. The heaviest snowstorms in Northern Europe are typically related to deep frontal cyclones approaching from the Atlantic Ocean. During such events, temperatures often vary rapidly in time and space, indicating that a rain–snow switch should be performed spatially and in real time.

Since December 1999, FMI has applied a spatially variable Z–R/Z_e–S relation adjustment at temporal measurement intervals of five minutes, which is applied to data from the seven Finnish radars in the national composite (Koistinen and Saltikoff, 1999). As these radars are not equipped with polarisation-diversity capability, we have to estimate the water phase of precipitation particles independently of radar measurements. Using 150 000 synoptic observations (SYNOP) of precipitation type in Finland and discriminant analysis, a simple model has been formulated for deriving the probability of rain as a function of temperature in degrees Celsius (T) and humidity in percent (RH) at the height of 2 m:

$$P_{(\mathrm{rain})} = \frac{1}{1 + e^{(22 - 2.7T - 0.20\mathrm{RH})}}. \qquad (3.6)$$

In dry conditions, it can be seen that evaporative cooling increases the probability of snow at relatively high temperatures. The FMI hydrometeor phase analysis at ground level is based on the assumption that the fraction of water in hydrometeors (PW), as a function of temperature and relative humidity, is, on average, the same as the probability of rain from (3.6). The most probable water phase of hydrometeors at ground level is determined from surface observations and orography information. Temperature (T) and relative humidity (RH) are interpolated from synoptical observations onto a 10-km grid using co-Kriging. The analysis is performed every three hours and interpolated to the one-km resolution of the real-time radar composite. The measured Z_e is transformed to precipitation intensity using the rain relation,

$$Z = 316 \ R^{1.5}, \tag{3.7}$$

if PW is greater than 0.8, and the snow relation

$$Z_e = 400 \ S^2, \tag{3.8}$$

is applied if PW is smaller than 0.2. Between these thresholds, coefficient A and exponent b are linearly interpolated between the values 400 and 316 and 2.0 and 1.5, respectively. Linear interpolation of A and b does not eliminate overestimation of precipitation intensity observed in melting layers (Sect. 3.4.2) but is a simple model to cover them continuously. In cases of rapidly moving fronts, the three-hourly time interval in the SYNOP-based phase analysis can lead to errors of 100–200 km in the location of the rain–snow borderline. Therefore, we have applied a real-time linear extrapolation of the observed rate of change of the PW value at each point in the analysis grid, based on the two latest SYNOP analyses. Thus each pseudo-CAPPI composite at radar measurement intervals of five minutes has an individual spatial precipitation phase pattern. Despite the limitations of the present method, a spatial phase analysis map is produced every three hours for the users. In the areas of observed precipitation at ground level, the agreement between the precipitation type in SYNOP code and in the PW-analysis has been good. The quantitative "improvement" of snowfall measurements, after applying the time-space variable reflectivity-precipitation intensity relation, will be dealt with in the next Sect. (3.3.2).

3.3.2 Accuracy of Snowfall Measurements

Despite the success gained in diagnostic and semiquantitative applications, the accuracies of radar measurements of solid precipitation are not well known. Work with rainfall has been much more extensive than that with snowfall. The introduction of a time-space variable hydrometeor water phase analysis (Sect. 3.3.1) has become a very useful diagnostic product in radar composites. A logical expectation would be the improvement of quantitative precipitation measurements in cases when rain and snow occur simultaneously in different parts of the coverage area of a radar network. In the following, the main results obtained by Saltikoff et al. (2000) are summarized. They prove that such expectations are not necessarily valid. The method applied was the comparision of precipitation accumulations from gauges and radar. Gauges were of the standard Finnish type

H&H-90, equipped with a Tretyakov wind shield (Yang et al., 1999). Radar data were selected from the one-km pixel nearest to the gauge location. Two sets of radar data were used:

- Accumulated 24-h precipitation derived using the constant Z–R relation (3.7) for rain, hereafter referred as rain only (R).
- Accumulated 24-h precipitation derived using both relations Z–R (3.7) for rain and Z_e–S (3.8) for snow, and their linearly interpolated average for partially melted precipitation, according to the time series of the water analysis, hereafter referred as precipitation with variable water phase (P).

Rain only (R) and variable phase radar precipitation estimates (P) were verified using gauge observations as independent reference measurements. The logarithmic ratio, hereafter referred to as log ratio (F), has been calculated when both instruments measured 0.2 mm or more precipitation:

$$F = 10 \ \log(G/R) \quad \text{or} \quad F = 10 \ \log(G/P). \tag{3.9}$$

Factor F is typically normally distributed in rain and in snowfall (Cain and Smith, 1976; Saltikoff et al., 2000; Koistinen and Michelson, 2002). Thus the average and standard deviation of F values give a reliable measure of the bias and of the random errors between gauges and radar, respectively. The data set consisted of 2939 gauge–radar pairs during January–April 2000, from which 68% contain snowfall or partially melted snowfall. Table 3.2 exhibits average F at all measurement ranges of 0–250 km from the radars. It can be seen that, on average, the radars tend to measure 2.0–2.5 dB lower precipitation accumulations than gauges. Tab. 3.2 shows that the real-time rain–snow adjustment does not bring the radar values closer to the gauge values, on average. The adjustment reduces the random scatter of F, but the effect is small, and its significance can be argued. Of course, the phase correction (P) in radar estimates of precipitation should become more clear in the 1990 cases when considerable snowfall during the 24-hour period has occurred at the gauge site. Even in such cases, the magnitude of the gauge–radar difference is the same as if all precipitation is assumed to be liquid. Application of the phase dependent relation implies that random errors

Table 3.2. Average gauge/radar log ratio (F) on a dB-scale and standard deviation of F values (in dB). All cases refer to the total sample of gauge–radar comparisons, and snow cases refer to the cases in which a considerable part of precipitation at the gauge site was solid. Rain only refers to constant use of (3.7) in radar measurements and variable phase to continuous selection and interpolation between (3.7) and (3.8) in radar measurements

Z–R/S function	Cases	F (average)	F(std)
Variable phase	All (2939)	2.05	3.6
Rain only	All (2939)	2.06	3.8
Variable phase	Snow (1990)	2.40	3.5
Rain only	Snow (1990)	2.46	3.9

decrease by only about 10%. Based on these experiences, it can be concluded that an optimal relation between radar reflectivity factor and precipitation intensity is hidden behind other, much larger, error sources.

The reason behind the large gauge–radar bias becomes evident when gauge–radar differences are classified by radar–gauge distance. The dominating feature is increasing radar underestimation at greater distances due to the effects of the vertical reflectivity profile (Sect. 3.4). The variability of F is also, at least partly, explained by wind errors in the gauge data. Indirect evidence of this can be seen in the average F values at 13 individual gauges, located 50 km or less from a radar. For such gauge measurements, all corresponding radar measurements are from the relatively low height of around 500 m, implicating small VPR effects. In the studied snowfall cases, the average value of F from gauge to gauge varied between -1.56 and 2.23 dB. Thus, the variability of F among these gauges is several times larger than the possible effect of the selection of Z–R or Z_e–S, as found in Table 3.2. One way of avoiding wind errors in gauge observations would be to apply surface snow depth as a reference value. Unfortunately, this variable depends on the density of snowflakes which is neither observable by radar nor can be statistically modelled in an easy way. Thus, radar reflectivity-snow depth algorithms have not been very succesfull (Super and Holroyd, 1997).

Gauge–radar comparisons from the winter period January–April 2000, containing snowfall in 68% of all cases, show that the application of water phase adjustment on radar measurements does not reduce the radar bias, and it reduces random errors only slightly if snowfall occurs at surface level. Even at closer ranges, the accuracy of radar measurements does not depend on the selection of the applied Z–R/Z_e–S relation. There is also evidence that the wind error in gauge observations may vary several decibels from gauge to gauge and between snowfall events. Thus, the use of gauges to diagnose relatively small error sources in radar measurements of solid precipitation seems doubtful or at least difficult. The main conclusion from these exercises is that the concept of "optimal Z_e–S relation" in winter precipitation is almost arbitary in operational radar measurements since other sources of error dominate both gauge and radar measurements. The result is parallel to the mainstream of conclusions from the last 10 years: the optimal Z–R and Z_e–S relationship is not the dominating factor affecting the accuracy of operational measurements of precipitation at ranges of 0–250 km from a radar (Zawadzki, 1984; Joss and Waldvogel, 1990). Obviously, this is even more valid in snowfall than in rainfall. We should first perform a vertical reflectivity profile correction (Sect. 3.4 and the previous chapter) and only after that search for possible improvement by applying optimal Z–R and Z_e–S relations.

3.4 Shallow Precipitation: the Main Limiting Factor

It had not been fully realised until the 1990s that the main factor introducing bias into radar estimates of surface precipitation is the vertical measurement geometry of weather radars (Joss and Waldvogel, 1990). Radar measurements

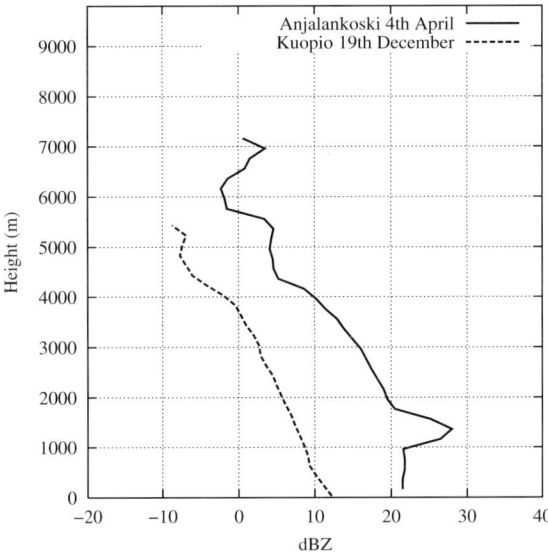

Fig. 3.8. Two vertical profiles of reflectivity averaged from single polar volumes at ranges of 2–40 km from the radar. The *solid line* represents rain and the *dashed line* snowfall

are made at increasing height and with an increasing measurement volume with increasing range, making them decreasingly representative of surface conditions. A radar measurement (Z_e), and possibly even R or S from (3.7)–(3.8), can be accurate aloft at the height of radar measurement, but it is not necessarily valid at the surface. This inaccuracy is not a measurement error but a sampling difference. The vertical profile of reflectivity (VPR) above each surface location can be denoted as $Z_e(h)$; where h is height above the surface. The shape of the VPR determines the magnitude of the sampling difference. Figure 3.8 shows an example of two measured VPRs. As we know the shape of the radar beam pattern f^2 and the height of the beam center h at each range r, it is easy to calculate from a VPR what the radar would measure at each range, $Z_e(h,r)$:

$$Z_e(h,r) = \int f^2(y) Z_e(y) \mathrm{d}y, \quad (3.10)$$

where the integration is performed vertically (y) from the lower to the upper edge height of the beam (Koistinen, 1991). The vertical sampling difference c (in decibels) is then

$$c = 10 \log \frac{Z_e(0)}{Z_e(h,r)} \quad (3.11)$$

where $Z_e(0)$ is the reflectivity at the surface in the VPR. Hence, by adding the sampling difference c to the measured reflectivity aloft (dBZ), we get the

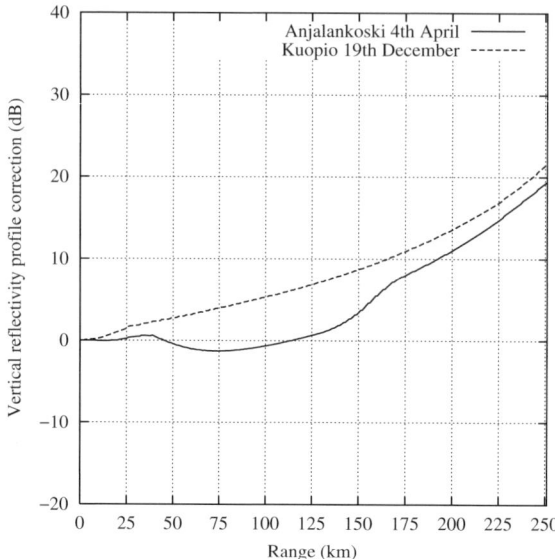

Fig. 3.9. Sampling difference (dB) between the measured reflectivity and actual reflectivity at surface level as a function of range. The graphs represent the vertical profiles of reflectivity in Fig. 3.8. The elevation angle is 0.4°, and one-way beam width is 0.95°

reflectivity at the surface, dBZ(0):

$$\mathrm{dBZ}(0) = \mathrm{dBZ} + c. \tag{3.12}$$

When we apply (3.11) to the reflectivity profiles in Fig. 3.8, the resulting sampling difference can be seen in Fig. 3.9. In snowfall, the difference increases monotonically as a function of range, indicating significant underestimation of surface precipitation already at close ranges. In rainfall, the radar measurement is relatively accurate up to the range 130–140 km. It should be noted that the overestimation introduced into the ground level precipitation estimate due to the bright band in Fig. 3.8 is very small in Fig. 3.9 (at 50–110 km). By comparing the two curves in Fig. 3.9, we can conclude the following: when the height of the bright band is more than approximately 1 km above the antenna, the overestimation due to the bright band will compensate for the underestimation effect of snow in the beam. As a result, a radar measurement is more accurate at longer ranges than it would be without a bright band. If the bright band is located at a low altitude (0–500 m), the resulting overestimation of surface precipitation will be much larger (typically 2–8 dB) but restricted to a short range interval close to the radar. Figures 3.8 and 3.9 show typical VPR examples in a cold climate. Figure 3.10 exhibits the average sampling bias c, i.e. the vertical reflectivity correction calculated from (3.11), applied to 96 000 measured precipitation profiles from the seven Finnish weather radars from March 2001 to February 2002. The lowest elevation angle is 0.4°, and the one-way half-power beam width is 0.95° in six systems. In the seventh system, the lowest elevation angle is 0.1°, and

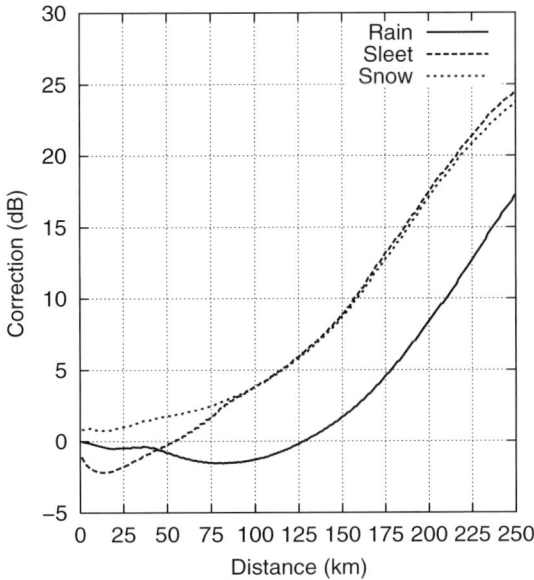

Fig. 3.10. Annual average of the sampling difference (dBZ) between the ground level reflectivity factor and the reflectivity measured at the 500 m pseudo-CAPPI level as a function of range from a radar. The data were obtained from the 7 Finnish weather radars and they consist of 96 000 measured VPRs of precipitation which were classified according to the hydrometeor phase at ground level to snow (*dotted*), melting snow (*dashed*), and rain (*solid*)

the beam width is 0.8°. The corrections are classified according to the precipitation type at ground level in each vertical profile of reflectivity to snow (56% of all precipitation profiles), melting snow (5%), rain with bright band (10%) and rain without bright band (39%). The curves should represent well the yearly average of the sampling differences brought about by the Finnish climate. It is strikingly evident that the sampling difference is by far a much more important factor in the accuracy of operational radar-based precipitation measurements than the effect of an "optimal" relation between the radar reflectivity factor and precipitation intensity.

Due to the shallow structure of precipitation with generally large negative reflectivity gradients from the surface level upward, especially in snowfall, this conclusion is even more valid in a cold climate. In such regions, the sampling bias is the only really outstanding factor influencing the accuracy of quantitative ground level precipitation measurements at 50–250 km ranges from a radar. In wintertime, the sampling bias at longer ranges is quite often so large, 20–40 dB, that the reflectivity signal is lost completely and the real detection range of precipitation is only 100–150 km. There are at least two solutions to correct the sampling biases (provided that some reflectivity level is measured). Gauge adjustment techniques can be applied for longer integration periods (12 hours and longer), since the amount of gauge data is usually representative only for longer periods (Sect. 3.4.1). For shorter periods, VPR correction schemes (3.11) based on measured or estimated VPRs can be applied (Sect. 3.4.2 and previous chapter).

3.4.1 Statistical Adjustment Using Gauge Data

Precipitation gauges are commonly viewed as providing accurate point measurements. Weather radar is commonly perceived as being able to capture precipitation's spatial distribution well in relative terms. Numerous studies over the past few decades have sought to integrate radar data with gauge observations to arrive at quantitatively accurate and spatially continuous radar-based precipitation measurements. Gauge adjustment techniques may be classified into those based on the gauge-to-radar (G/R) ratio and "sophisticated" techniques which can involve probability matching of radar reflectivity and rain rate, statistical interpolation methods, or Kalman filters (Barbosa, 1994).

G/R-based techniques are generally well suited for operational real-time use since they are robust and generate results which are more quantitatively useful than unadjusted radar data. We have applied such a gauge adjustment technique (Michelson et al., 2000; Michelson and Koistinen, 2000), for the purposes of generating precipitation data sets for the Baltic Sea Experiment (BALTEX) (Brandt et al., 1996; Raschke et al., 2001), which is a further development of a technique for use with Finnish single-site radar data and a high density gauge network (Koistinen and Puhakka, 1981). Their technique, in turn, is based on improvements to that developed in the United States (Brandes, 1975) and an application of the improved Barnes analysis technique (Barnes, 1973).

In short, the gauge adjustment technique involves the derivation and application of the logged gauge-to-radar ratio $F(\mathrm{dB}) = \log(G/R)$. This variable is used to derive a uniform distance-dependent relation between radar and gauge data, and a spatially analyzed F field. The final adjustment factor applied to a given radar-based value is a weighted combination of the uniform and spatial adjustment factors, based on observation density; the spatial adjustment is given that proportion of the total weight which the local observations can support. The primary innovations of this adjustment technique are

- Radar sums, and the distance information, are managed as composites.
- Individual radar sums are subjected to a preliminary adjustment to normalize their content to a common level throughout the network, as described by Gekat et al. (2003), this book. This step is also designed to minimize the bias between radar and gauge sums at each radar site due to systematic errors such as differences in calibration.
- G/R point pairs are collected in a moving time window to ensure a large enough sample such that the risk of overfitting in the relation with distance is minimized. Also, the use of a quality control routine is used whereby individual point pairs are assigned a variable quality weight [0–1] within an accepted quality interval.

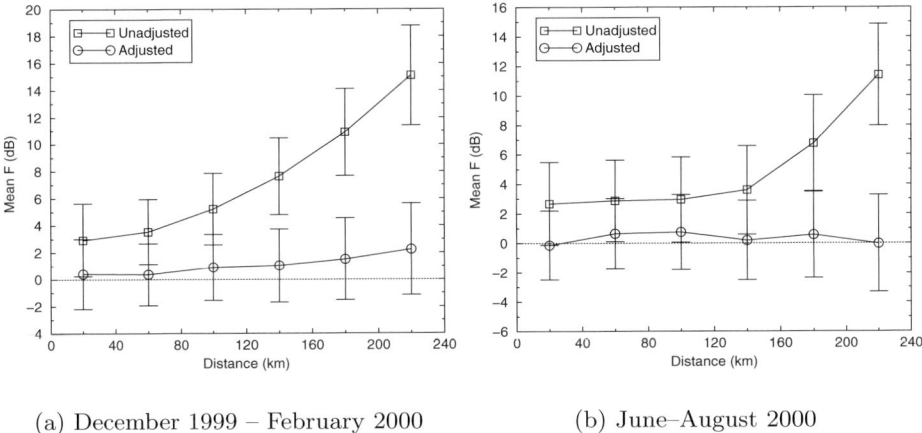

(a) December 1999 – February 2000 (b) June–August 2000

Fig. 3.11. Bias as a function of distance with daily accumulations before and after gauge adjustment. Error bars denote one standard deviation. Note the difference in Y-axis scales

- A second-order polynomial between $\log(G/R)$ and ground distance from the radar is the basis for the adjustment.
- The gauge adjusted radar sums are integrated with results of an optimal interpolation of corrected gauge sums in areas without radar coverage.

An additional innovation is the use of systematically corrected gauge observations through the application of a statistical model driven by MESAN (Häggmark et al., 2000) fields (Førland et al., 1996; Michelson et al., 2000).

The gauge adjustment technique was evaluated using the three-month winter and summer periods and independent climate station gauge data, described in Gekat et al. (2003), this book, and Koistinen and Michelson (2002). This was done by binning the G/R point pairs into 40-km strata and deriving averages, standard deviations and histograms of F for each stratum (Figs. 3.11–3.12).

The histograms in Fig. 3.12 show that the mean biases are minimized, while the variabilities in all but the most proximate 40-km stratum are also reduced. In all but the most distant couple of strata for the winter evaluations, the mean bias is minimized to within one dB which is around a 25% loss. Only in the most distant winter stratum does the bias exceed 2 dB, which is roughly a 60% loss, where the corresponding unadjusted bias exceeds 3000%! Standard deviations are lower for adjusted data in all but a couple of strata. This means that significant improvements to the accuracy of radar-derived accumulated precipitation are gained out to full operational range resulting from the use of this gauge adjustment procedure. Results from previous similar work in the UK (Collier, 1986) report roughly similar performance but do not go beyond ranges of 75 km. Results from northern Australia using the Window Probability Matching Method (Rosenfeld et al., 1995) are also roughly compatible but only extend to ranges of 130 km.

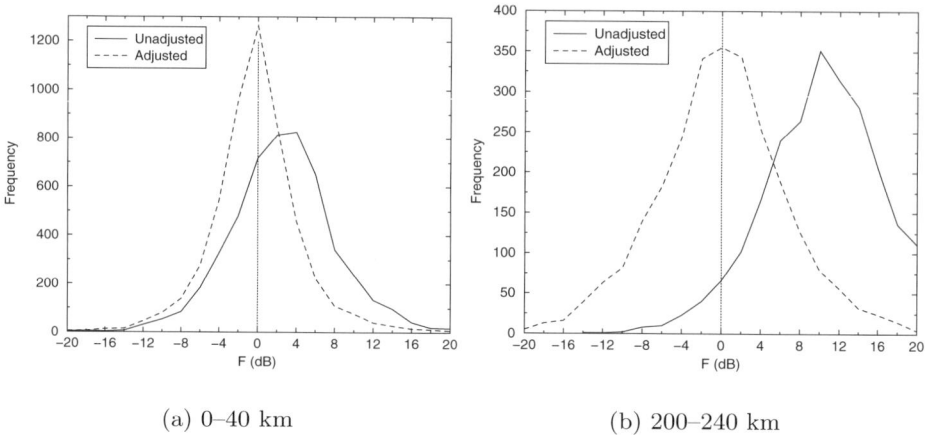

Fig. 3.12. F(dB) histograms for the period June–August 2000 for the most proximate and the most distant of the six 40-km wide distance strata

These results show the value of gauge adjustment in improving the quantitative value of radar-based accumulated precipitation estimates by addressing the issue of bias against gauge observations. Further improvements may be gained, in the form of reduced scatter in the comparison with gauges, if precipitation phase type-dependent Z–R relations and VPR corrections are applied. This combination of strategies has been identified (Collier, 1996) as a means of improving the accuracy of radar data; it is the subject of ongoing research, and interesting examples can be found in Sects. 3.3 and 3.4 and elsewhere in this book.

3.4.2 Real-Time Approach Using the Vertical Profile of Reflectivity

The lack of accurate and densely telemetered gauge networks for operational measurements, especially of snowfall, prevents the application of gauge adjustment methods to short-term radar data (instantaneous reflectivities to 12 hourly accumulations), unless gauge and radar observations are integrated in time (Sect. 3.4.1) and considered representative for periods which extend beyond their integration periods. One way to improve radar-based surface measurements of precipitation is to estimate the sampling bias c (3.11), i.e. to apply a correction based on the VPR. Despite the fact that several VPR correction schemes have been proposed and tested (Koistinen, 1991; Joss and Lee, 1995; Andrieu and Creutin, 1995; Vignal et al., 2000; Marzano et al., 2002), operational solutions are not widely established (Collier, 2001). The main difficulty is that the measurement geometry of a radar system, together with the shielding topography, prevents us from seeing the actual VPR at longer ranges above each geographical location at each moment, as discussed by Germann and Joss (2003), this book. A further problem appears when we try to apply VPR corrections in a radar network. The magnitude of the correction factor at longer ranges is quite sensitive to small spatial variations in the vertical shape of the VPR. This is most pronounced in cold climates, where precipitation is shallow and where a strong negative reflectivity gradient is dominant in the snowfall part of a precipitating system (Figs. 3.8 and 3.10). Although straightforward with a single radar system, slightly different VPR corrections, applied independently at neighbouring radars, easily amplify reflectivity discrepancies where data from two neighbouring radars meet in a radar network composite product. In the following, we describe a solution to these problems, i.e. the VPR correction scheme implemented with the Finnish radar network in November 2002. Although it is only one of several proposed schemes, the following describes problems and their solutions which are inherent in any VPR correction scheme.

A common correction approach is to apply a VPR derived at close range to data acquired at long range. The measured profiles should be obtained from polar volumes with high vertical angular resolution and many elevation angles up to 30–60°. In Finland, the VPRs are derived from three-dimensional polar volumes at 2–40 km ranges with 10 elevation angles (commonly 0.4–45°) and with a vertical resolution of 200 m.

A diagnostic tool is required to separate precipitating VPRs from other, inapplicable VPRs such as clear air echoes. In a one year data set, from March 2001 to February 2002, consisting of 240 000 VPRs from Finnish radars, 40% of all VPRs were classified as precipitation reaching the surface, 20% as overhanging precipitation (not reaching the surface) and 40% as clear air echos (Pohjola and Koistinen, 2002). The most important information for the VPR classification, in addition to the profile itself, is the freezing level height, which can be used to diagnose the hydrometeor water phase at the surface and occurrence of the bright band. Since the Finnish radars are not polarimetric systems, the freezing level height must be estimated from time–space interpolated radiosonde data (or from NWP model data). In the classification scheme, a local reflectivity maximum is diagnosed as bright band when its height is within ±500 m of the estimated freezing level height. In cases where a bright band reaches the surface and partially melted snowfall is found there, it is important that the surface reference reflectivity is not taken from the measured reflectivity in the lowest, bright band layer but the reflectivity interpolated to the surface from the first dry snow layer above the bright band. In such cases, the correction scheme will produce negative VPR correction factors at close ranges from the radar and thus, will properly correct overestimation due to the bright band close to ground level (Fig. 3.10). The vertical temperature profile is also very useful to separate shallow precipitation from clear air echoes; it is assumed that all precipitation in Finland is initiated in the ice crystal process in temperatures of $-6°C$ or less. This means that the echo top of a profile should be at least one km above freezing level to be diagnosed as precipitation. This assumption implies that we are not able to diagnose drizzle from clear air echoes or clutter.

Prior to the application of a measured VPR from the surface reaching precipitation aloft, quality control is needed to guarantee that the vertical structure of the profile is physically reasonable. Each layer should contain enough measurement bins to be at least representative for it. An additional quality test requires that reflectivity gradients between layers should not be too steep outside the bright band. Despite efficient Doppler filtering of the reflectivity data, some very strong clutter targets close to radars may partly remain and introduce a pronounced maximum in the VPR close to the surface. The height of the freezing level helps to separate real bright bands at the surface from spurious bright bands due to residual clutter. In the latter case, "clutter cutter" is applied, i.e. the vertical reflectivity gradient close to ground level is limited to at most -1 dBZ/200 m.

In case the quality of a measured VPR is poor or there is no precipitation within a 40-km range, we should apply estimated profiles of reflectivity. We apply a so-called climatological VPR which has a fixed shape relative to the varying bright-band height. The height of the freezing level is obtained from time–space extrapolated radiosonde data (a linear trend is assumed based on the latest two soundings). As has been shown (Koistinen, 1991; Joss and Lee, 1995), climatological or average VPRs remove the major part of the sampling bias.

Even when a high quality measured VPR exists, the instantaneous VPR correction factors c for a single polar volume as a function of range are always derived as the weighted mean of the corrections derived from both the quality weighted measured VPR (weight 0–1) and from the climatological VPR (weight 0.2). Further averaging is needed to make the corrections more representative in the whole single radar measurement area up to a 250-km range. Assuming an average speed of 10 m/s for a precipitating system, it takes approximately six hours for a VPR to move across a single radar's coverage area. Therefore, the corrections based on climatological and measured instantaneous VPRs are further averaged in time at each range.

The spatial representativity of a VPR correction in a network is obtained by spatial averaging of the time-averaged correction factors obtained independently from each radar. In a network composite pixel, the reflectivity value $dBZ(h, r)$ is typically taken from a single radar: the one which has the best visibility to the pixel, see also Germann and Joss (2003), this book. However, the time-averaged VPR correction from only the same radar is not necessarily representative enough, especially if the pixel is located almost as close to two or more radars. We apply spatially weighted time-averaged correction factors from all neighbouring radars closer than 300 km from the pixel to be corrected. The weight of each radar is inversely proportional to the distance to each radar squared. An important detail in the method is that the resulting spatial correction factor field will not introduce border effects or amplify possibly existing ones (e.g., due to calibration differences or due to elevation angle errors, as described by Gekat et al., 2003, this book) along the borders which separate data from neighbouring radars in the given composite.

A comprehensive validation of the network composite VPR scheme has not yet been performed. However, a composite image pair example from the Finnish

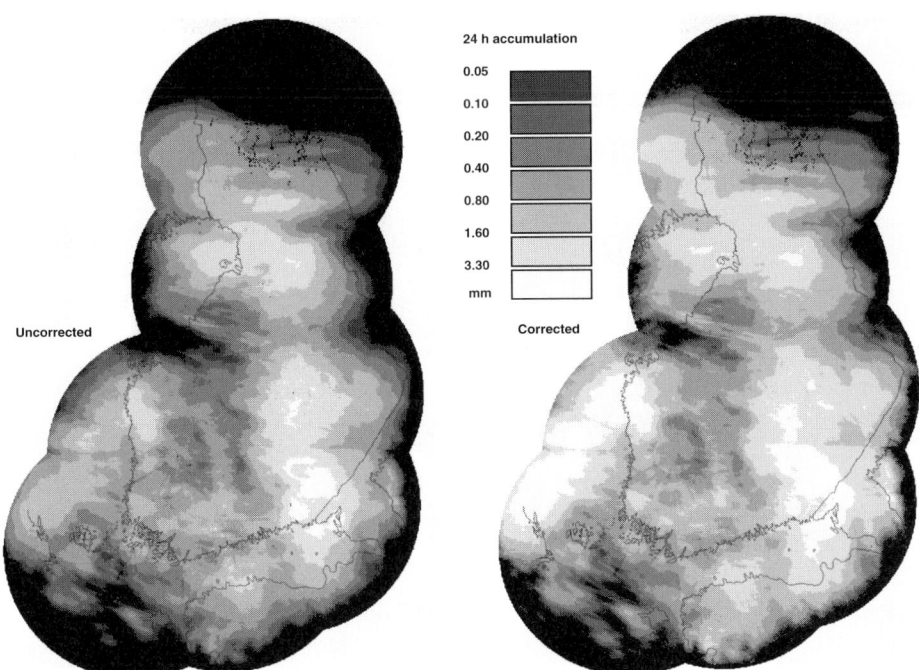

Fig. 3.13. Accumulated 24-h precipitation in the Finnish radar network on December 20, 2002 at 8 UTC. The outer range of the measurement range is 250 km at each radar. Left: accumulation without any VPR corrections. Right: accumulation after application of the composite VPR correction scheme. Structures appear more clearly in the color Fig. 10 on page 320

network shows clearly the effect of the VPR correction. Although the correction is performed separately for each reflectivity composite, the resulting effect is most clearly visible in the accumulated precipitation fields. Figure 3.13 exhibits the 24-h accumulated precipitation both with and without a VPR correction. The strong underestimation has disappeared at longer ranges. On the other hand, the correction seems stable such that reflectivities at long ranges have not increased to unrealistic levels. It can also be seen in the example that total beam overshooting occurs at the longest ranges (200–250 km), which means that the VPR correction cannot help in such areas. A final threshold is applied to prevent the corrected reflectivity values from exceeding selectable limits (45 dBZ in snowfall and 60 dBZ in rainfall).

A general conclusion of this section is that a VPR correction or a range dependent gauge adjustment is an extremely important application in cold climates. Without such corrections, quantitative precipitation estimates at the surface will be biased due to the sampling difference, by 1–5 dB at 50–150 km ranges and 5–30 dB at 150–250 km ranges. After a VPR correction has been applied, it still remains unclear if the application of an "optimal" Z–R or a Z_e–S relation would improve the accuracy of measurements. If such improvement is desired, the reflectivity–precipitation intensity relation should be a time–space variable quantity applied to data from the whole radar network, according to the analysed water phase of hydrometeors, such as that described in Sect. 3.3.1.

3.5 Nowcasting

NWP models are incapable of accurately forecasting the very near future. This is an especially fatal shortcoming if we are thinking about the weather phenomenon with probably the greatest public interest: precipitation. One way to predict precipitation a couple of hours into the future is to examine the time series of radar data to determine the motion and development of precipitating areas and development of precipitating areas. Several tracking and nowcasting techniques have been developed recently for hydrological and warning applications: RADVIL (Boudevillain et al. 2000), TRACE3D (Handwerker at al. 2000), COTREC (Li et al. 1995; Schmid et al. 2000), GANDOLF (Hand 1996), NIMROD (Golding 1998) and GSF (Toussaint et al. 2000). Some of these have been compared in the World Weather Research Programme (WWRP) Sydney 2000 Forecast Demonstration Project (Keenan et al., 2001). An example of an operationally implemented advective method is presented in this section which is based on deriving the motion vectors of precipitating areas from successive radar composite images. Trajectories and their inaccuracies computed from the vector field are used to predict the movement of precipitation up to five hours into the future.

The most important thing when deriving motion vectors is to use a method capable of producing not only the vectors themselves but also analyzing their general qualities and inaccuracies in speed and direction. These inaccuracy parameters are crucial when determining the uncertainties of the forecast. There are several methods for deriving the motion vector field. Ours is based on autocorrelation between successive radar images. It is a simplified version of the atmospheric motion vectors (AMV) system used by EUMETSAT to compute vectors from geostationary satellite data (Holmlund, 1998).

There are some general requirements to be fullfilled before radar data can be used as a basis for precipitation nowcasting:

- Spatial and temporal resolutions of radar data must be adequate enough. One kilometer spatial and five minute temporal resolutions are used here. Tests with 15-minute temporal resolution in the data generated significantly worse motion vectors than the use of radar scans every five minutes.
- Coverage of the radar composite must be noticeably larger than the forecast area. Otherwise boundary errors can be advected over larger parts of the forecast area.
- Proximity of neighbouring radars must be sufficient to detect precipitation at all measurement ranges, so that no arbitrary precipitation gaps exist in the composite product. In cold climates in winter, this typically means that

the distance between neighboring radars must be less than approximately 150 km (see Sect. 3.4).
- Data transfer from radars should be fast enough, as should processing computers, to complete the nowcasting process within a few minutes after radar data are available.
- The radar data must be free from spurious, nonprecipitating echoes like clutter and radiation from external transmitters. Otherwise, these are forecasted as precipitation.

The last requirement is the most difficult one to fullfill, as can be seen from Sect. 3.2. The other requirements are more or less economic constraints.

3.5.1 Nowcasting Method

With the advection method, we can schematically divide the local rate of change of the reflectivity factor $\partial Z/\partial t$ into the advection part $-\boldsymbol{c} \cdot \boldsymbol{\nabla} Z$ and a development part following the motion, dZ/dt:

$$\partial Z/\partial t = -\boldsymbol{c} \cdot \boldsymbol{\nabla} Z + dZ/dt \qquad (3.13)$$

where $\boldsymbol{c}$ is the displacement or motion vector (also called the steering wind). In our application, the development term dZ/dt is assumed to be zero or at least negligible. This is a reasonable assumption when used in nowcasting of widespread precipitation. In convective cases, this leads to unqualified point forecasts but still gives decent forecasts of accumulated precipitation over larger areas.

The nowcasting process begins with the derivation of the precipitation motion vectors. This is done on the 16-km resolution subgrid of a radar composite containing data from seven Finnish radars. As a first guess, the derivation is done using unfiltered reflectivity data. The time series of the first-guess vector fields (Fig. 3.14a) are used to analyze vector qualities and components along with their spatial and temporal variations.

The result of the time series analysis can be used as an input to the anomaly detection and removal process (Sect. 3.2). As part of this process, the vector field characteristics are used to determine the probability of echoes belonging to "naturally moving" precipitation. As a side effect, this also helps the classification of nonprecipitating echoes exhibiting often more or less chaotic motion vector patterns.

Finally, the quality controlled motion vector field of precipitating echoes can be obtained directly from the analysis of the first-guess fields, or it can be done separately using anomaly-free data. Before using it as an input to the nowcasting process, the vector field has to be interpolated to nonprecipitating areas. This is necessary, as the five-hourly trajectories have to be extended outside the precipitating areas. In the current system, a modified Barnes interpolation scheme (Holmlund, 1999) is used to obtain the smoothed and quality controlled motion vector field covering the whole forecast grid (Fig. 3.14b).

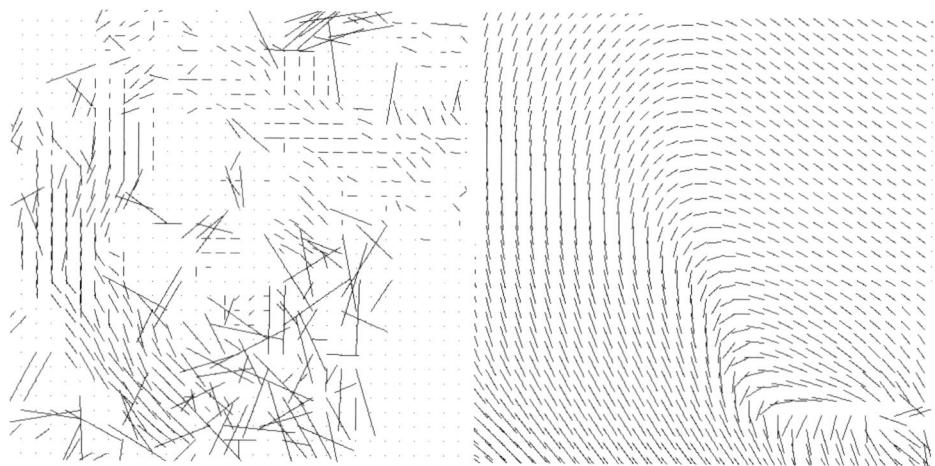

(a) Example of raw motion vector field in a 16-km grid. Only the most qualified vectors are accepted for further processing, for example, interpolation

(b) Corresponding interpolated vector field using the modified Barnes interpolation scheme

Fig. 3.14. Examples of precipitation motion vectors on a 16-km grid covering a region of 500 × 500 km. The vectors are calculated from the movement of the composited reflectivity field of seven radars

The nowcasting process applies the initial reflectivity data from the lowest four elevation angles at five-minute intervals. Nonmeteorological echoes have been filtered from the data (Sect. 3.2). The goal of the nowcasting process is to predict precipitation intensity (R) at ground level. Therefore, the compositing system not only merges reflectivities from single radars, but it also applies a vertical profile correction (Sect. 3.4) and performs the appropriate Z–R conversion (Sect. 3.3.1). Thus, the forecast field is the ground level precipitation intensity.

3.5.2 Calculation of Point Forecasts

In addition to the motion vector field, we obtain quality weights and error variances of speed and direction for each vector, together with the initial intensity data. To get a prediction of intensity at a chosen point at a chosen time in the future, we have to calculate a trajectory which ends at that point and starts somewhere in the initial intensity field. To find out where the trajectory starts, we must calculate a reversed trajectory backward in time from the end point by applying the vector field. Forward calculation will easily lead to spurious "divergence" and breakup of uniformly precipitating patches. Inaccuracies in the speed and direction of vectors along the trajectory route will be cumulative during the trajectory calculation as a function of time. This means that around

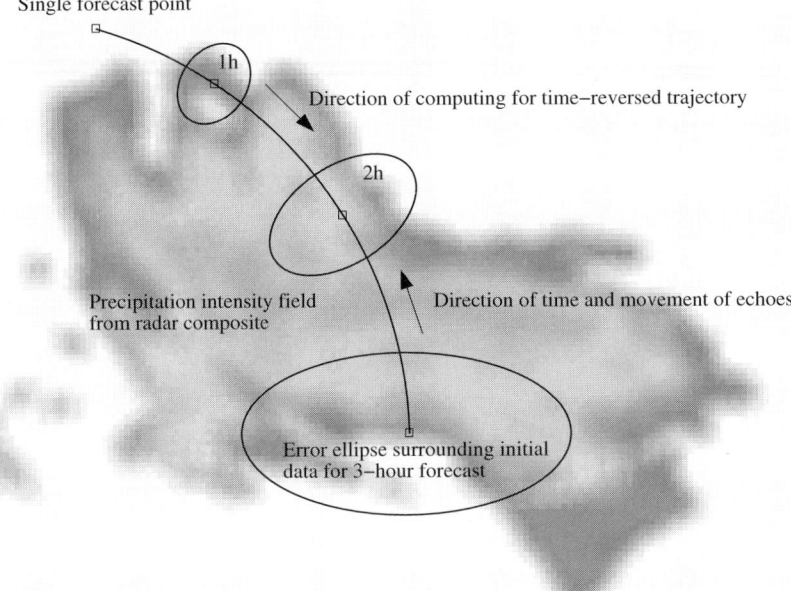

Fig. 3.15. Advective nowcasting method. The trajectory plume is a result of inaccuracies of the vector field and is approximated by the growth of an error ellipse as a function of forecast lead time

each starting point of a trajectory, an "error ellipse" will grow as a function of forecast time (Fig. 3.15). The length of the ellipse axis along the motion is proportional to the speed errors, and the length of the axis perpendicular to the motion is proportional to the direction errors of the motion vectors. The most probable precipitation intensity value (Fig. 3.16a) is then calculated by averaging the intensity content of the error ellipse by assuming a two-dimensional Gaussian distribution of probability within it. The content of an error ellipse is also used to forecast intensity class probabilities of precipitation (Fig. 3.16b).

This procedure is repeated for each forecast point in the one-km grid for each time step. The order of computing being a "spatial loop" inside a "temporal loop" allows us to choose when to output the forecasts for the whole area. The error ellipses are calculated, and their contents averaged at these times. Only trajectory start point coordinates and ellipse parameters (length of axes and orientation) are calculated at each time step. For intensity forecasts, an interval of 15 minutes might be sufficient. For good forecasts of accumulated precipitation in highly convective cases, even the time interval of intensity forecasts should be five minutes. Otherwise a moving convective cell will create accumulated precipitation patterns resembling a chain of pearls (the so-called stroboscope effect).

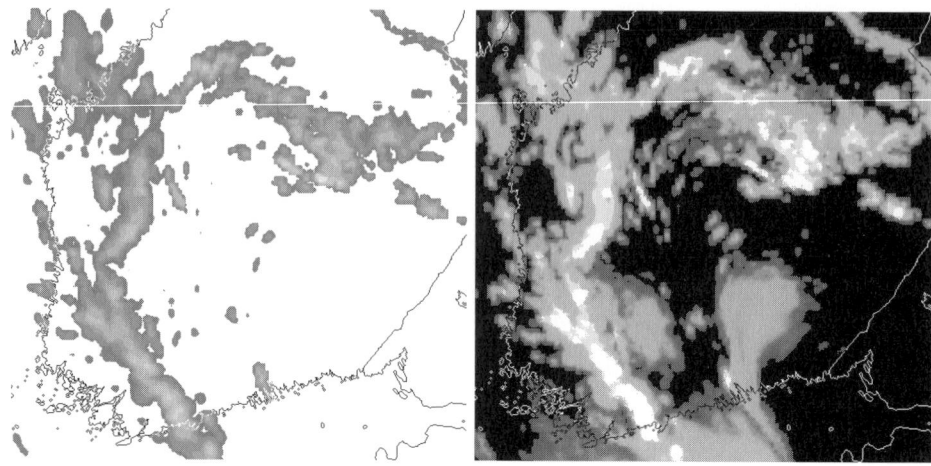

(a) Example of three-hour forecast of precipitation intensity. Shading is from low (*dark*) to high (*light*) intensities

(b) Example of three-hour forecast of intensity class probabilities. Darker grey shadings indicate probabilities of light rain. Light grey and white indicate increasing probabilities of moderate and heavy rain

Fig. 3.16. Examples of three-hour forecasts of precipitation intensity and precipitation type probability

Trajectory calculation takes minimal processing time compared to the averaging of the content of error ellipses, which means that the time step of trajectory calculation can be short, typically one minute. The direction of forecast time during computing defines that we get shorter forecasts out of the system first, which is useful from the end user's point of view. On the other hand, the growth of the error ellipses leads to rapid increase of computing time as a function of forecast time and especially as a function of the density of output times. This can be a major problem if computational resources are limited. We have solved this problem using time-dependent grid space. An alternative strategy is to split up the forecast area into subareas and process each separately in a parallel computing environment.

3.5.3 Verification

The verification of forecast precipitation areas can be tricky, and the use of only traditionally qualitative measures like threat score, POD and FAR (Sect. 3.2.3) can dramatically penalize what may visually, and in practise, be useful forecasts. This is especially valid for convective rainfall with variable, scattered precipitation patterns. In case of a small solitary cell, the magnitude or shape of the rainfall pattern being incorrect or the forecast pattern being displaced from the

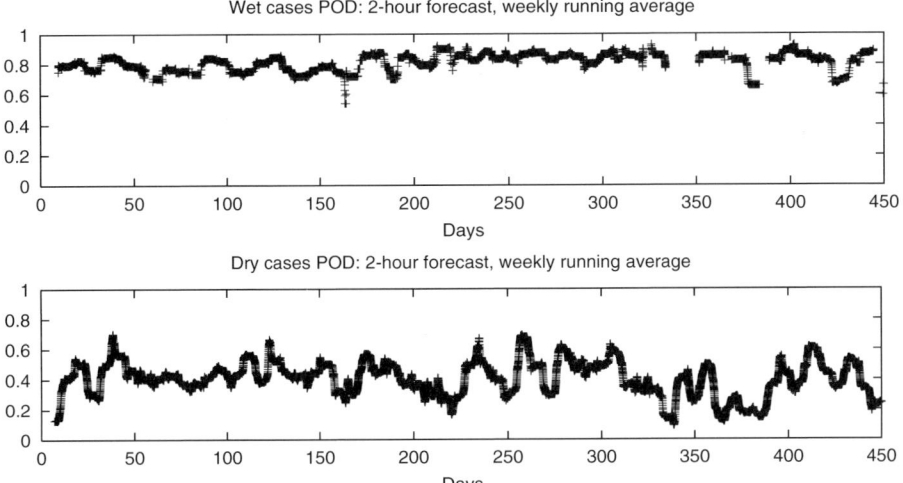

Fig. 3.17. Probability of Detection (POD) in wet and dry cases of a verification experiment in a 105 000 km² area with full radar coverage

observed location would result in a very low skill. However, for smooth large-scale stratiform precipitation areas, there are typically no verification problems.

As an example, a result from a verification experiment is presented in Fig. 3.17. In this experiment, we selected all forecasts during a year starting with the summer of 2001, and we verified rain coverage against observed radar images. The size of the area used was 300 ×350 km in southern Finland. The data, which comprised over one billion forecast intensities, was split into "wet" and "dry" cases based on the number of rainy pixels in the observed fields: if less than 33% of all pixels were rain, the case was classified dry, otherwise wet. Cases with rain coverage over 66% were rejected, as these cases are rare, and forecasts are equal to persistence in these cases. In the example, two-hour "wet" and "dry" forecasts are compared. The difference between POD values in dry and wet cases demonstrates the verification problems when this traditional quantity is used. The verification issue is, nevertheless, a very important component in the assessment and further development of the present forecasting tool.

3.6 Future Outlook

We have shown how Nordic efforts have been devoted toward improving the accuracy and use of radar precipitation estimates in our cold climates. We now have a suite of methods with which quality control of radar data may be conducted, and their quantitative accuracy raised. So where do we go from here? Some of the methods will become operationally implemented in the second generation NORDRAD network, as a part of the common Nordic research and development collaboration. Some will inevitably require improvement, and improvements in radar hardware and signal processing technologies will be invaluable in providing systems with greater sensitivity and the ability to more confidently identify and treat nonprecipitation echoes. Other improvements will follow as a result of more comprehensive use of multisource methods employing higher quality satellite observations (e.g., from MSG), and information from NWP and analysis systems (Hardaker et al., 1999).

Radar hydrology, presented and discussed by Collier and Hardaker (2003), this book, is relatively uncommon in the Nordic countries. However, the higher quality radar-based precipitation estimates resulting from our method developments will lead to added value to hydrological forecasting across our radar network.

NASA's forthcoming Global Precipitation Mission (GPM) is an undertaking which will put European weather radar activities to the test in terms of their ability to provide a network capable of supporting validation of spaceborne precipitation estimates from both passive and active microwave systems, as discussed by Testud (2003), this book. One of many valuable issues is how accurately spaceborne sensors retrieve snow estimates in cold climates. The activities presented in Chapter 1 on radar network inhomogeneities and those presented in this chapter hopefully provide the foundation necessary for achieving such a ground-based infrastructure in NORDRAD.

3.7 Acknowledgements

The authors gratefully acknowledge the work, both past, present and future, performed by Heikki Pohjola (Sect. 3.4), and Elena Saltikoff (Sect. 3.3.1) of FMI, and by Tage Andersson (retired, Sect. 3.2.2), Günther Haase, and Daniel Sunhede (Sect. 3.2.3) of SMHI. We also gratefully acknowledge support from the Nordic Council of Ministers for funding a bilateral exchange between our two institutes. A great proportion of this work has been performed under the umbrella of the Baltic Sea Experiment, and much is also relevant to the ongoing COST Action 717 (use of weather radar observations in hydrological and NWP models). Support from the European Commission, through Framework IV projects "DARTH" (ENV4-CT96-0261) and "PEP in BALTEX" (ENV4-CT97-0484) along with Framework V projects "CLIWA-NET" (EVK2CT-1999-00007) and "CARPE-DIEM" (EVG1-CT-2001-0045), has been highly beneficial and is also gratefully acknowledged.

References

1. Alberoni, P. P., T. Andersson, P. Mezzasalma, D. B. Michelson, and S. Nanni, 2001: Use of the vertical reflectivity profile for identification of anomalous propagation. Meteorol. Appl., **8**(3), 257–266.
2. Andrieu, H. and J. D. Creutin, 1995: Identification of vertical profiles of radar reflectivity for hydrological applications using an inverse method: Parts 1 and 2 – Formulation, sensitivity analysis and case study. J. Appl. Meteorol., **34**, 225–259.

3. Atlas, D. (ed.), 1990: *Radar in Meteorology*. AMS, Boston.
4. Barbosa, S., 1994: Brief review of radar-raingauge adjustment techniques. In M. E. Almeida-Teixeira, R. Fantechi, R. Moore, and V. M. Silva (eds.), *Advances in Radar Hydrology*, European Commission, Brussels, EUR 14334 EN, pp. 148–169.
5. Barnes, S. L., 1973: *Mesoscale Objective Map Analysis Using Weighted Time-Series Observations*. Technical Report NOAA Technical Memorandum ERL NSSL-62, National Severe Storms Laboratory, Norman, Oklahoma.
6. Battan, L. J., 1973: *Radar Observations of the Atmosphere*. University of Chicago Press, Chicago.
7. Boudevillain, B., J. Thielen and H. Andrieu, 2000: Definition of the characteristics of an urban hydrological radar: Interest in the vertically integrated liquid water content. Phys. Chem. Earth (B), **25**, 1311–1316.
8. Brandes, E. A., 1975: Optimizing rainfall estimates with the aid of radar. J. Appl. Meteorol., **14**, 1339–1345.
9. Brandt, R., C. Collier, H.-I. Isemer, J. Koistinen, B. Macpherson, D. Michelson, S. Overgaard, E. Raschke, and J. Svensson, 1996: *BALTEX Radar Research – A Plan for Future Action*. Publication No. 6, International BALTEX Secretariat, GKSS Research Center, Geesthacht, Germany.
10. Cain, D. E. and P. L. Smith, 1976: Operational adjustment of radar estimated rainfall with rain gage data: A statistical evaluation. *Proc. 17th Conf. Radar Meteorol.*, AMS, pp. 533–538.
11. Campos, E. and I. Zawadzki, 2000: Instrumental uncertainties in Z-R relations. J. Appl. Meteorol., **39**, 1088–1102.
12. Collier, C. G., 1986: Accuracy of rainfall estimates by radar, Part I: Calibration by Telemetering Raingauges. J. Hydrology, **83**, 207–223.
13. Collier, C. G., 1996: *Applications of Weather Radar Systems. A Guide to Uses of Radar Data in Meteorology and Hydrology*. Praxis/John Wiley and Sons, Chichester/London, 2nd ed.
14. Collier, C. G., 1998: Observations of sea clutter using an S-band weather radar. Meteorol. Appl., **5**, 263–270.
15. Collier, C. G. (ed.), 2001: *COST Action 75 - Advanced Weather Radar Systems – 1993-97. Final report*. European Commission, Luxembourg, EUR 19546.
16. Daley, R., 1991: *Atmospheric Data Analysis*. Cambridge University Press, New York.
17. Ebert, E. E. and G. T. Weymouth, 1999: Incorporating satellite observations of "no rain" in an Australian daily rainfall analysis. J. Appl. Meteorol., **38**, 44–56.
18. Fabry, F. and I. Zawadzki, 1995: Long term radar observations of the melting layer of precipitation and their interpretations. J. Atmos. Sci., **52**, 838–851.
19. Førland, E. J., P. Allerup, B. Dahlström, E. Elomaa, T. Jónsson, H. Madsen, J. Perälä, P. Rissanen, H. Vedin, and F. Vejen, 1996: *Manual for operational correction of Nordic precipitation data*. Report nr. 24/96, DNMI, P.O. Box 43, Blindern, Oslo, Norway, 66 pp.
20. Fulton, R. A., J. P. Breidenbach, D.-J. Seo, and D. A. Miller, 1998: The WSR-88D rainfall algorithm. Weather and Forecasting, **13**, 337–395.
21. Golding, B.W., 1998: NIMROD : A system for generating automated very short range forecasts. Meterol. App., **5**, 1–16.
22. Häggmark, L., S. Gollvik, K.-I. Ivarsson, and P.-O. Olofsson, 2000: Mesan, an operational mesoscale analysis system. Tellus, **52A**(1), 2–20.
23. Hand, W.H., 1996: An object-oriented technique for nowcasting heavy showers and thunderstorms. Meterol. App., **3**, 31–41.

24. Handwerker, J., J. Ressing and K.D. Beheng, 2000: Tracking convective cells in the upper Rhine valley. Phys. Chem. Earth (B), **25**, 1317–1322.
25. Hardaker, P. J., B. Macpherson, and P. R. A. Brown, 1999: Weather radar and Numerical Weather Prediction models. In *COST 75 Advanced weather radar systems. International seminar*, European Commission, EUR 18567 EN, pp. 451–459.
26. Harrison, D. L., S. J. Driscoll, and M. Kitchen, 2000: Improving precipitation estimates from weather radar using quality control and correction techniques. Meteorol. Appl., **7**, 135–144.
27. Holmlund, K., 1998: The utilization of statistical properties of satellite-derived atmospheric motion vectors to derive quality indicators. Weather and Forecasting, **13**, 1093–1104.
28. Holmlund, K., 1999: The use of observation errors as an extension to Barnes interpolation scheme to derive smooth instantaneous vector fields from satellite-derived atmospheric motion vectors. In *Proc. 1999 EUMETSAT Meteorol. Satellite Data Users Conf.*, EUMETSAT, EUM P26, pp. 633–637.
29. Joe, P. and S. Lapczak, 2002: Evolution of the Canadian operational radar network. In *Proc. ERAD (2002)*, EMS, Copernicus GmbH, pp. 370–382.
30. Joss, J. and R. Lee, 1995: The application of radar–gauge comparisons to operational precipitation profile corrections. J. Appl. Meteorol., **34**, 2612–2630.
31. Joss, J. and A. Waldvogel, 1990: Precipitation measurement and hydrology: A review. In *In Battan Memorial and Radar Conference. Radar in Meteorology*, AMS, Boston, Chap. 29a, pp. 577–606.
32. Källén, E., 1996: *HIRLAM Documentation Manual, system 2.5*. Technical report, SMHI, S-601 76 Norrköping, Sweden.
33. Keenan, T.D., J.W. Wilson, P.I. Joe, C.G. Collier, B. Golding, D.W. Burgess, R. Carbone, A. Seed, P.T. May, L. Berry, J. Bally and C.E. Pierce, 2001: The World Weather Research Programme (WWRP) Sydney 2000 Forecast Demonstration Project: Overview. *Proc. 30th Conf. Radar Meteorol.*, AMS, pp. 474–476.
34. Kerr, D. E. (ed.), 1951: *Propagation of Short Radio Waves*. Dover Publications, New York.
35. Kitchen, M. and R. B. Blackall, 1992: Representiveness errors in comparisons between radar and gauge measurements of rainfall. J. Hydrology, **134**, 13–33.
36. Koistinen, J., 1991: Operational correction of radar rainfall errors due to vertical reflectivity profile. *Proc. 25th Conf. Radar Meteorol.*, AMS, pp. 91–94.
37. Koistinen, J., 1997: Clutter cancellation and the capabilities of modern Doppler radars. *Proc. COST 75 Workshop on Doppler Weather Radar*. European Commission, Luxembourg, pp. 7–11.
38. Koistinen, J. and D. B. Michelson, 2002: BALTEX weather radar-based precipitation products and their accuracies. Boreal Env. Res., **7**(3), 253–263.
39. Koistinen, J. and T. Puhakka, 1981: An improved spatial gauge-radar adjustment technique. *Proc. 20th Conf. Radar Meteorol.*, AMS, pp. 179–186.
40. Koistinen, J. and T. Puhakka, 1986: Can we calibrate radar with raingauges? Geophysica (Helsinki), **22**, 119–129.
41. Koistinen, J. and E. Saltikoff, 1999: Experience of customer products of accumulated snow, sleet and rain. In *COST 75 Advanced Weather Radar Systems. International seminar*, European Commission, EUR 18567 EN, pp. 397–406.
42. Li, L., W. Schmid and J. Joss, 1995: Nowcasting of motion and growth of precipitation with radar over a complex orography. J. Appl. Meteorol., **34**, 1286–1300.

43. Liao, L. and R. Meneghini, 1999: A study of effective dielectric constant of ice-water spheres where fractional water content is prescribed as function of radius. *Proc. 29th Conf. Radar Meteorol.*, AMS, pp. 699–702.
44. Marzano, F. S., E. Picciotti, and G. Vulpiani, 2002: Reconstruction of rainrate fields in complex orography from C-band radar volume data. In *Proc. ERAD (2002)*, EMS, Copernicus GmbH, pp. 227–232.
45. Matrosov, S. Y., 1992: Radar reflectivity in snowfall. *IEEE Trans. Geoscience Remote Sensing*, **30**, 454–461.
46. Michelson, D. B., T. Andersson, J. Koistinen, C. G. Collier, J. Riedl, J. Szturc, U. Gjertsen, A. Nielsen, and S. Overgaard, 2000: *BALTEX Radar Data Centre Products and their Methodologies*. Reports Meteorology and Climatology RMK 90, SMHI, SE-601 76 Norrköping, Sweden.
47. Michelson, D. B. and J. Koistinen, 2000: Gauge-radar network adjustment for the Baltic Sea Experiment. Phys. Chem. Earth (B), **25**(10–12), 915–920.
48. Michelson, D. B., J. Koistinen, R. Bennartz, C. Fortelius, and A. Thoss, 2002: BALTEX radar achievements at the end of the main experiment. In *Proc. ERAD (2002)*, EMS, Copernicus GmbH, pp. 357–362.
49. Pamment, J. A. and B. J. Conway, 1998: Objective identification of echoes due to anomalous propagation in weather radar data. *J. Atmos. Oceanic Technol.*, **15**, 98–113.
50. Peura, M., 2002: Computer vision methods for anomaly removal. In *Proc. ERAD (2002)*, EMS, Copernicus GmbH, pp. 312–317.
51. Pohjola, H. and J. Koistinen, 2002: Diagnostics of reflectivity profiles at the radar sites. In *Proc. ERAD (2002)*, EMS, Copernicus GmbH, pp. 233–237.
52. Raschke, E., J. Meywerk, K. Warrach, U. Andræ, S. Bergström, F. Beyrich, F. Bosveld, K. Bumke, C. Fortelius, L. P. Graham, S.-E. Gryning, S. Halldin, L. Hasse, M. Heikinheimo, H.-J. Isemer, D. Jacob, I. Jauja, K.-G. Karlsson, S. Keevallik, J. Koistinen, A. van Lammeren, U. Lass, J. Launianen, A. Lehmann, B. Liljebladh, M. Lobmeyr, W. Matthäus, T. Mengelkamp, D. B. Michelson, J. Napiórkowski, A. Omstedt, J. Piechura, B. Rockel, F. Rubel, E. Ruprecht, A.-S. Smedman, and A. Stigebrandt, 2001: The Baltic Sea Experiment (BALTEX): A European contribution to the investigation of the energy and water cycle over a large drainage basin. Bull. Amer. Meteorol. Soc., **82**(11), 2389–2413.
53. Rosenfeld, D., E. Amitai, and D. B. Wolf, 1995: Improved accuracy of radar WPMM estimated rainfall upon application of objective classification criteria. J. Appl. Meteorol., **34**, 212–223.
54. Saltikoff, E., J. Koistinen, and H. Hohti, 2000: Experience of real time spatial adjustment of the Z-R relation according to water phase of hydrometeors. Phys. Chem. Earth (B), **25**(10–12), 1017–1020.
55. Schmid, W., S. Mecklenburg, and J. Joss, 2000: Short-term risk forecasts of severe weather. Phys. Chem. Earth (B), **25**, 1335–1338.
56. Seed, A. W., J. Nicol, G. L. Austin, C. D. Stow, and S. G. Bradley, 1996: The impact of radar and raingauge sampling errors when calibrating a weather radar. Meteorol. Appl., **3**, 43–52.
57. Sekhon, R. S. and R. C. Srivastava, 1970: Snow size spectra and radar reflectivity. J. Atmos. Sci., **27**, 299–307.
58. Smith, P. L., 1984: Equivalent Radar Reflectivity Factors for Snow and Ice Particles. J. Climate Appl. Meteorol., **23**, 1258–1260.
59. Sonka, M., V. Hlavac, and R. Boyle, 1993: *Image Processing, Analysis and Computer Vision*. Chapman and Hall Computing.

60. Steiner, M. and J. A. Smith, 2002: Use of three-dimensional reflectivity structure for automated detection and removal of nonprecipitating echoes in radar data. J. Atmos. Oceanic Technol., **19**, 673–686.
61. Sugier, J., J. Parent du Châtelet, P. Roquain, and A. Smith, 2002: Detection and removal of clutter and anaprop in radar data using a statistical scheme based on echo fluctuation. In *Proc. ERAD (2002)*, EMS, Copernicus GmbH, pp. 17–24.
62. Super, A. B. and E. W. Holroyd, 1997: Snow accumulation algorithm for the WSR88D radar. Second Annual Report of the Bureau of Reclamation R-97-5, Denver.
63. Toussaint, M., B. Jacquemin, I. Donet, A. Carlier, and M. Malkomes, 2000: GSF – A Doppler weather radar based tracking tool. Phys. Chem. Earth (B), **25**, 1339–1442.
64. Vignal, B., G. Galli, J. Joss, and U. Germann, 2000: Three methods to determine profiles of reflectivity from volumetric radar data to correct precipitation estimates. J. Appl. Meteorol., **39**(10), 1715–1726.
65. Watson, R. J., 1996: *Data comparisons for spatially separated meteorological radars*. Ph.D. Thesis, Dept. Electronic Systems Engineering, University of Essex, UK.
66. Wessels, H. R. A. and J. H. Beekhuis, 1992: Automatic suppression of anomalous propagation clutter for noncoherent weather radars. Scientific reports; WR 92-06, kNMI, PO Box 201, 3730 AE De Bilt, The Netherlands.
67. Wilson, C., 2001: Review of current methods and tools for verification of numerical forecasts of precipitation. COST 717 Working Document WDF_02_200109_1.
68. Wilson, J. W. and E. A. Brandes, 1979: Radar measurement of rainfall – A summary. Bull. Am. Meteorol. Soc., **60**(9), 1048–1058.
69. Yang, D., B. E. Goodison, J. R. Metcalfe, P. Louie, G. Leavesley, D. Emerson, C. I. Hanson, V. S. Golubev, E. Elomaa, T. Gunther, T. Pangburn, E. Kang and J. Milkovic, 1999: Quantification of precipitation measurement discontinuity induced by wind shields on national gauges. Water Resource Res., **35**, 491–508.
70. Zawadzki, I., 1984: Factors affecting the precision of radar measurements of rain. *Proc. 22nd Conf. Radar Meteorol.*, AMS, pp. 251–256.

4 Using Radar in Hydrometeorology

Christopher Collier[1] and Paul Hardaker[2]

[1] Telford Institute of Environmental Systems, University of Salford, UK
[2] 2. Met Office, Bracknell, UK

4.1 The Role of Radar in Flood Forecasting and Water Resources Management

It is now over 40 years since digital weather radar data became available in significant quantities. Whilst the exciting possibilities for improved flood forecasts were quickly recognised, it is only in recent years that the development of spatially distributed hydrological models has provided a basis to utilise effectively these new data. Nevertheless, for some time, a few operational hydrologists have made use of radar data qualitatively to define the start and cessation of rainfall events and there have been some attempts to demonstrate the potential and difficulty of using radar data as input to isolated event and similar lumped models.

Unlike practical rain-gauge networks, radar provides estimates of surface precipitation at high resolution over wide areas from a single radar location in near realtime. Unfortunately, the error characteristics of these measurements can be very variable both in space and time. Consequently, river flow forecasts using such data can be very accurate at one time and very inaccurate only a short time later (Fig. 4.1). To some extent, the same can be said of predictions using rain-gauge network data, although the use of sometimes complex quality control procedures may mask significant errors in the rainfall field derived from a network of rain-gauges, particularly for rainfall fields having large spatial and temporal gradients. Nevertheless, it has been known for some time that radar measurements of areal rainfall can outperform rain-gauge network measurements. We discuss this further later.

Radar data are well suited to real-time flood forecasting applications, but they have also been used for assessing water resources and providing a basis for engineering design. This is particularly true in remote areas where rain-gauges are sparse, and satellite techniques for estimating rainfall are too inaccurate (see, for example, Georgakakos and Kavvos, 1987). For these applications, very carefully quality controlled data sets are required. Often radar data are discounted as not providing a statistically sound basis upon which to extract design criteria. However, by using rain-gauge and radar data sets in a complementary fashion, improved analysis can be achieved. We will discuss this in the context of dam and dam spillway design later.

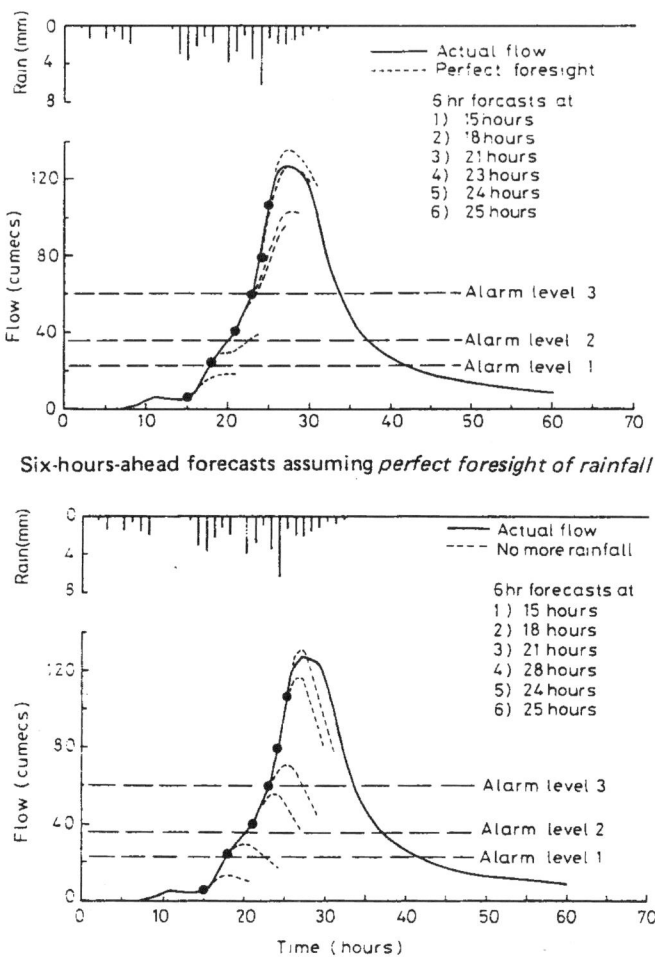

Fig. 4.1. Six hour-ahead forecasts for 14 March 1982 made for Blackford Bridge, assuming perfect foresight of rainfall and no further rain will occur (Cluckie et al., 1987)

4.2 Radar Data Quality Control

In earlier chapters, the problems of interpreting radar data in terms of rainfall have been addressed. The introduction of Doppler technology, enabling radars to measure the radial motion of the target hydrometeors as well as the signal backscattered from them, has led to significant improvements being made in the removal of ground clutter echoes, including anomalous propagation echoes. Operationally, Doppler clutter cancellation is most effective when combined with techniques based upon either maps locating ground echoes and/or statistically based procedures (see, for example, Grecu and Krajewski, 2000; Germann and Joos, 2003, this book).

Polarisation-diversity, as touched upon in Gekat et al. (2003), this book and discussed by Illingworth (2003), this book, is also a new area of development for operational radar systems, although the technology has been available for some time for research studies. Whilst there are still important operational decisions to make regarding the appropriate implementation of polarisation-diversity at operationally used wavelengths, it is clear that this technology offers considerable benefit to quality control procedures, with more limited benefits to enhancing rainfall estimation techniques directly (Collier, 2002; Illingworth, this book).

Whilst the use of S-band (10-cm wavelength frequency, as in the USA and in Spain), removes problems associated with the attenuation of the radar beam as it passes through precipitation, the use of C-band (5-cm wavelength) frequency by much of the rest of the world, particularly in Europe, to gain the advantage of improved sensitivity continues to be a source of measurement problems. This is a particularly important problem in situations of very heavy convective rainfall which often leads to flash flooding. The occurrence of very heavy rainfall often accompanied by hail in such storms may cause radar estimates of the surface rainfall at ranges beyond the storm cores to be very significant underestimates (Fig. 4.1). On the other hand, where the radar beam does intersect hail the rainfall estimates may be very significant overestimates. The next chapter discusses the potential for alleviating these problems using polarisation-diversity techniques. It may be that operational improvements in accuracy are possible using such technology. The attenuated area indicated in Fig. 4.2 has been derived by using a polarisation-based algorithm.

Equally important problems, particularly in midlatitudes, arise from the error in estimates of surface precipitation which may occur when converting radar reflectivity to precipitation as the radar beam intersects the brightband (the region where snow melts to form rain) or when evaporation or growth occur below the radar beam. This led to operational attempts to get around these problems using raingauge adjustment techniques (see, for example, Collier et al., 1983). However, more recently, it was recognised that a more reliable approach was one based upon the derivation and adjustment of the Vertical Profile of Reflectivity (VPR), as demonstrated and discussed in depth by Germann and Joss (2003), this book, and Koistinen et al. (2003), this book.

Over the last ten years or so, models of the physical processes involved in generating the VPR have been developed (Smith, 1986; Kitchen et al., 1994; Hardaker et al., 1995; Andrieu et al., 1995). It is now generally accepted that the VPR must be adjusted before retrieving surface estimates of precipitation. The appropriateness of subsequent rain-gauge adjustment of the radar estimates remains uncertain, although Vignal et al. (2000) point out that the improvement in radar–rain-gauge agreement achieved by adjustment using groups of rain-gauges (increasing the representativeness) can be similar to that of the VPR correction alone.

In conclusion, removal of bright-band effects can significantly improve the quality of radar measurements of rainfall, as shown in Fig. 4.3. It seems clear that daily and longer period rainfall can be measured by radar after careful

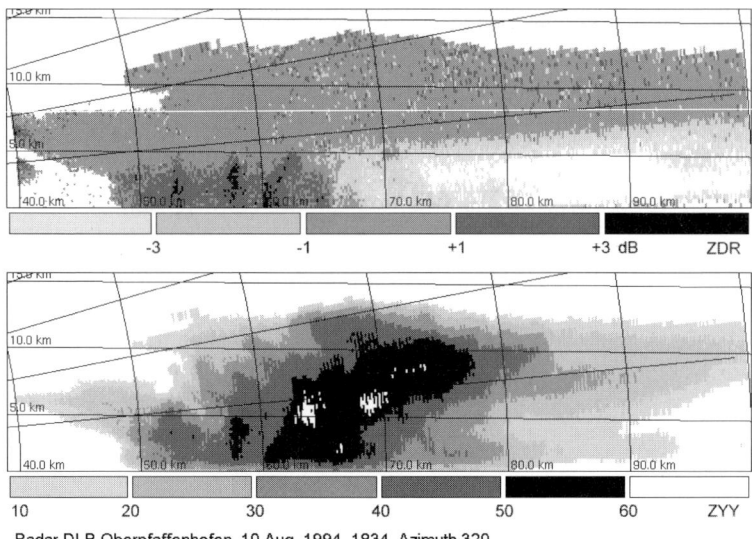

Fig. 4.2. RHI radar images of a severe thunderstorm. The area of high reflectivity causes severe attenuation of the radar beam such that the signal strength is much reduced beyond the area of high reflectivity. The lower image is reflectivity and the upper image is differential reflectivity, (courtesy DLR Germany). The differential reflectivity becoming smaller than about -10.5 dB is caused by more intense attenuation of horizontally polarised waves than vertically polarised waves. See also color Fig. 11 on page 320

quality control within about 100 km of the radar site to an accuracy of around 10%. For shorter period rainfall estimates, the accuracy achieved can be 25–30% which is comparable with the accuracy achieved using quite dense rain-gauge networks. Fortunately, operational radars are now very reliable, a necessary condition for use by operational hydrologists, as illustrated in Table 4.1. Downtime can be planned to coincide with no rainfall situations, and a number of methods now exist to undertake necessary calibration activities in realtime without taking the radar out of operations, for example, Manz et al. (2000) and Goddard et al. (1994). Such a level of performance is vital for real-time flood forecasting applications.

4.3 Flood Forecasting

As already stated in this chapter, the availability of radar measurements of rainfall greatly increased the availability of data for input to hydrological models. Of particular importance are the short-period forecasts of rainfall derived by using extrapolation techniques. Extensive reviews of this work have been published, see, for example, Collier (2000) and the papers contained in Collier and Krzysz-

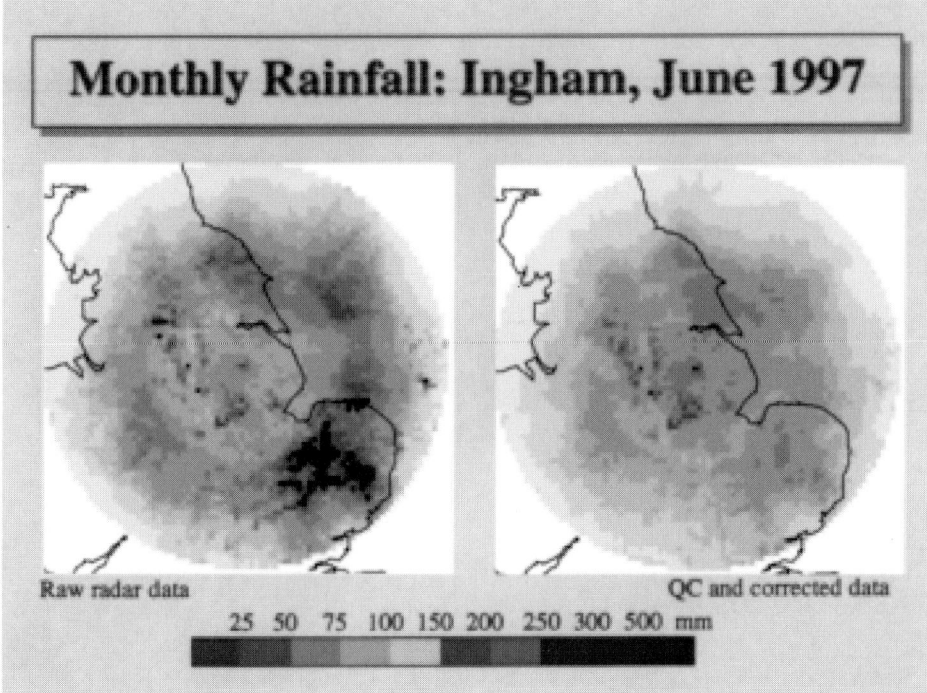

Fig. 4.3. Monthly integration in June 1997 before and after VPR adjustment for the Ingham radar in the UK (courtesy Met. Office, UK). The coastline of England is shown by the black line. The grey scale gives the monthly rainfall measured with the rain radar data (left panel) and the quality controlled radar data (right panel). See also color Fig. 12 on page 321

togowicz (2000). The level of performance achieved using the UK Met Office Nimrod system using the advections of radar echoes and a mesoscale numerical model is shown in Fig. 4.4. The data indicate that purely advection forecasts only outperform forecast from the model for the first 2–3 hours ahead.

Distributed models, originally developed for use with rain-gauge network data, are ideally suited to the use of radar data. Some hydrologists argue that this complicates the procedure for deciding whether a model generates reliable predictions, and even go as far as to query whether it is worth using data which may be unreliable (Bevan, 1996). However, others argue that if a model is structured to cope with distributed data such as radar data, which are not reliant on calibration processes, but rather on error feedback procedures which make use of the error structure of the radar rainfall measurements, then distributed hydrological models can perform satisfactorily (Refsgaard et al., 1996).

In recent years, whatever type of model is adopted, the importance of associating model predictions with estimates of predictive uncertainty is being increasingly recognised. Two approaches to estimating flow prediction uncer-

Table 4.1. Availability of radars in the UK and Irish weather radar networks and for the States of Jersey radar for December 2001, courtesy Met. Office, UK

Radar (denoted by radar site name)	Availability (%)
Dublin	95.4
Cobbacombe Cross	93.8
Lewis	99.5
Dudwick	99.8
Corse Hill	98.4
Jersey	95.6
Wardon Hill	92.2
Crug-y-Gorllwyn	94.5
Ingham	98.6
Predannack	99.7
Caster Bay	91.8
Shannon	94.8
Chenies	99.8
Hameldon	99.6
Clee Hill	98.7
Average network availability	96.8

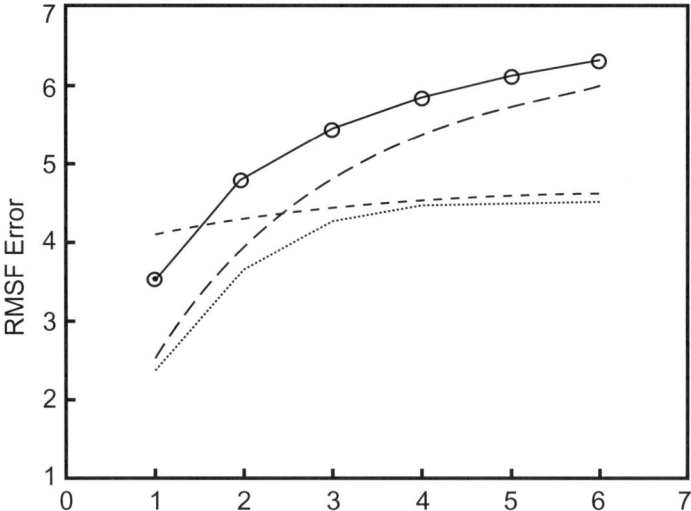

Fig. 4.4. RMSF hourly precipitation accumulation errors computed from 15 km x15 km averages and verified against quality controlled radar actuals over the area of reliable radar cover for the period September 1998 to August 1999 as a function of lead time from the time of availability. *Full line:* Nimrod forecasting system; *dashed line:* advection component of Nimrod forecasts; *dotted line:* mesoscale model forecasts; *light line with circles persistence*, RMSF = 1.0 corresponds to a perfect forecast (Golding, 2000)

tainty are being pursued: First, procedures based upon the assumption that the uncertainty derives from that in model parameter values and boundary conditions, with maximum likelihood or some other parameter estimation technique being used to find the optimal estimates of model parameters (for example, Bell and Moore, 2000). Second, procedures based upon Monte Carlo simulation using randomly chosen sets of parameter values (for example, Bevan and Binley, 1992) are applied.

Hydrologists have utilised the idea of error feedback to improve future flow predictions. Model flow is constantly compared with measured flows, and the measured difference (error) is used to provide a correction factor to the input data used for the next model iterations (for example, Storm et al., 1988). However, almost universally, the models have generated deterministic forecasts. Meteorologists, on the other hand, have not employed real-time error feedback procedures, except in so much as model error analyses have been used to guide model development. They have though recently increasingly invested research effort in the development ensemble prediction schemes leading to forecasts with a defined level of uncertainty (for example, Trevisan et al. 2001). It is likely that hydrological models will make use of ensemble predictions of rainfall to provide ensemble predictions of river flow.

Operationally, many flood forecasters are reluctant to consider model output couched in terms of an uncertainty distribution, whether generated by ensemble prediction or through the use of stochastic physics within the model (see Hardaker et al. 1995). This must change, as flow forecasts will always have errors arising both from the models used and the propagation, through the models, of error in the input data. Model errors need to be combined with errors in input data to produce flows which have an error range associated with them. The use of Baysian statistics to achieve this is one possible approach. It will also be necessary to blend model output uncertainty with decision and risk analyses. This will involve the associations of the statistical distributions of errors in modelled flows with cost–loss functions derived for specific flood risk regions of river catchments.

Figure 4.5 shows the impact of rainfall data input quality on the hydrograph generated, using a hydraulic model of a combined sewer system for a small (30 km^2) urban catchment in Bolton, North West England. The random errors (not shown) in the flow predictions are approximately the same for each observing system (rain-gauge networks having 1, 2, 3 and 4 rain-gauges per 20 km^2 and rainfall inferred from the attenuation caused by the rainfall along a microwave communication link). However, the bias errors in the flow predictions change with time through the event in different ways for each observing system. The bias error in the modelled flows is used to modify the input to each time step to the model to improve the flows derived by the next model iteration.

Real-time assessment of these errors enables the error distribution characteristics at each time step to be modified to adjust with the model input, enabling improved predictions to be made. Unfortunately if the error changes rapidly with time, then the updating procedure will lag reality. It is not necessarily straight-

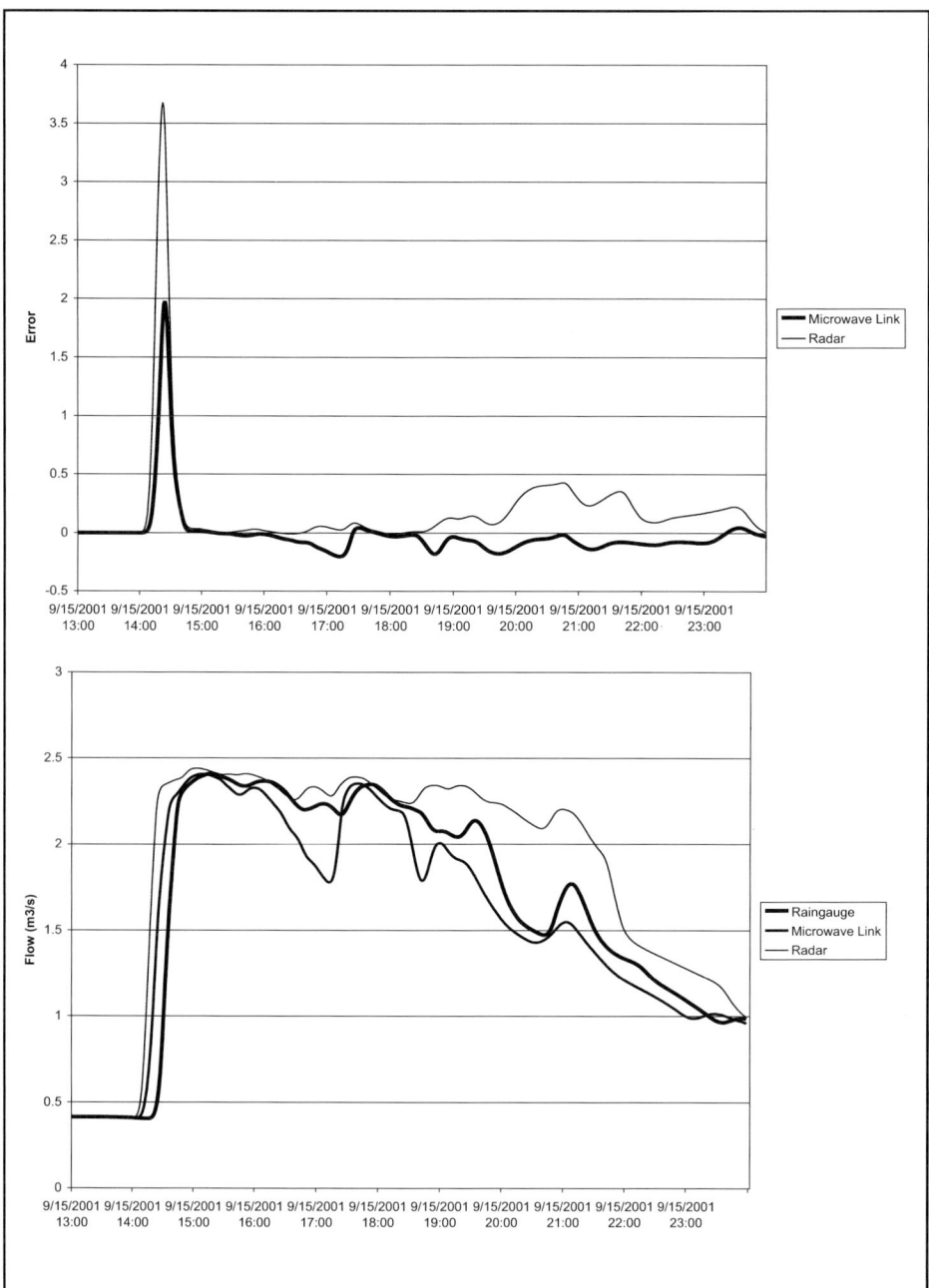

Fig. 4.5. Bias errors in model flow predictions made using different observing systems for a combined sewer system in Bolton, North West England. Also shown are hydrographs using rainfall input from three rain-gauges, two microwave links and weather radar, courtesy G.L. Robbins, University of Salford. The ordinate on the lower graph is flow in m^3 s^{-1} and in the upper graph is bias error. The abscissa in both graphs is time

forward to allow for such changes in a way which prevents wildly varying model output. In the next section we discuss prospects for combining atmospheric and hydrological models.

4.4 Coupled Atmospheric and Hydrologic Numerical Models

Three basic types of numerical models have been used to forecast rainfall, namely, stochastic, dynamical and statistical-dynamical models. Following early work on the development of viewing the rain process as organised two-dimensional turbulence operating on scales about, or larger than, 5 km (see, for example, Krajewski and Georgakakos, 1985; Georgakakos and Kavvos, 1987), recent work has characterised rainfall variability using multifractal analysis (Lovejoy and Schertzer, 1990) or using four-dimensional statistical models (Cowpertwait et al., 2002). Radar data have been used to assess the validity of such approaches and whether or not the models can be used to downscale the output from dynamical models (Wheater et al., 1999). Collier (1993) proposed an approach to using radar data to distribute numerical model derived rainfall across a model grid square. A log normal frequency distributions was defined using parameters whose values were derived from radar data for NW Europe, as shown in Fig. 4.6.

The simplest type of dynamical model is based upon the estimation of topographically-forced vertical motion (Collier, 1975; 1977). The introduction of cloud physical processes leads to improved rainfall forecasts (for example, Bader and Roach, 1977; Sinclair, 1994). However, to capture fully small-scale weather systems and, therefore, improve operational reliability necessitates the use of four-dimensional models having fine spatial and temporal resolution. Attempts to avoid the complexity of these models have been described. These so-called statistical dynamical models develop state estimators with conservation of mass in an atmospheric column (see, for example, Lee and Georgakakas, 1990). Unfortunately, these models are not reliable enough for operational use.

The current operational performance of dynamical models in the United Kingdom has been discussed by Golding (2000). Experiments are being carried out in several countries with very high resolution compressible nonhydrostatic models, a recent example is described by Benoit et al. (1997).

Radar data are used to verify dynamical models, but are also assimilated into them (Jones and Macpherson, 1997), as discussed by Macpherson et al. (2003) this book. However, there is growing activity aimed at coupling atmospheric and hydrological models. This enables streamflow predictions made using observational input, including radar data, to be compared with those using numerical model input. Benoit et al. (2000) have demonstrated that a distributed hydrological model can be sufficiently sensitive and accurate to diagnose both numerical model and radar errors. The potential of using radar with a coupled atmospheric–hydrological model was stressed. These types of research studies aim to provide a basis for the operational use of coupled models. However, as pointed out by Johnson (2000), given a mean absolute error of radar or atmo-

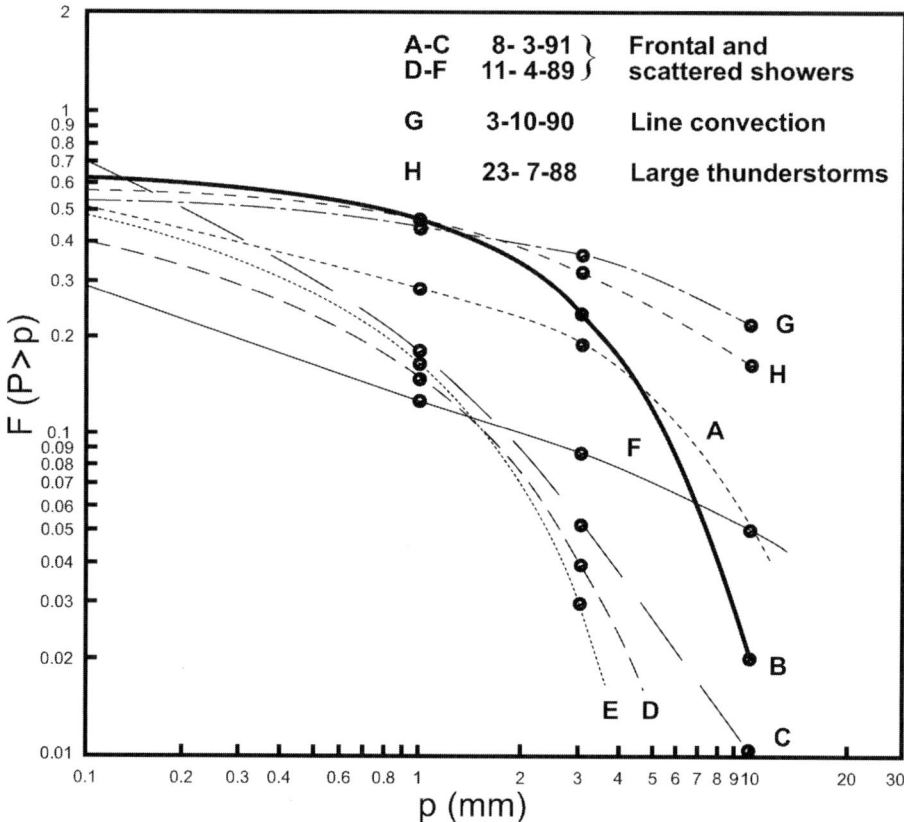

Fig. 4.6. A log-log plot of observations of the fraction of a $100\ km \times 100\ km$ numerical model grid square that has rainfall accumulations greater than given values. The letters refer to the particular cases and weather types shown. Data from radars in NW Europe were used to derive the rainfall distributions (Collier, 1993)

spheric rainfall estimates of approximately 25%, the land surface response error would be of the same magnitude. By coupling the rainfall input uncertainty to that of infiltration and runoff routing uncertainties, large prediction uncertainties in flood peak magnitude and timing are to be expected. Hence, it is likely to be necessary to use categorical Monte Carlo approaches to model parameter calibration and output flow assessment, as discussed in the previous section, plus comprehensive data assimilation procedures.

The advent of real-time radar and high resolution model data offered flood forecasters the comprehensive data sets needed for distributed hydrological modelling. However, if significant (greater than 20% over 100 km^2) errors remain in these data from time to time, then, as pointed out by Bevan (1996) and Johnson (2000), fast simple hydrological models may be adequate for operational use in providing fewer, as opposed to more, site-specific forecasts.

4.5 Engineering Design

The use of radar data for engineering design purposes, that is, the estimation of probable maximum precipitation (PMP) (the theoretically maximum precipitation that can occur over a specific time period and area), probable maximum flood (PMF) (the theoretically maximum flood that can occur at a specific locations on a river) and return period analysis is often dismissed. The reason for this is twofold; firstly, the length of record provided by operational radars is usually very much shorter than that available from rain gauges, and secondly, the radar error characteristics cast doubt on the quantitativeness of the rainfall record. Looking firstly at the error characteristics, radar makes measurements of rainfall over areas, not at a point. These areas may be quite small (100 m × 100 m), but nevertheless one must be aware of this difference from raingauge measurements since rainfall can vary by an order of magnitude over a very small distance. One should not expect radar estimates to exactly coincide with rain gauge estimates (see Kitchen and Blackall, 1992). The two devices measure rainfall in very different ways, and the different sampling techniques by definition will lead to different respective measures of rainfall, even with perfect observations.

It may be possible to reanalyse a long radar data set to ensure that quality control procedures are up-to-date and have been uniformly applied. However, the amount of work involved should not be underestimated. Given that it is usual in carrying out depth duration frequency (DDF) analyses (the river depth is related to the duration of the rainfall and the frequency of occurrence of the flow) using rain-gauge data, to work where possible with at least 30 year records for daily data and 20 year records for hourly data, one must consider the consequences of working with short period records. Nevertheless, the UK Flood Estimation Handbook concludes that a minimum record length of 9 years is acceptable. If this is so for rain-gauges, then the same should apply to radar records, although for these short record lengths, it is unwise to investigate return periods much in excess of 1 in 50 years. The validity of the extrapolation of DDF models, based upon relatively short rainfall records, to very rare return periods of 1 in 10 000 years equivalent to the PMP has recently been challenged by MacDonald and Scott (2001). Therefore, radar records, however accurate, may as yet be of inadequate length for use in engineering design statistical frequency analyses for very rare events.

Is there any alternative to creating long highly quality controlled radar data sets for engineering design purposes? Collier and Hardaker (1996) use Lagrangian analysis techniques to estimate maximum storm rainfall totals, from which the efficiencies of observed storms were derived and used in the context of a storm model. This approach has some success in leading to realistic estimates of PMP, for example Collier and Hardaker (1995, 1996) and Hardaker (1996). Such an approach still requires highly quality controlled radar data, but from individual storms rather than a continuous long sequence of data, providing rainfall data that is as accurate as possible.

So far we have discussed using rainfall data sets derived from radar reflectivity on their own. Faulkner and Prudhomme (1988) consider the application

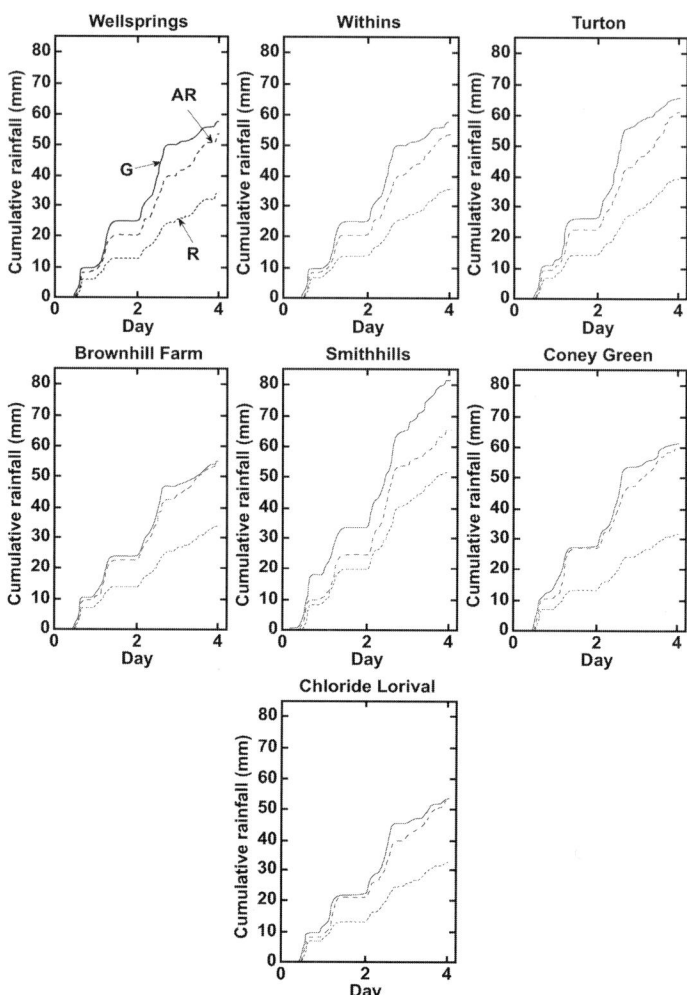

Fig. 4.7. Four-day storm accumulated rainfall totals for seven rain-gauges and the radar pixel overlying the rain-gauge location (Day 0 = 28 February 1999, Day 1 = 1 March 1999, etc; raingauges (G) radar, data adjusted using a probability density function technique (AR) and raw radar (R) are indicated in the first frame (after Tilford et al., 2002)

of georegression techniques for improving the interpolation of rainfall extremes. Such techniques involved the use of topography and lightning strike data, although they were less successful for shorter durations (one hour). It is possible to apply, off-line, an adjustment technique based upon the probability density functions (PDFs) for radar and rain-gauges, as applied by Tilford et al. (2002). Figure 4.7 shows the success of such a technique in modifying rainfall amounts measured by radar when compared to contemporaneous rain-gauge totals.

Given that a weather radar can generate, within a range of 75 km in a single scan, data from the equivalent of over 4400 rain-gauge locations, it is clear that there is enormous potential to increase record lengths for frequency analyses so enabling returns period and PMP analyses to be derived more accurately in areas where there are no long rain-gauge records.

4.6 Future Prospects

New algorithm development coupled with improvements in radar technology such as polarisation-diversity promise significant improvements in radar data accuracy which will be sustained throughout a range of differing precipitation events. Nevertheless, the only certainty we have is that all errors will not be removed all the time, but what is important is that quality control of all model inputs, particularly radar rainfall are optimised, and that assimilation schemes can actively account for uncertainty in rainfall inputs. Further, while work on improving the accuracy of rainfall measurements is key (and radar data are now as accurate as rain-gauge data when like is compared to like), it is the sensitivity of the output of the forecast model to the quality of the input data that is important, not directly the absolute accuracy of rainfall estimates.

The development of flood forecasting models and engineering design procedures should be undertaken using radar data because of the benefits of their spatial information content. It may not be necessary to accrue very long periods of highly quality controlled data to achieve this aim, due to the extensive spatial coverage provided by radar.

References

1. Andrieu, H. and J.D. Creutin, 1995: Identification of vertical profiles of radar reflectivity for hydrological applications using an inverse method. Part I: Formulation. J. Appl. Meteorol., **34**, 225–239.
2. Bader, M.J. and W.T. Roach, 1977: Orographic rainfall in warm sectors of depressions. Q. J. R. Meteorol. Soc., **103**, 269–280.
3. Bell, V.A. and R.J. Moore, 2000: The sensitivity of catchment runoff models to rainfall data at different spatial scales. Hydrology and Earth Syst. Sci., **4**, 653–667.
4. Benoit, R., P. Pellerin, N. Kouwen, H. Ritchie, N. Donaldson, P. Joe, and E.D. Soulis, 2000: Toward the use of coupled atmospheric and hydraulic models at regional scale. Mon. Weather Rev., **128**, 1681–1706.
5. Benoit, R., M. Desgagne, P. Pellerin, S. Pellerin, Y. Chartier and S. Desjardins, 1997: A semi-Lagrangian, semi-implicit wide-band atmospheric model suited for finescale process studies and simulation. Mon. Weather Rev., **125**, 2382–2415.
6. Bevan, K.J., 1996: A discussion of distributed hydrological modelling. Chapter 13A in *Distributed Hydrological Modelling*. Editors M.B. Abott and J.C. Regsgaard, publ. Kluwer Academic, pp. 255-278.
7. Bevan, K.J. and A.M. Binley, 1992: The future of distributed models: model calibration and uncertainty prediction. Hydro Process **6**, 279–298.

8. Cluckie, I.D., P.F. Ede, M.D. Owens, A.C. Bailey and C.G. Collier, 1987: Some hydrological aspects of weather radar research in the United Kingdom. Hydrol. Sci. J., **32**, 328–346.
9. Collier, C.G., 2002: Developments in radar and remote sensing methods for measuring and forecasting rainfall. Philos. Trans. Ser. A, **360**, 1345–1361.
10. Collier, C.G., 2000: Precipitation estimation and forecasting. WMO Operational Hydrology Report No. 46, WMO-No. 887, Geneva.
11. Collier, C.G. and R. Krysztogowicz, 2000: Special Edition of J. Hydrology on Quantitative Precipitation Forecasting. Vol. **239**, Nos. 1–4.
12. Collier, C.G. and P.J. Hardaker, 1996: Estimating probable maximum precipitation using a storm model approach. J. Hydrology, **183**, 277–306.
13. Collier, C.G. and P.J. Hardaker, 1995: Radar and storm model-based estimation of probable maximum precipitation in the tropics, dams and reservoirs, **5**, 19–21.
14. Collier, C.G., 1993: The applications of a continental – scale radar database to hydrological process parametrization within atmospheric general circulation models. J. Hydrology, **142**, 301–318.
15. Collier, C.G., P.R. Larke and B.R. May, 1983: A weather radar correction procedure for real-time estimation of surface rainfall. Q. J. R. Meteorol. Soc., **109**, 589–608.
16. Collier, C.G., 1977: The effect of model grid length and orographic rainfall efficiency on computed surface rainfall. Q. J. R. Meteorol. Soc., **103**, 247–253.
17. Collier, C.G., 1975: A representation of the effects of topography on surface rainfall within moving baroclinic disturbances. Q. J. R. Meteorol. Soc., **101**, 407–422.
18. Cowpertwait, P.S.P., C.G. Kilsby and P.E. O'Connell, 2002: A Spatial-temporal Neyman–Scott model of rainfall: Empirical analysis of multisite data (to appear in Water Res. Res.).
19. Faulkner, D.S. and C. Prudhomme, 1988: Mapping an index of extreme rainfall across the UK. Hydrology Earth Sys. Sci., **2**, 183–194.
20. FEH, 1999: *Flood Estimation Handbook.* NERC/Institute of Hydrology.
21. Georgakakos, K.P. and M.L. Kavvos, 1987: Precipitation analysis modelling and prediction in hydrology. Rev. Geophys., **25**, 163–178.
22. Goddard, J.W.F., J. Tan and M. Thurai, 1994: Technique for calibration of meteorological radars using differential phase. Electron. Lett., **30**, 166–167.
23. Golding, B.W., 2000: Quantitative precipitation forecasting in the UK. J. Hydrology, **239**, 286–305.
24. Grecu, M. and W.F. Krajewski, 2000: An efficient methodology for detection of anomalous propogation echoes in radar reflectivity data using neural networks. J. Atmos. Oceanic Technol. **117**, 121–129.
25. Hardaker, P.J., A.R. Holt and J.W.F. Goddard, 1998: Comparing model and measured rainfall rates obtained from a combination of remotely-sensed and in-situ observations, Radio Sci., **32**, 1785–1796.
26. Hardaker, P.J., 1996: Estimating probable maximum precipitation for a catchment in Greece using a storm model approach. Meteorol. Appl., **3**, 137–145.
27. Hardaker, P.J., A.R. Holt and C.G. Collier, 1995: A melting layer model and its use in correcting for the bright band in single polarisation radar echoes. Q. J. R. Meteorol. Soc., **121**, 495–525.
28. Johnson, L.E., 2000: Assessment of flash flood warning procedures. J. Geophys. Res., **105**, D2, 2299–2313.
29. Jones, C.D. and B. Macpherson, 1997: A latent heat nudging scheme for assimilation of precipitation data into an operational mesoscale model. Meteorol. Appl., **4**, 269–277.

30. Kitchen, M., R. Brown and A.G. Davies, 1994: Real-time correction of weather radar data for the effects of bright-band range and orographic growth in widespread precipitation. Q. J. R. Meteorol. Soc., **120**, 1231–1254.
31. Kitchen, M. and R.M. Blackall, 1992: Representativeness error in comparisons between radar and gauge measurements. J. Hydrology, **134**, 13–33.
32. Krajewski, W.F. and K.P. Georgakakos, 1985: Synthesis of radar rainfall data. Water Resources Res., **21**, 764–768.
33. Lee, T.H. and K.P. Georgakakos, 1990: A two-dimensional stochastic-dynamical quantitative precipitation forecasting model. J. Geophys. Res., **95**, D3, 2113–2126.
34. Lovejoy, S. and D. Schertzer, 1990: Multifractals, universality classes and satellite and radar measurements of cloud and rain. J. Geophys. Res., **95**, D3, 2021–2034.
35. MacDonald, D.E. and C.W. Scott, 2001: FEH vs FSR rainfall estimates: An explanation for the discrepancies identified for very rare events. Dams & Reservoirs, December.
36. Manz, A., A.H. Smith and P.J. Hardaker, 2000: Comparison of different methods of end-to-end calibration of the UK weather network. Phys. Chem, Earth, **25**, 1157–1162.
37. Refsgaard, J.C., B. Storm and M.B. Abbott, 1996: Comment on 'A discussion of distributed hydrological modelling' by K. Bevan. Chapter 13B in *Distributed Hydrological Modelling*, editors M.B. Abbott and J.C. Regsgaard, Kluwer Academic, pp. 279–287.
38. Sinclair, M.R., 1994: A diagnostic model for estimating orographic precipitation. J. Appl. Meteorol., **33**, 1163–1175.
39. Smith, C.S., 1986: The reduction of errors caused by bright-bands in quantitative rainfall measurements made using radar. J. Atmos. Oceanic Technol., **3**, 129–141.
40. Storm, B., K.H. Jensen and J.C. Refsgaard, 1988: Estimation of catchment rainfall uncertainty and its influence on runoff prediction. Nordic Hydrology, **19**, 77–88.
41. Tilford, K.A., N.I. Fox and C.G. Collier, 2002: Application of weather radar data for urban hydrology. Meteorol. Appl., **9**, 95–104.
42. Trevisan, P., F. Pancotti and F. Molteni, 2001: Ensemble prediction in a model with flow regimes. Q. J. R. Meteorol. Soc., **127**, 343–358.
43. Vignal, B., G. Galli, J. Joss and U. Germann, 2000: Three methods to determine profiles of reflectivity from volumetric radar data to correct precipitation estimates. J. Appl. Meteorol., **39**, 1715–1726.
44. Wheater, H.S., T.J. Jolley, C. Onof, N. Mackay and R.E. Chandler, 1999: Analysis of aggregation and disaggregation for grid-based hydrological models and the development of improved precipitation disaggregation procedures for GCMs. Hydrology Earth System Sci., **3**, 95–108.

5 Improved Precipitation Rates and Data Quality by Using Polarimetric Measurements

Anthony Illingworth

Dept of Meteorology, University of Reading, Reading, RG6 6BB, UK

5.1 Introduction

The additional information provided by polarisation-diversity radar has the potential to remove many of the ambiguities and uncertainties present when only the conventional reflectivity (Z) and Doppler information are available. In this chapter we emphasise the application in an operational environment with a typical one-degree beam-width radar and a dwell time of about one-sixth of a second, so that it completes a PPI every minute and provides rainfall estimates with a spatial resolution of 2 km or better. An excellent and very comprehensive book on polarimetric Doppler weather radar has recently appeared by Bringi and Chandrasekar (2001), and we refer the reader to this book for detailed derivations and analysis which we will simply quote. Essentially, polarimetric radar provides information on the shape and orientation of the radar targets. In Sect. 5.2, we define the polarisation parameters, summarise their applications and discuss the theoretical and practical limits to the accuracy with which they can be estimated. The variation of raindrop shape with size and the typical raindrop size spectra are reviewed in Sect. 5.3. The following sections consider how to exploit the shape and orientation information from a polarimetric radar which can transmit and receive horizontally and vertically polarised radiation, as described by Gekat et al. (2003), this book.

- *Recognition of anomalous propagation and ground clutter (Sect. 5.4).*
  Precipitation particles are essentially an array of quasi-spherical scatterers which are usually small enough to Rayleigh scatter and so can be distinguished from ground echoes. Ground clutter and 'anaprop' are essentially non-Rayleigh scatterers and are far from spherical, having very different scattering cross sections for horizontally and vertically polarised radiation.
- *Improved rainrate estimates (R) when rain alone is present (Sect. 5.5).*
  Empirical $Z(R)$ relationships of the form $Z = aR^b$ result in errors in R of up to a factor of two. They arise because of the variability in the raindrop size spectrum: $Z = \sum ND^6$ for rain, where N is the concentration and D the raindrop diameter, but $R \simeq \sum ND^{3.67}$. Because raindrops are oblate to a degree which depends upon their size, polarisation observations have the potential to estimate mean raindrop size and so provide better rain rates.
- *The use of combined and integral polarisation parameters (Sect. 5.6).*
  High resolution polarisation parameters tend to be very noisy. An alternative approach is to exploit their path integrated properties which can be

estimated much more accurately and use them as a constraint to provide, for example, absolute calibration of the radar reflectivity, an estimate of the constant 'a' in $Z = aR^b$ and a measure of the attenuation.
- *Improved rainfall estimates when ice is present (Sect. 5.7).*
  Large hailstones lead to very high values of Z which can be interpreted as spuriously high rainfall rates but may be identified because they are oblate and tend to tumble as they fall. Melting snowflakes are responsible for the enhanced radar return at the melting layer known as 'the bright band' and can easily be recognised because they are wet, oblate and fall with a characteristic rocking motion.
- *Correction for attenuation (Sect. 5.8).*
  Attenuation for the C-band (5.6 cm wavelength) radars used in Europe and Japan is a serious problem in heavy rain. Polarisation parameters which are immune to attenuation can be used to correct for such effects.
- *Identification of hydrometeors (Sect. 5.9).*
  For ice hydrometeors other than hail and melting snow, identification using polarisation is more difficult, but polarisation can yield useful information.

In Sect. 5.10, the potential for implementation of polarisation parameters is summarised for an operational environment both for the 5.6 cm wavelength (C-band) as generally used in Europe and Japan and 10 cm (S-band) used in the USA. The use of S-band for polarisation observations has two major advantages: firstly, we have Rayleigh scattering for nearly all meteorological targets, and secondly, there are normally negligible propagation or attenuation problems. Conventional radar networks yield estimates of Z and rainfall rate with a spatial resolution of about 2 km and an update time of 5 (or perhaps two and a half) minutes. If polarisation techniques require longer dwell times, then the poor sampling could negate any increased accuracy of specific rainfall estimates.

5.2 The Polarisation Parameters

5.2.1 Introduction

We shall restrict the discussion to the use of linear polarisation in which the radar can transmit pulses which are alternately polarised in the horizontal (H) and vertical (V), and can measure both the two copolar returns Z_H, Z_V; and, if the radar is Dopplerised, the phase of the horizontally and vertically polarised returns, ϕ_h and ϕ_v. We shall consider the following four parameters: the differential reflectivity (Z_{DR}), the copolar correlation (ρ_{hv}), specific differential phase shift (K_{DP}), and the linear depolarisation ratio (LDR). In principle, these four parameters can also be derived from measurements made with circular polarisation, but nearly all recent work has used linear polarisation. We will also briefly discuss transmission of pulses polarised at 45° with reception at H and V, which has the advantage that no rapid polarisation switching between transmitted pulses is needed.

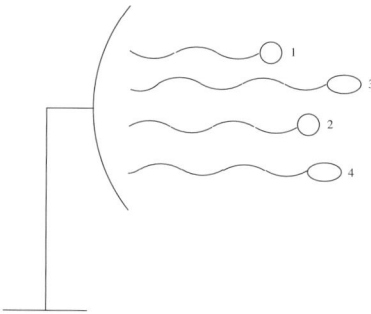

Fig. 5.1. Signals backscattered from precipitation particles arrive at the antenna with random phase

Recall from Fig. 5.1 that the amplitudes of the backscattered waves from individual hydrometeors arrive at the antenna with random phase, so that as the scatterers move relative to one another, the observed intensity Z will fluctuate. The fractional standard error of the mean Z estimate is given by $1/\sqrt{n}$ for large n, where n is the number of independent samples of Z. The time to independence is given by $\lambda/(4\sigma_v \sqrt{\pi})$ where σ_v is the Doppler width of the target. Assuming that at a low beam elevation the precipitation has a Doppler width of 0.5m s^{-1}, then the time to independence at C-band is about 16 ms, and at S-band about 28 ms. If we need one PPI a minute then the dwell time is 0.166 s and averaging over six adjacent gates, each typically 150 m long (1 μs transmitted pulse), provides a range resolution of 1 km to match the one-degree beam-width azimuthal resolution at a range of 57 km. This leads to n of about 63 at C-band and 36 at S-band, so the statistical sampling error of Z will be about 13% (0.53 dB) and 17% (0.68 dB), respectively. System errors are discussed by Gekat et al. (2003, this book).

5.2.2 Z_{DR} - Differential Reflectivity

Z_{DR} [$= 10 \log(Z_H/Z_V)$] is a measure of mean particle shape. Z_{DR} is particularly useful for rain, because small raindrops are spherical, but larger ones become increasingly oblate. An improved rainfall rate, $R(Z, Z_{DR})$, can be obtained from Z and Z_{DR}. Assuming an exponential Marshall–Palmer (1948) raindrop size distribution

$$N(D) = N_0 \exp(-3.67 D/D_0), \tag{5.1}$$

where D_0 is the median volume drop diameter, then the value of D_0 can be estimated from Z_{DR}. The advantage of Z_{DR} is that it is a ratio which is independent of the concentration parameter, N_0. Once D_0 is known, then the value of N_0 is fixed by the observed value of Z. An empirical Z–R relationship is approximately equivalent to assuming we have a Marshall–Palmer distribution with a variable D_0 but N_0 constant and equal to 8000 m^{-3} mm^{-1}. The use of Z and Z_{DR} to fix both N_0 and D_0 should result in more accurate rainfall estimates

(Seliga and Bringi, 1976). For a given value of Z_{DR} (fixed D_0), the rainfall rate and the value of Z both scale linearly with N_0, so we expect relationships of the form,

$$Z = cRf(Z_{\text{DR}}). \tag{5.2}$$

Ice particles have a lower dielectric constant than liquid water, and so even if they are oblate, they tend to have low values of Z_{DR}, particularly if, as is the case of snow, they are a low density mixture of air and ice. If oblate ice particles become wet, the value of Z_{DR} increases, and so the melting snow in the bright band is associated with high values of Z_{DR}. In vigorous convection, supercooled raindrops can be recognised by narrow vertical columns of positive Z_{DR} extending above the freezing level (Illingworth et al., 1987). Hail usually tumbles as it falls and so should be associated with a Z_{DR} of 0 dB. At low altitudes where rain is to be expected, a Z_{DR} value of 0 dB accompanied by a high value of Z can be used to identify hail. Unfortunately, hail is often mixed with rain – so the result will be an intermediate value of Z_{DR}, which is lower than it should be for the rain, but the presence of hail will lead to higher than expected values of Z. In Sect. 5.5, we shall show that for improved rainfall estimates using $R(Z, Z_{\text{DR}})$, Z must be accurately calibrated, and Z_{DR} estimated to within 0.2 dB. Z_{DR} can be calibrated to 0.1 dB by observing precipitation at vertical incidence which we know has a value of 0 dB, or, less satisfactorily, looking at the statistics of Z_{DR} values in very light drizzle. Bringi et al. (1983) show that for individual observations to be accurate to 0.2 dB requires about 60 independent samples of Z and a correlation coefficient of the H and V time series (see next subsection) above 0.98. These values should be satisfied by an operational radar scanning in rain with a resolution of about 1–2 km, but in practice, other factors may limit the accuracy of Z_{DR}:

- *Reflectivity gradients.*
  In the presence of reflectivity gradients, a significant fraction of the power may be received through the side lobes which are mismatched in their polarisation characteristics. Herzegh and Carbone (1984) report spurious values of Z_{DR} of up to 10 dB in low reflectivity regions adjacent to intense echoes. The sidelobes of operational radars may well quite commonly lead to the introduction of errors in Z_{DR} of 0.2–0.5 dB.
- *Mismatched beams.*
  If the H and V beam patterns are not well matched, then the two beams will not be sampling the same volume of precipitation, and the correlation may fall below 0.98.
- *Triple scattering echoes.*
  Triple scattering echoes (Illingworth and Caylor, 1988; Hubbert and Bringi, 2000) can occur in a weak echo region behind an intense storm because some of the radiation scattered by the main echo is reflected by the ground and then scattered a third time by the main echo back to the antenna. No vertically polarised power is scattered to the ground, so such echoes are associated with anomalously high values of Z_{DR}.

- *Differential attenuation.*
  This leads to a much more severe problem at C-band. The large oblate raindrops attenuate the horizontally polarised radar beam more than the vertical one, and this differential attenuation leads to increasingly negative values of Z_{DR} with range. One example is shown by Collier and Hardaker (2003), this book. To demonstrate the severity of this problem, note that at C-band, a differential attenuation of 0.2 dB results from only 1 km of rain at 65 mm hr^{-1} or 2 km of rain at 40 mm hr^{-1}. Upton and Fernandez-Duran (1999) report values as low as -5 dB behind intense echoes, such attenuation must be corrected to an accuracy of 0.2 dB.
- *Hail.*
  Finally, as described above, the presence of hail mixed with the rain will lead to a reduction in Z_{DR}, so that application of the $R(Z, Z_{DR})$ algorithm will lead to the inference of a spuriously high rainfall rate.

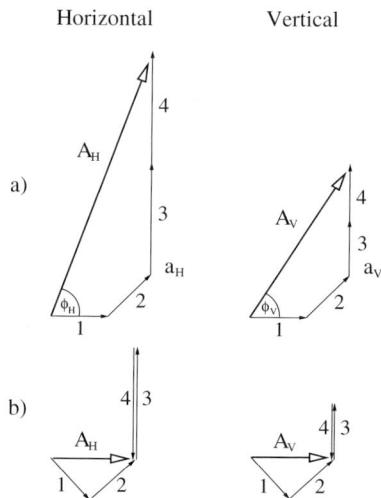

Fig. 5.2. Argand diagram showing that the addition of backscattered amplitudes from the mixture of particles shapes in Fig. 5.1; (a) at time $t = t_1$ and (b) at a later time $t = t_2$, leads to a lowering of the correlation of the H and V returns and fluctuations in the differential phase

5.2.3 Copolar Correlation Coefficient

The copolar correlation coefficient (ρ_{hv}) is the correlation of the time series of successive estimates of Z_H and Z_V and is a measure of the variety of shapes of hydrometeors present. The importance of this parameter is demonstrated in Fig. 5.2 which shows schematically how the amplitude and phase of the signal returned from four targets, two spheres and two oblate spheroids (see Fig. 5.1), add to give a fluctuating return, as the targets reshuffle in space between two times, $t = t_1$, and $t = t_2$. In the figure, the spheres and the oblate spheroids all have the same backscatter amplitude (a_v) for vertical polarisation, but for horizontal polarisation, the spheres have $a_h = a_v$ whilst for the oblate spheroids $a_h = 2a_v$. The figure shows that as the particles reshuffle, the relative phases change and the resultant amplitudes A_h and A_v at the two polarisations will fluctuate. In this case because the ratio of a_h to a_v is not the same for all the scatterers, then as the particles reshuffle, the ratio of A_h to A_v will fluctuate and so the time series of Z_H and Z_V (proportional to A_h^2 and A_v^2) will not be perfectly correlated. As the correlation falls, then the estimate of Z_{DR} also becomes less accurate. The correlation coefficient, (ρ_{hv}), will only be unity if the ratio of a_h to a_v is the same for all the particles, that is to say, the particles all have the same shape and orientation in space. The values of correlation can be used to identify targets as follows:

- *Raindrops.*
  Raindrops are nearly spherical, and if oblate highly oriented, and so the correlation is at least 0.98.
- *Hail and the bright band.*
  The correlation drops to about 0.9 when a wide mixture of hydrometeor shapes is present. This occurs in the bright band or when hail particles are large enough to result in non-Rayleigh scattering.
- *Clutter and anaprop.*
  Returns from the ground, whether as clutter or anomalous propagation, have a correlation which is essentially zero because the returns in the two polarisations have almost random amplitudes and phases.

In practice, the observed values of ρ_{hv} will be lowered somewhat because the H and V beams can never be perfectly matched in space and also because H and V are not sampled simultaneously. The effect of staggered H and V samples can be corrected by either assuming a Gaussian Doppler width of the target or, more accurately, by interpolating the two time series. For very precise estimates of ρ_{hv} the dwell times must be much longer than those allowed in an operational system (Illingworth and Caylor, 1991).

5.2.4 K_{DP} – Specific Differential Phase

The velocity of a horizontally polarised radar wave propagating through a region containing oblate raindrops is slightly less than that of a vertically polarised wave; so, as shown in Fig. 5.3, the phase of the horizontal return (ϕ_h) lags progressively behind the phase of the vertical polarised return (ϕ_v). As a result, the differential phase, $\phi_v - \phi_h = \phi_{DP}$, normally increases monotonically with range and K_{DP}, the rate of change of ϕ_{DP} with range (measured in $°$ km^{-1}), should be positive. K_{DP} should not be affected by tumbling hailstones but should increase with rain rate. $R(K_{DP})$ relationships (Sachidananda and Zrnic, 1986, 1987) have been proposed of the form,

$$R = aK_{DP}^b, \qquad (5.3)$$

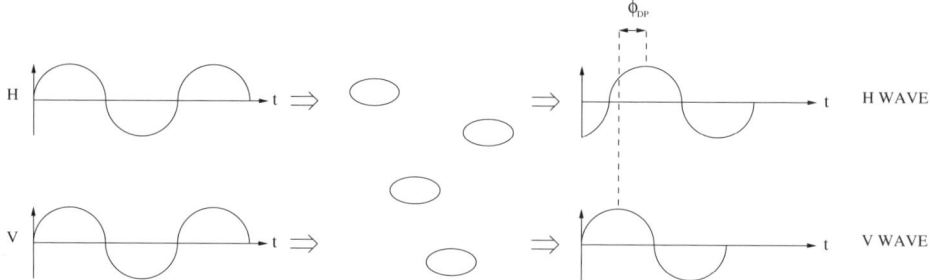

Fig. 5.3. Differential phase shift, ϕ_{DP}, introduced because the H wave propagates more slowly than the V wave through a region of oblate raindrops

with a value of $b = 0.866$. This technique appears to be very attractive and has the following advantages (Blackman and Illingworth, 1995; Ryzhkov and Zrnic, 1996:)

- *Calibration and attenuation.*
  The measurement of phase is immune to calibration problems and is unaffected by attenuation.
- *Hail.*
  The phase measurement should not be affected by hail.
- *Linearity.*
  If the value of 'b' in (5.3) is 0.866, then K_{DP} has a more linear dependence on R than conventional $R(Z)$ relations. Bringi and Chandrasekar (2001) show that K_{DP} should be less dependent upon the precise form of the drop shapes and, assuming Rayleigh scattering, is approximately proportional to the product of the liquid water content times D_0, or to the fourth moment of the drop size spectrum and so should be closer to R than Z.
- *Integrated rainfall over a catchment.*
  If (5.3) is nearly linear, then the total integrated differential phase shift along the path of the radar beam (Φ_{DP}) will be a measure of the integrated rainfall along the path.
- *Use of partially blocked beams.*
  If ϕ_{DP} can be observed in low elevation beams which are partially blocked so that Z values are considerably reduced, it may be possible to estimate R from K_{DP}, and so use these low elevations beams, which will remain in the rain, out to greater distances.

A difficulty arises because the phase shifts are quite small. If these advantages are to be realised, then ϕ_{DP} should be estimated to 1° or better, but unfortunately the phase measurement can be quite noisy. Firstly, there is a theoretical limit to the accuracy of the ϕ_{DP} estimate:

- *Sequential pulses.*
  ϕ_{DP} is usually measured from sequential pulses transmitted with alternate

horizontal and vertical polarisation, but the phase itself is continually changing due to the mean Doppler velocity of the targets, and so interpolation is required to estimate $\phi_{\rm DP}$. The accuracy of this interpolation is limited by the Doppler width of the target. Ryzhkov and Zrnic (1998b) show that for 60 pulse pairs and a normalised spectral width ($= 2\sigma_{\rm v} t_{\rm s}/\lambda$, where $t_{\rm s}$ is the time between pulses) of 0.1, the standard error of $\phi_{\rm DP}$, (σ_ϕ), should be about $1°$.

- *'Hybrid' transmission at 45°.*
  Transmission of a pulse polarised at 45° and simultaneous reception at H and V (Bringi and Chandrasekar, 2001) avoid the interpolation problem, but the correlation of the target is slightly less than unity, and this introduces noise into the $\phi_{\rm DP}$ estimate which still leads to a σ_ϕ of about 1°. The mechanism by which fluctuations in the estimate of $\phi_{\rm DP}$ are introduced by the low correlation can be seen in Fig. 5.2.

This theoretical accuracy of $\phi_{\rm DP}$ is not achieved in practice. Ryzhkov and Zrnic (1995) cite a typical σ_ϕ at S-band of 3°; Hubbert et al. (1993) and Keenan et al. (1998) quote similar figures for their C-band radars. This degradation is probably caused by the following factors:

- *Gradients of reflectivity.*
  Gradients in reflectivity across the beam cause one part of the beam to experience a larger $\phi_{\rm DP}$ than another (Ryzhkov and Zrnic, 1998a) and can give rise to negative $K_{\rm DP}$. Gradients along the beam lead to biases in inferred $K_{\rm DP}$ (Gorgucci et al., 1999a) whereby, for example, a sudden change in Z of 30 dBZ can bias the value of $K_{\rm DP}$ by up to 30%.
- *Extreme sensitivity to clutter.*
  If the random phase of a ground clutter signal is added to a precipitation return which has an amplitude ten times larger, then the resultant signal will have a phase noise of 5°. This will render the value of $\phi_{\rm DP}$ virtually useless, even though the ground clutter will have a Z (i.e. an intensity) which is 20 dB below (i.e. 1% of) the Z of the precipitation.
- *Side lobes.*
  $\phi_{\rm DP}$ is much more sensitive to signals from mismatched side lobes than $Z_{\rm DR}$. If the phase of the side lobe signal is mismatched in H and V, then again all we need is a side lobe signal which is 20 dB below the precipitation return from the main lobe to introduce a 5° noise in $\phi_{\rm DP}$.
- *Differential phase shift on backscatter.*
  If the particles or rain or hail are large enough for non-Rayleigh scattering, then a differential phase shift on backscatter, δ, will be introduced as a transient superposed on the monotonic increase in $\phi_{\rm DP}$ due to propagation and can lead to apparent negative values of $K_{\rm DP}$. For rain, Testud et al. (2000) show that the effect is negligible at S-band, but at C-band, δ is about 1°, 4.5°, and 9.5° for $Z_{\rm DR} = 2$ dB, 3 dB, and 4 dB, respectively. Ryzhkov and Zrnic (1996) suggest removing the δ transient by 'light' filtering over a distance of 2.4 km for heavy rain and 'heavy' filtering over 7.2 km for light rain.

The result of the above effects is a noisy trace of ϕ_{DP} not always increasing monotonically with range. Filtering is difficult because the extent of the δ region may not be known and it leads to an unacceptable loss of spatial resolution. Ryzhkov and Zrnic (1996) suggest using the modulus of K_{DP} in (5.3) to avoid the problem of negative values, but this does not seem physically justifiable.

5.2.5 Linear Depolarisation Ratio (LDR)

LDR is defined as $10 \log (Z_{VH}/Z_H)$, where Z_{VH} is the cross-polar return at horizontal polarisation for a vertically polarised transmission. Only oblate particles falling with their major axis at an angle to the vertical or horizontal direction yield a cross-polar return. LDR is able (Frost et al., 1991; Straka et al., 2000) to identify melting snowflakes associated with the bright band; on occasion it can differentiate ice from rain. Finally, ground clutter has anomalous LDR signals.

Figure 5.4a demonstrates that raindrops result in negligible depolarisation because they fall with their major axis aligned in the horizontal; the horizontal field excites a single dipole along the major axis which reradiates with horizontal polarisation. Contrast this with the oblate nonaligned hydrometeor in Fig. 5.4b; the incident field excites two dipoles along the major and minor axes, and if the polarisabilities along the two axes are not the same, then the resultant induced dipole is no longer parallel to the induced field, and so the reradiated field has an orthogonal or cross-polar component.

The difference in polarisability is largest for wet oblate particles, so the highest values of LDR of about -15 dB are associated with melting snowflakes because such particles rock and roll as they fall. The LDR of melting hail and dry high density ice crystals is in the range -20 to -26 dB. Rain has an LDR generally below -30 dB but the observed value is often limited by the antenna isolation. The cross and copolar signals are essentially uncorrelated, so if the error in Z is about 0.5 dB, then the value of LDR can be estimated at 0.7 dB. Masking out of echoes, where LDR > -10 dB, removes much of the ground clutter (see Sect. 5.4).

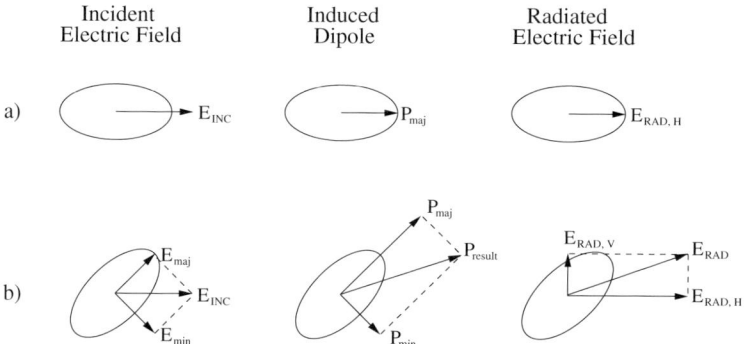

Fig. 5.4. Linear depolarisation ratio. (**a**) No depolarisation of incident wave by aligned raindrops. (**b**) Depolarisation by nonaligned oblate spheroids

5.2.6 An Example of a PPI Scan Through Vigorous Convection

An S-band low elevation PPI scan though vigorous convective showers which illustrates the signatures of Z, Z_{DR} and ϕ_{DP} is displayed in Fig. 5.5 which also appears as a colour plate. The observations were made with the Chilbolton 0.28° beam-width radar in the UK on 28 July 2000 at 1308 UT at 0.5° elevation, so that the beam is dwelling in the rain. Because of the narrow beam-width, we believe that this image is not affected by any of the artifacts discussed above such as side lobes and reflectivity gradients either across or along the main beam. From Fig. 5.5 we see that once Z is above about 35 dBZ, then Z_{DR} reaches about 1 dB, but that the high phase shifts generally result from regions of Z above 45 dBZ extending over a few kilometres. Two such regions producing large phase shifts are clearly visible: one at 90 km E and −12 km N; the other at 115 km E and about −8 km N. Even for such large phase shifts, the estimates of K_{DP} from differentiating the phase are quite noisy and have a spatial resolution which is much less than the 300 m of the Z and Z_{DR} and observations. Note that the highest values of Z_{DR} are about 4 dB, but are quite localised and often not coincident with the highest Z; they do not produce any phase shift, and we believe that they result from low concentrations of anomalously large raindrops.

5.3 Raindrop Shapes and Size Spectra

5.3.1 Introduction

Both Z_{DR} and K_{DP} essentially provide information on the mean shape and hence the size of raindrops. If this is to lead to improved rainfall rate estimates, then the shape of raindrops as a function of size must be known precisely, but there is currently some uncertainty over these shapes. In addition, if the size sensed by the radar is to be related to the appropriate mean drop size for the rainfall rate, then we need to know the range of size spectra for naturally occurring rainfall. Again, there is some uncertainty over these spectra.

5.3.2 Raindrop Shape Model

Until very recently, a simple 'linear' relationship between axial ratio, r, and drop diameter, D, in mm, was in widespread use (Pruppacher and Pitter, 1971):

$$r = 1.03 - \beta D, \qquad (5.4)$$

with β having a value of 0.062 mm^{-1} for drops larger than 0.5 mm diameter; drops smaller than 0.5 mm are assumed to be spherical. However, in some of the earliest observations, Goddard et al. (1982) found that observed values of Z_{DR} in the range 0 dB to 1.5 dB were from 0.5 to 0.2 dB lower than the values of Z_{DR} computed using the linear shapes from a raindrop disdrometer at the ground 200 m below the radar beam. They could only obtain agreement if they empirically adjusted the shape of drops smaller than 2.5 mm to be more spherical

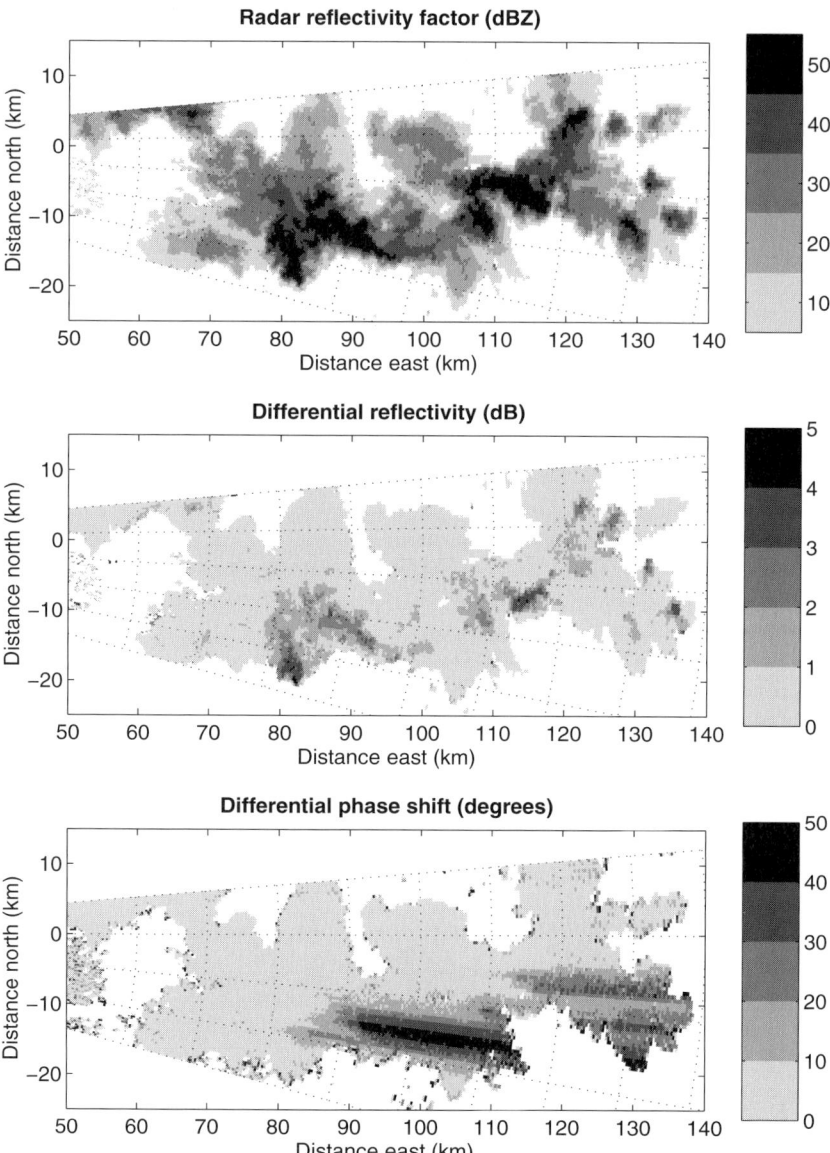

Fig. 5.5. An example of Z, Z_{DR} and ϕ_{DP} observed with the narrow beam S-band Chilbolton radar in the UK on 28 July 2000. The data are from a low elevation (0.5°) PPI scan, so the beam is dwelling in the rain. Even for these large values of ϕ_{DP} the derived values of K_{DP} are rather noisy with much poorer range resolution than Z and Z_{DR}. Figure courtesy of R.J.Hogan (U of Reading). Radar data kindly supplied by RCRU, Rutherford Appleton Laboratory. This figure also appears as color Fig. 13 on page 322

than given by (5.4). They proposed (Goddard et al., 1995) a new drop shape model:
$$r = 1.075 - 0.065D - 0.0036D^2 + 0.0004D^3 \qquad (5.5)$$
for drops larger than 1.1 mm with smaller drops again assumed spherical. Confirmation of such drop shapes is not simple. If the raindrop shape is measured close to the ground, then the drops have just experienced atypically high shear as they fall through the boundary layer. Similarly, drop shapes measured with in-situ aircraft probes have been subjected to unusual stresses as they are deflected around the aircraft wing. It is only recently that careful experiments in long wind tunnels, such as those carried out by Andsager et al. (1999), and references therein, have proposed the following polynomial:
$$r = 1.012 - 0.01445D - 0.01028D^2 \qquad (5.6)$$
for D in the range 1–4 mm. The three drop shape models are compared in Fig. 5.6. Equations (5.5) and (5.6) are very similar and predict virtually the same values of K_{DP} and Z_{DR}. This remarkable independent confirmation of the empirical adjustment gives us confidence that the 'linear' shapes are indeed an oversimplification and that these 'new' drop shapes are probably correct. The last two equations lead to negligible differences in the predicted polarisation parameters; when we refer to 'new' drop shapes, we will be using those of (5.5), as opposed to the 'linear' shapes of (5.4). Equation (5.6) is equally valid.

The linear drop shapes used in the past lead to two effects. Firstly, the linear drop shapes are more oblate than those occurring naturally, leading to predicted values of Z_{DR} and K_{DP} which are too high, so that values of c in (5.2) and a in (5.3) are too small; this may account for the underestimates of rainfall using K_{DP} reported, for example, by May et al. (1999) and Petersen et al. (1999)

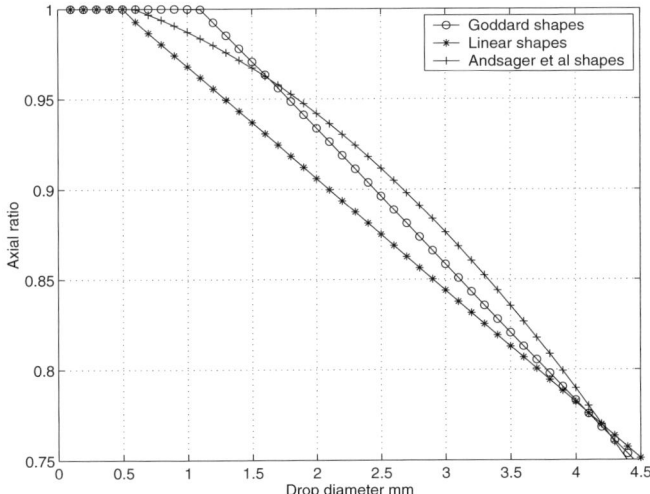

Fig. 5.6. Different raindrop shape models

who used the 'linear' raindrop, as discussed by Bringi and Chandrasekar (2001). Secondly, for drops in the range 1–3 mm, the 'new' drop shapes predict a larger value of β than the 0.062 in (5.4); this means that the increase in the values of $Z_{\rm DR}$ and $K_{\rm DP}$ as the rainfall rate increases, will be more rapid than predicted from the linear drops, so that in reality the value of b in the $R(K_{\rm DP})$ exponent will be lower than the value of 0.866 predicted using linear shapes (Illingworth and Blackman, 2002).

5.3.3 Raindrop Size Spectra

To gauge the performance of polarisation techniques in improving rainfall estimates we need to know how variations in the drop size spectra affect the representativeness of the mean drop size inferred from the polarisation parameters. Rather than the simple exponential representation in (5.1), natural raindrop size spectra are well represented by a normalised gamma function of the form,

$$N(D) = N_{\rm w} f(\mu) \left(\frac{D}{D_0}\right)^\mu \exp\left[-\frac{(3.67+\mu)D}{D_0}\right], \qquad (5.7)$$

where

$$f(\mu) = \frac{6}{(3.67)^4} \frac{(3.67+\mu)^{\mu+4}}{\Gamma(\mu+4)}, \qquad (5.8)$$

which, when $\mu = 0$, reduces to the simple exponential form of (5.1) with $N_{\rm w} \equiv N_0$. This form has the advantage that the three variables are independent, with D_0 being the median volume drop diameter, μ being the shape of the drop spectrum, and $N_{\rm w}$ is a normalised drop concentration, so that the liquid water content remains constant even if μ changes, and is preferable (Illingworth and Blackman, 2002) to the nonnormalised gamma function of Ulbrich (1983),

$$N(D) = N_0 D_0^\mu \exp\left[-\frac{(3.67+\mu)D}{D_0}\right]. \qquad (5.9)$$

We have followed the convention of Bringi and Chandrasekar (2001); $N_{\rm w}$ is the same as the $N_{\rm L}$ used by Illingworth and Blackman (1999, 2002) and the N_0^* used by Testud et al. (2001, 2003 - this book), with 'w' and 'L' both indicating normalisation with respect to liquid water content. These equations are sometimes expressed in terms of $D_{\rm m}$, the mass weighted mean diameter, in which case $D_{\rm m}$ is equal to D_0 times the factor $(4+\mu)/(3.67+\mu)$.

Ulbrich (1983) derived the expected range of values of N_0 and μ in the nonnormalised (5.9) in natural rainfall by comparing expressions for R and Z from the appropriately weighted integral of (5.9) with the 69 $Z(R)$ relationships of Battan (1973); he deduced, the range of μ was from -1 to 5. The mathematical validity of this approach for deriving the 'Ulbrich' range of values of N_0 and μ has been questioned by Illingworth and Blackman (2002), but it has become common practice to derive relationships between R and the polarimetric variables by cycling over these 'Ulbrich' values of N_0 and μ, calculating Z, $Z_{\rm DR}$, $K_{\rm DP}$ and R and performing a nonlinear regression.

We now explore other methods of deriving the range of values of μ, N_w and D_0 in naturally occurring rainfall. One approach is to use observed raindrop spectra. A least squares fit is not appropriate because it gives equal weight to small and large rain drops, whereas for radar and rainfall, the larger drops are much more important. Kozu and Nakamura (1991) and Illingworth and Johnson (1999) both equated the sixth, fourth and third moments of observed raindrop size distributions to the appropriately weighted integral of the gamma function and deduced values of μ in the range 0 to 15 with a mean value of about 5 or 6. Illingworth and Johnson (1999) found that in the UK the mean value of the normalised raindrop concentration, N_w, was close to the Marshall–Palmer value of 8000 m^{-3} mm^{-1} with a standard deviation of a factor of three. Bringi and Chandrasekar (2001) found a similar spread of N_w when they analysed an entire season of rainfall spectra from Darwin, Australia.

These high values of μ derived from fitting the higher moments have been criticised because they are very dependent upon maximum drop size in the spectra which are poorly sampled by disdrometers. The values obtained depend upon the moments chosen for the fit. Higher moments are appropriate for relationships between Z and R and lead to higher values of μ. Testud et al. (2001) and Bringi and Chandrasekar (2001) infer μ values closer to unity; this may be because they first derive N_0 and D_0 in an exponential spectrum by fitting moments, but then choose μ to minimise a least squares fit to the observed spectrum. This procedure for fixing μ assigns equal weight to drops of all sizes and may lead to values of μ which are inappropriate for the higher moments involved in radar studies. Ulbrich and Atlas (1998) used truncated moments by limiting the experimental spectra to $D_{\max}$ and found that this reduced μ from about 4 to a median value of 0 with a mean of 1.6.

The disadvantage with the above techniques is the use of a disdrometer which has poor sampling of the larger drops which are important for the higher moments. This can be avoided by appealing to radar measurements themselves; because of the much larger radar sampling volume, the large drops are sampled satisfactorily. Wilson et al. (1997) reported values of μ derived from the 'Differential Doppler Velocity' (DDV), the difference in the Doppler velocity of rain for horizontally and vertically polarised radiation with the radar beam dwelling at a finite elevation. They showed that the value of DDV as a function of Z_{DR} depends upon the value of μ and found a range of μ between 2 and 10 with a mean value of 5. There is also some evidence from observations of the Doppler spectral width of rainfall at vertical incidence which indicate (pers. comm., D. Bouniol) that the width is significantly narrower than would be observed for an exponential and is consistent with a μ of around 5.

5.3.4 Implications for $Z = aR^b$ Relationships

It appears that the range of naturally occurring raindrop size spectra is accurately described by the normalised gamma function in (5.7) with values of μ near 5 and values of N_w which have a standard deviation of a factor of three. We now follow the treatment of Bringi and Chandrasekar (2001) and show that the normalised gamma function leads to $Z(R)$ relationships of the form $Z = aR^{1.5}$ with different values of a reflecting changes in N_w. There is, however, an additional dependence of a upon the value of μ. Integrating the appropriately weighted normalised gamma function produces expressions of the form,

$$Z = F_Z(\mu) N_w D_0^7, \qquad (5.10)$$

and assuming the terminal velocity varies as $D^{0.67}$,

$$R = F_R(\mu) N_w D_0^{4.67}; \qquad (5.11)$$

eliminating D_0 gives

$$\frac{Z}{N_w} = H(\mu) \left(\frac{R}{N_w}\right)^b \qquad (5.12)$$

where b = $7/4.67 \approx$ 1.5 or

$$Z = H(\mu) N_w^{1-b} R^b = \frac{H(\mu) R^{1.5}}{\sqrt{N_w}}. \qquad (5.13)$$

It is usually assumed that $H(\mu)$ is a slowly varying function of μ. This is so if we restrict μ to the range 0 to 2, but earlier we argued that μ has an average value of 5 and can vary between 0 and 10. As μ increases from 0 to 5 and 10, the value of $H(\mu)$ falls by a factor of 1.55 (2 dB) and 1.79 (2.5 dB).

Accordingly we can associate changes in the multiplicative factor a in $Z = aR^b$ not only with changes in N_w but also with μ. For example, a standard deviation of N_w of a factor of three will lead to a change in a of a factor of 1.7 and a change in rainfall rate derived from Z of ±40%. To this, we must add a similar change in a and Z from the range of μ; if the two are uncorrelated, then the error in R will be about 64%. If the range of values of N_w is a factor of ten, this will lead to changes in a of a factor of three and a factor of two error in the rain rate. This gives us a good estimate of the error expected from a simple $Z = aR^b$ relationship because of variability in drop spectra. In Sect. 5.6, we will discuss a simple polarimetric method of estimating N_w and hence gauging the value of a, leaving only the variability due to the unknown μ.

5.4 Identification of Ground Clutter and Anomalous Propagation

Anomalous propagation can be one of the most difficult signals to reject using conventional radar (Pamment and Conway, 1994), but fortunately their identification via polarisation returns is unambiguous. Both clutter and 'anaprop' give rise to non-Rayleigh scattering, so that the both the amplitudes and phases of the H and V returns are essentially uncorrelated. Accordingly, these signals can be identified by the following:

- The most reliable technique relies upon a copolar correlation of essentially zero as opposed to values above 0.9 for precipitation echoes (Caylor and Illingworth, 1992; Ryzhkov and Zrnic, 1998b).
- A noisy ϕ_{DP} profile (Ryzhkov and Zrnic, 1998b).
- Higher values of LDR than occur in natural precipitation (Wilson et al., 1995; Hagen, 1997).
- Noisy estimates of Z_{DR} which fluctuate between ± 3 dB from gate to gate (Hall et al., 1984).

5.5 Improved Rainfall Rates using Polarisation Parameters

5.5.1 Introduction

In this section, we discuss the improved accuracy in rainfall rates by using Z_{DR} or K_{DP}. The error in R derived from Z alone can be up to a factor of two; the goal is to reduce this to $\pm 25\%$. We shall initially consider cases when we know that the beam is dwelling in rain and there is no ice present. This will apply to summertime situations in the midlatitudes out to a range of about 75 km when the melting layer is above 2 km and no hail is present. The situation is more common in the tropics because the melting level is much higher and large hail is rarer.

5.5.2 $R = f(Z, Z_{DR})$

The procedure outlined in Sect. 5.2.2 for calculating rainfall from Z and Z_{DR} seems straightforward. Bringi and Chandrasekar (2001) suggest that at S-band $D_0 = 1.529 Z_{DR}^{0.467}$ (with Z_{DR} in dB) and that we should be able to estimate D_0 from Z_{DR} to within 0.1 mm; but plots of D_0 as a function of Z_{DR} computed for normalised gamma functions (Fig. 5.7) with the 'new' drop shapes are initially discouraging, showing great sensitivity to μ. From Fig. 5.7, if $Z_{DR} = 1.5$ dB and $\mu = 0$, then we have $D_0 = 1.5$ mm, but for the narrower spectrum with $\mu = 5$ and the same Z_{DR}, the absence of larger drops in the tail of the spectrum means that we need $D_0 = 2.2$ mm. However, the goal is to measure R not D_0, and R is a function of both D_0 and μ. Fortunately, for $D_0 = 2.2$ mm and $\mu = 5$, both the rainfall rate and the value of Z are about five times higher than for $D_0 = 1.5$ mm and $\mu = 0$. As we shall see in the next paragraph, this means that the value of Z per unit R as a function of Z_{DR} for the normalised gamma functions has a much reduced dependence upon μ, with the curves for $\mu = 0$ and $\mu = 5$ only 1 dB apart.

In Sect. 5.2.2, we argued that Z/R should be a function of Z_{DR} (5.2). Values of Z/R as a function of Z_{DR} for the 'linear' (5.4) and 'new' (5.5) drop shapes are plotted in Fig. 5.8. These curves can be interpreted as the Z which would give R of 1 mm hr^{-1} for a particular value of Z_{DR}. The rain rate scales with Z, so, if for a given Z_{DR}, the observed value of Z is 'x'dB above the line, then the rain rate is 'x'dB above 1 mm hr^{-1}.

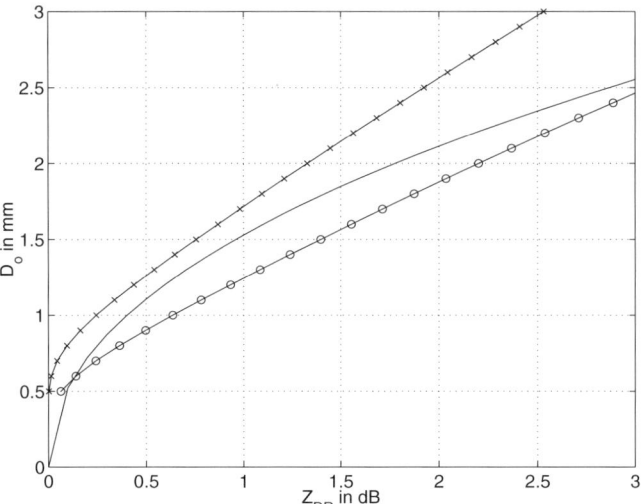

Fig. 5.7. Plot of D_0 against Z_{DR} for the new drop shape models at S-band with $\mu = 0$ (o) and $\mu = 5$ (x). The *solid line* is $D_0 = 1.529 Z_{\mathrm{DR}}^{0.467}$ suggested by Bringi and Chandrasekar (2001)

It was argued in Sect. 5.3.3 that a μ of 5 is the average value for natural rainfall. A third-order polynomial fit of Z/R as a function of Z_{DR} for $\mu = 5$ gives the following expression for (5.2) (Illingworth and Blackman, 2002):

$$Z/R = 21.48 + 8.14 Z_{\mathrm{DR}} - 1.385 (Z_{\mathrm{DR}})^2 + 0.1039 (Z_{\mathrm{DR}})^3 \qquad (5.14)$$

at S-band which is accurate to 0.5 dB. The equivalent expression at C-band is

$$Z/R = 21.50 + 8.35 Z_{\mathrm{DR}} - 1.89 (Z_{\mathrm{DR}})^2 + 0.1976 (Z_{\mathrm{DR}})^3 \,. \qquad (5.15)$$

This assumes that the values of Z_{DR} have been corrected for any attenuation to an accuracy of 0.2 dB.

An alternative approach has been widely adopted in the literature (see discussion in Bringi and Chandrasekar, 2001) in which the requirement for Z to scale with R for constant Z_{DR} is dropped and a relationship of the form,

$$R = c_1 Z_{\mathrm{H}}{}^{a_1} 10^{0.1 b_1 Z_{\mathrm{DR}}} \,, \qquad (5.16)$$

is used. The coefficients are chosen by allowing the N_{w} (or N_0), D_0 and μ to cycle over the 'Ulbrich' (Sect. 5.3.3) range of values, calculating Z, Z_{DR} and R and performing a nonlinear regression. The curve proposed by Bringi and Chandrasekar (2001) with $c_1 = 0.0067, a_1 = 0.93, b_1 = -3.43$ at S-band using the 'new' drop shapes for $R = 1$ mm hr^{-1} is also plotted in Fig. 5.8 and predicts rainfall rates 2 dB higher than (5.14), once Z_{DR} is above 1 dB. For a higher rain rate of 10 mm hr^{-1} the nonlinearity with Z, introduced by $a_1 = 0.93$, means that the curve would be 0.7 dB higher, but the rainfall overestimate would be

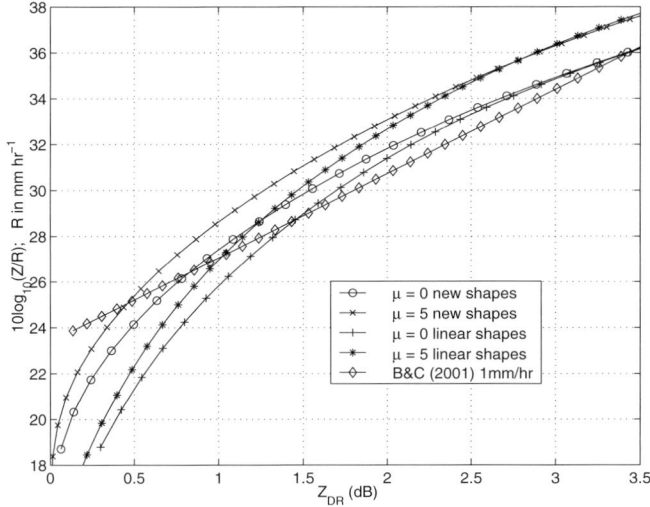

Fig. 5.8. Plot of Z/R against Z_{DR} for the linear and new drop shape models at S-band

still over 1 dB. Physically, it is difficult to justify using a value of a_1 which is not unity, unless the heavier rainfalls are systematically associated with higher values of μ. Earlier relationships based on (5.16) (Gorgucci et al., 1994; Chandrasekar and Bringi, 1988; Chandrasekar et al., 1990) used the linear drop shapes and the 'Ulbrich' range of N_0 and μ (Illingworth and Blackman, 2002) leading to curves about 1–2 dB lower than the values of Bringi and Chandrasekar (2001) and a probable overestimate of the rainfall rate by up to a factor of two.

If we seek a value of R accurate to $\pm 25\%$ (1 dB) then from Fig. 5.8 we note:

- *Accuracy of Z_{DR} estimate.*
  A 25% accuracy in R implies that Z/R should be known to 1 dB, which from the slope of the curves imposes a limit of 0.2 dB for the accuracy of the Z_{DR} estimate for R > 10 mm hr^{-1}. For a rainfall rate of 3 mm hr^{-1}, an accuracy of 0.1 dB is needed. As we saw in Sect. 5.2.2, this is very difficult to achieve.
- *Calibration of Z.*
  R is only as accurate as the value of Z, so a 25% error in R implies that Z must be calibrated to 1 dB. This is a serious constraint and will be discussed further in Sect. 5.6.
- *Sensitivity to drop shapes.*
  The correct choice of drop shapes is crucial. The 'new' drop shapes will lead to rainfall rates from 1 or 3 dB lower than the traditional 'linear' shapes. Using the Andsager shapes (5.6) rather than Goddard shapes (5.5) changes the values of Z/R by less than 0.5 dB.
- *Theoretical limit to accuracy of R estimate.*
  Supposing that the drop shapes are known, then the naturally occurring variability of μ limits the fundamental accuracy of the technique. If the

mean value of μ in rain is 5 but varies between 2 and 10, then this will introduce an uncertainty in the R estimates of about 0.5 dB or $\pm 12\%$. A greater variability of μ with values down to 0 implies an error of 1 dB or 25%.

- *Sensitivity to truncation of drop spectrum.*
  A value of $\mu = 5$ in rain introduces a natural truncation of the drop spectrum so that the curves in Fig. 5.8 are insensitive to the choice of maximum drop size, providing it is larger than 8 mm.

5.5.3 $R = f(K_{DP})$

The potential advantages of deriving R from K_{DP} were summarised in Sect. 5.2.4, tempered by remarks on the noisy character of observed values of K_{DP}. Plots of the values of K_{DP} (one-way) as a function of R at S-band for the 'linear' and 'new' drop shape models using the normalised gamma function with $\mu = 0$ and 5 with constant $N_w = 8000$ m^{-3} mm^{-1} are displayed in Fig. 5.9 and demonstrate:

- *Sensitivity to drop shape model.*
  For a value of K_{DP} of 0.55° km^{-1}, the linear drop shapes with $\mu = 0$ predict a rainfall of about 20 mm hr^{-1} rather than about 25 and 30 mm hr^{-1} for the 'new' drop shapes with $\mu = 0$ and 5, respectively. This may account for the persistent underestimates of rainfall using K_{DP} with these drop shapes (e.g., most recently, Brandes et al., 2001) from the widespread use of the Equation:

$$R = 40.56 K_{DP}^{0.866}. \tag{5.17}$$

Bringi and Chandrasekar (2001) suggest that using a coefficient of 50.7 rather than 40.56 is more appropriate for the 'new' drop shapes.

- *Increased nonlinearity and sensitivity to drop concentration and μ.*
  We have argued that $\mu = 5$ is more appropriate than $\mu = 0$; over the range 10–100 mm hr^{-1} with the new drop shapes at S-band (9.75 cm), this yields

$$R = 50.1 K_{DP}^{0.7} \quad or \quad K_{DP} = 0.00417 R^{1.4}. \tag{5.18}$$

Note that the index for R of 1.4 in (5.18) is almost as large as the value of 1.5 derived for a $Z(R)$ relationship in Sect. 5.3.4 and identical to the index for the NEXRAD default Z(R) relationship ($Z = 300R^{1.4}$) quoted in Brandes et al. (1999). This suggests that, apart from the immunity to hail, many of the advantages claimed for the $R(K_{DP})$ approach may be difficult to realise. Increases of N_w by a factor of ten will lead to the inferred rainfall rate being underestimated by a factor of two for all values of R. Changing μ from 5 to 0 reduces R by about 30% for values of R of about 20 mm hr^{-1}.

The $R(K_{DP})$ dependence at C-band is similar to that shown in Fig. 5.9 but with the values of K_{DP} approximately scaled by the increased frequency. The appropriate equation for the 'new' drop shapes and $\mu = 5$ at C-band (5.35 cm) is

$$R = 31.4 K_{DP}^{0.7} \quad or \quad K_{DP} = 0.00802 R^{1.4}. \tag{5.19}$$

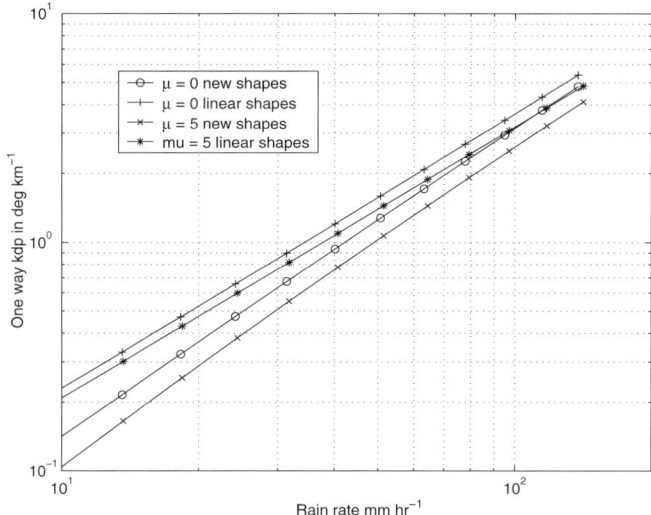

Fig. 5.9. Plot of R against one-way K_{DP} at S-band for the linear and new drop shape models with $N_w = 8000$ m^{-3} mm^{-1} and $\mu = 0$ and 5

Matrosov et al. (1999) present differential phase observations at X-band (3.2 cm) and suggest an equation similar to (5.19) for rain rates up to 15 mm hr^{-1} with a coefficient of 14.2 and an exponent of 0.85 for the linear shapes in (5.4) with β of 0.062 mm^{-1}; but they found better agreement with rain gauges if they used much less oblate drops with a rather unrealistic β of 0.044 mm^{-1} which gave a coefficient of 20.5. For these rain, rates they calculated that the maximum value of δ was an insignificant 1°. Although ϕ_{DP} is larger at X-band, in heavy rain, the bigger drops will Mie scatter, so δ will be larger and the correlation lower; both effects lead to an increasingly noisy ϕ_{DP}.

Thus far, we have only discussed the form of the $R(K_{DP})$ dependence. The next stage is to consider the error in the derived values of R, σ_R, resulting from the error in the differential phase estimate σ_ϕ which can in theory be as low as 1° but in practice is at least 3°. The value of the one-way K_{DP} is derived by a least squares fit through the N gates each of length $\Delta(r)$ to give a range resolution of $L = N\Delta(r)$ and has an error given by

$$\sigma_{KDP} = \frac{\sigma_\phi}{L} \sqrt{\frac{3}{N - (1/N)}}. \tag{5.20}$$

The errors in K_{DP} predicted from (5.20) appear quite encouraging (Bringi and Chandrasekar, 2001); with a gate length of 150 m, a resolution of 2.4 km and σ_ϕ of 1 and 3°, the value of σ_{KDP} from a fit over the 16 gates is 0.18 and 0.54° km^{-1}, respectively. However, Blackman and Illingworth (1997) extended the analysis to the context of an operational radar and derived the error in the rainfall rate, σ_R, with an azimuthal resolution L to match the range resolution, for a radar with a beam-width of θ_b at a range r, and derived an expression:

$$\sigma_R = \frac{\sigma_\phi}{abR^{b-1}L^2}\sqrt{12\theta_b r \Delta(r)}, \qquad (5.21)$$

where θ_b is in radians, R is the rainfall rate and a and b are the constants in the expression $K_{DP} = aR^b$. This leads to the following errors in rain-rate estimates:

- *Resolution of 2.4 km, range of 25 km, and σ_ϕ of 1°.*
  An R of 10 mm hr^{-1} can be estimated at 60% and $R=50$ mm hr^{-1} at 6%.
- *For a more realistic $\sigma_\phi = 3°$ and range = 100 km.*
  The accuracy for $R = 10$ and 50 mm hr^{-1} is 361% and 40%.
- *Performance at C-band.*
  At C-band, the value of b is about half that at S-band, so these percentages would be reduced by a factor of two, provided low values of σ_ϕ can be achieved at C-band.
- *Performance at 4.8 km resolution rather than 2.4 km.*
  The fractional errors are reduced by a factor of four, but at a range of 100 km, the S-band error at 10 mm hr^{-1} would be still 90% and 10% for 50 mm hr^{-1}.

The remaining possibility is to multiply the dwell time by four which would halve the phase error, σ_ϕ and the rain rate error, σ_R, but lead to an unacceptably slow scanning rate. The use of the integrated differential phase shift, Φ_{DP}, along a path length of 30 km (Ryzhkov et al., 2000) and 10 km (Bringi et al., 2001a) and exploiting the assumed linearity of the $R(K_{DP})$ relationship to provide an integrated average rainfall over a catchment has been proposed but, for most applications, this approach leads to an unacceptable loss of spatial resolution. We are forced to conclude that the noisy characteristics of ϕ_{DP} mean that K_{DP} cannot provide sufficiently accurate rainfall rates at the scales of 2 km required in an operational environment.

5.6 Improved Rainfall Rate Using Integrated Polarisation Parameters

In this section, we analyse what information can be retrieved from the three parameters Z, Z_{DR} and K_{DP}. First, we look at their use in an operational environment and conclude that their accuracy is probably insufficient to provide an improved rainfall rate using all three parameters on the 2 km scale but that their integrated properties can provide a valuable constraint. We then consider how the nonindependence of the three parameters can be used to provide an autocalibration of Z to within 0.5 dB every time there is heavy rain. Next we discuss an alternative approach, whereby the radar is assumed calibrated but the nonindependence of the parameters is used to derive a drop shape model. Finally, we consider the ZPHI technique, whereby the value of the integrated phase shift is used to fix the value of a in the $Z = aR^b$ relationship.

5.6.1 R from Z, Z_{DR} and K_{DP}

In the previous section, deriving R from Z and Z_{DR} or from K_{DP} were discussed. A natural progression would seem to be to use all three parameters, Z, Z_{DR} and K_{DP}, from which one might derive the three parameters of the normalised gamma function, N_w, D_0 and μ, and hence an improved rainfall rate. Several authors have published equations of the form:

$$R(K_{DP}, Z_{DR}) = cK_{DP}{}^a Z_{DR}{}^b \tag{5.22}$$

with the values of a, b, and c obtained by a regression analysis of values obtained by scanning over the 'Ulbrich' values of N_0, D_0 and μ. Bringi and Chandrasekar (2001) provide a table of the coefficients in (5.22) with, for example, at S-band, $c = 90.8$, $a = 0.89$ and $b = -1.69$ (for Z_{DR} in linear units). Four comments may be made concerning this approach:

- *Physical arguments would lead us to expect that $a = 1$.*
  For a given value of Z_{DR}, both R and K_{DP} should scale linearly with N_w.
- *The Z information is not used at all.*
  Z can be estimated more accurately than either Z_{DR} or K_{DP}, but of course, Z must be calibrated, whereas Z_{DR} and K_{DP} are 'self-calibrating.'
- *Insensitivity to hail.*
  Although K_{DP} is very noisy, it has the unique advantage that it is insensitive to hail, but in the above equation this advantage is lost. Hail will lead to a reduced value of Z_{DR} and an erroneous inferred rainfall rate.
- *Nonindependence of Z, K_{DP} and Z_{DR} in rainfall.*
  Once Z and Z_{DR} are known, then the value of K_{DP} can be derived.

5.6.2 Autocalibration of Z using Polarisation Redundancy.

Calibration of Z is sometimes neglected but is crucial if accurate rainfall rates are to be derived from Z alone or from both Z and Z_{DR}. Traditional methods rely on characterising the link-budget or comparisons with rain gauges but are often only good to a factor of two. Goddard et al. (1994) showed (Fig. 5.10) that for rain, K_{DP}/Z_H is a unique function of Z_{DR}, which is virtually independent of μ, and proposed that this redundancy could be used to provide an automatic calibration of Z to within 0.5 dB each time there was moderately heavy rainfall. The technique is as follows: The observed value of Z and Z_{DR} at each gate along a ray are used to predict the value of K_{DP} at that gate, and so the predicted phase shift at each gate can be calculated and added up to give the theoretical total phase shift along the ray, Φ_{DP}. This total phase shift can be compared with the observed total phase shift and the value of Z scaled until the computed value agrees with the observed value, as illustrated in Fig. 5.11. The technique has the following advantages:

- *The total phase shift is a robust observation.*
  The technique avoids deriving K_{DP} by differentiating an observed noisy ϕ_{DP} profile to give an even noisier gradient, but instead uses the two almost constant values of ϕ_{DP} before and after the heavy rain.

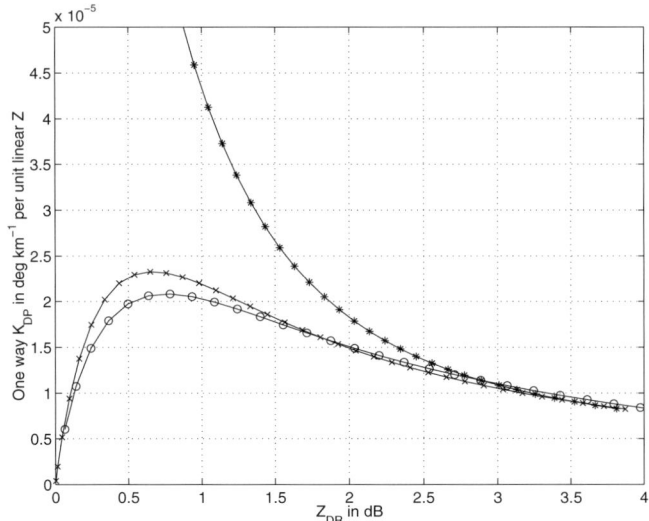

Fig. 5.10. Autocalibration technique for Z at S-band. Values of one-way K_{DP}/Z for the normalised gamma function as a function of Z_{DR}. 'New' shapes: 'o' $\mu = 0$: 'x', $\mu = 5$. 'Linear' shapes: '$\star$' $\mu = 5$

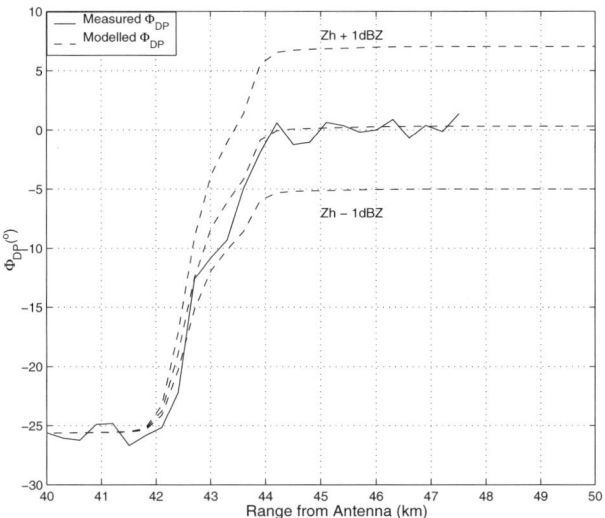

Fig. 5.11. Example of autocalibration. Different calibrations of Z lead to thee different traces for Φ_{DP}. Comparison with the observed Φ_{DP} in rain fixes the calibration of Z to 0.5 dB (10%). The initial value of ϕ_{DP} is an arbitrary system offset

- *Calibration accuracy of 10% or 0.5 dB.*
  This can be achieved if an observed Φ_{DP} of $10°$ can be estimated to $1°$ accuracy. The calibration can be repeated for many rays, so that any anomalous rays with very noisy Z_{DR} or ϕ_{DP} can be recognised and rejected.
- *Insensitivity to errors in Z_{DR}.*
  Most of the phase shift occurs with Z_{DR} of between 1 and 2 dB; so from the slope of the curve in Fig. 5.10, we only need estimate Z_{DR} to about 0.4 dB to achieve the required calibration accuracy of 0.5 dB.
- *The drop shape model provides a limit to the attainable accuracy.*
  The use of the 'linear' drop shapes rather than the new ones changes the calibration typically by about 1 dB (see Fig. 5.10).
- *The effect of attenuation.*
  Le Bouar et al. (2001) express concern over attenuation of Z and Z_{DR}. The technique does indeed fail at X-band, but if Φ_{DP} at C-band is limited to $10°$, then the Z attenuation is less than 0.5 dB and does not pose a problem.

Gorgucci et al. (1999b) propose a rather similar calibration technique based upon examination of the consistency between the two polarimetric rain-rate estimates, one using Z and Z_{DR} the other using K_{DP}. Because this method involves deriving K_{DP} directly from the observations, the noisy value of K_{DP} introduces considerable scatter into the data, making the comparison rather difficult.

5.6.3 Use of Polarisation Redundancy to Derive Drop Shape Model

A rather different approach in exploiting the nonindependence of the polarisation variables in rainfall is to assume that the radar is independently calibrated but allow the raindrop shape versus size relationship to be a free variable which is fixed by the observations. Gorgucci et al. (2000) consider the raindrop spectrum to be described as a gamma function and calculate values of Z, Z_{DR}, and K_{DP} by cycling randomly over the 'Ulbrich' (Sect. 5.3.3) range of N_0, D_0 and μ to be expected and also allow a variation of the values of the slope, β, in (5.4), for the linear model of drop shape. A nonlinear regression analysis yields:

$$\beta = 2.37 \ Z_H{}^{-0.377} \ K_{DP}{}^{0.396} \ 10^{0.093 Z_{DR}} \ . \tag{5.23}$$

They suggest that the β can be estimated to 10%. Inferred values in one storm range from 0.04 to 0.08, with average values falling from 0.061 to 0.053 for the Z ranges 40–45 dBZ to above 53 dBZ which they ascribe to raindrop oscillations. Gorgucci et al. (2001) use this approach and report rainfall estimates which are immune to changes in the raindrop shape–size relation. Recently, Matrosov et al. (2002) report that at X-band, the use of a variable β leads to improved rainfall rates than from K_{DP} alone. The immunity to drop shape models provided by this approach is attractive, but the following caveats apply:

- *Apparent change of β with rainfall rate.*
  Equation (5.23) is virtually identical to the Goddard et al., 1994) calibration

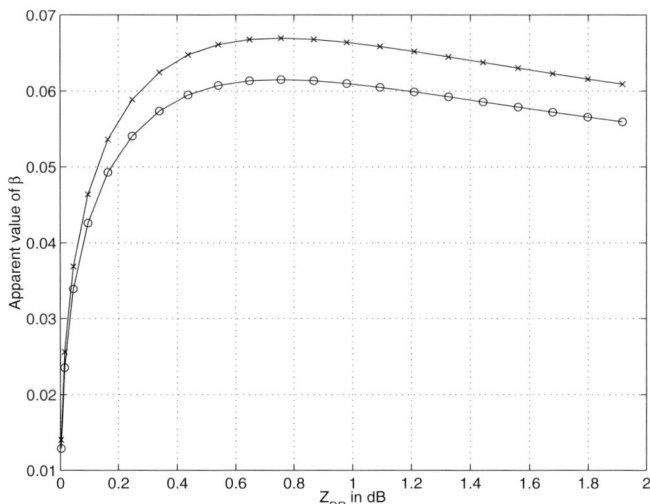

Fig. 5.12. The apparent value of β as a function of Z_{DR} if the rain drop spectrum is described as a normalised gamma function with $\mu = 5$ [*lower curve* (o)]. In the heavier rain, D_0, and hence Z_{DR}, increases and leads to an apparent fall in β. The *upper curve* (x) shows the change in apparent β if the Z calibration changes by 1 dB

curve; the indices of Z_{H} and K_{DP} are nearly equal and opposite, so for a given β (5.23) reduces to a single curve of $Z_{\mathrm{H}}/K_{\mathrm{DP}}$ against Z_{DR}. Rather than use a nonlinear regression, the values of Z_{H}, K_{DP} and Z_{DR} as D_0 varies have been calculated for a normalised gamma function with $\mu = 5$, and the apparent value of β in (5.23) plotted out in Fig. 5.12 as a function of Z_{DR} for the 'new' drop shapes. From Fig. 5.12, we now see that the apparent fall of β in heavy (high Z_{DR}) rain arises naturally from the fall in the K_{DP}/Z ratio with increasing Z_{DR} in the calibration curve (Fig. 5.10), rather than appealing to drop oscillations (Gorgucci et al., 2000).

- *Sensitivity to calibration.*
  Because the redundancy is being used to fix β, it is assumed that Z is perfectly calibrated by some independent means. A small error of 1 dB in Z calibration will change β by about 10% (Fig. 5.12).
- *Drop shape model.*
  The use of a single free variable β may not be sufficient to define the drop shape model; the intercept is also a variable (see Fig. 5.6) and the curve is not necessarily a straight line.
- *Sensitivity to noisy K_{DP}.*
  This approach involves the implicit computation of K_{DP} from the observations with its associated uncertainty, rather than the approaches above which use the more accurate integrated phase shift.

5.6.4 The ZPHI Technique

The ZPHI technique is described by Testud et al. (2000), and encouraging results at C-band in heavy tropical rainfall are reported by Le Bouar et al. (2001). This method uses the total phase shift, Φ_{DP}, recorded along a path, along which Z is observed, to provide a constraint and fix the value of a in the $Z = aR^b$ relationship. We now describe the principle of the technique at nonattenuating wavelengths such as S-band. At shorter wavelengths, C-band and below, attenuation can be appreciable, and the full technique which also uses the phase shift to correct for the attenuation is described in Testud et al. (2003), this book. In Sect. 5.3.4, it was shown that if naturally occurring raindrop spectra can be described by a normalised gamma function, then the value of b should be 1.5. Recall (5.12):

$$\frac{Z}{N_{\text{w}}} = H(\mu) \left(\frac{R}{N_{\text{w}}}\right)^{7.67/4.67} = H(\mu) \left(\frac{R}{N_{\text{w}}}\right)^b. \tag{5.24}$$

Figure 5.13 shows that, for a normalised gamma function, a log-log plot of $K_{\text{DP}}/N_{\text{w}}$ against Z/N_{w} with varying D_0 is also very nearly a straight line which is only weakly dependent upon μ. Thus, we may write,

$$K_{\text{DP}}/N_{\text{w}} = f(Z/N_{\text{w}})^g \quad \text{or} \quad K_{\text{DP}} = f N_{\text{w}}^{1-g} Z^g \tag{5.25}$$

where f and g are constants. Integrating along the path, r, we have:

$$\Phi_{\text{DP}} = f N_{\text{w}}^{1-g} \int Z^g \text{d}r. \tag{5.26}$$

Accordingly, since f and g are known (Fig. 5.13) the value of N_{w} in Eqn 5.26 can be derived from the observed total phase shift, Φ_{DP} and the path integral of the reflectivity raised to the power 'g'; the value of N_{w} then fixes 'a' in $Z = aR^{1.5}$. Le Boar et al. (2001) report much improved rainfall rates at C-band, both as a result of the attenuation correction and from using the appropriate values of N_{w}. The technique has the following characteristics:

- *The total phase shift is a robust observation.*
  As for the calibration method described above, the technique avoids the use of a noisy value of K_{DP} obtained by differentiating the ϕ_{DP} phase profile, but instead reduces the noise by integration. Note the sensitivity to the drop shape model. For the new drop shapes, when $Z_{\text{H}}/N_{\text{W}}$ is below 1 (e.g., $Z = 39$ dBZ and $N_{\text{w}} = 8000$ m^{-3} mm^{-1}), g increases and approaches unity, but this region does not correspond to the regions of heavy rainfall with large phase shifts.
- *Need for accurate calibration of Z.*
  Any calibration error in Z will directly affect the derived value of N_{w} and the value of a in the $Z = aR^b$ relationship. Le Bouar et al. (2001) suggest that the method can be used to calibrate Z, based on the 'climatological' stability of N_{w}. They also propose a calibration technique which relies on the consistency

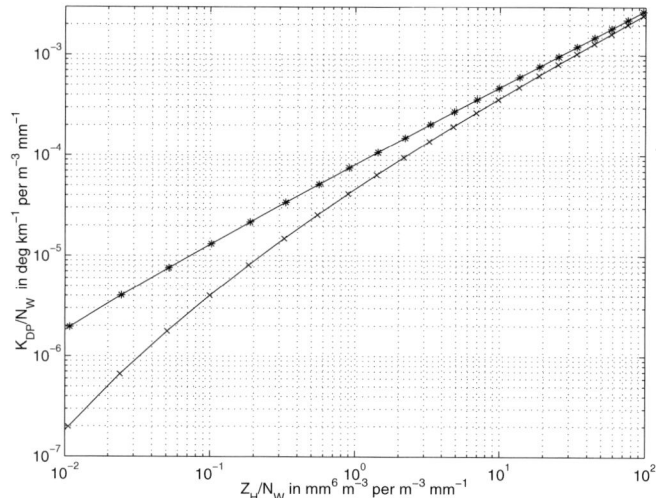

Fig. 5.13. K_{DP}/N_w for rain at S-band as a function of Z_H/N_w for a normalised gamma function with $\mu=5$: x – New shapes. ⋆ – Linear shapes. The ZPHI technique relies on this curve being a straight line with constant slope g which is not unity

of deriving the rainfall rate from ZPHI and from an attenuation corrected Z_{DR} and the inferred attenuation coefficient; in other words, the method exploits the integrated phase information, Z_{DR} and Z, and is equivalent to the calibration techniques in Sect. 5.6.2 in another guise.

- *Sensitivity and operation in moderate rain rates.*
  At C-band, Φ_{DP} must be above 6° if R is to be estimated to 32% and above 32° for 21% error. At this frequency, 10 mm hr^{-1} has a one-way K_{DP} of 0.2° km^{-1} which would need to extend over 15 km to produce a total differential phase shift of 6°. For 50 mm hr^{-1}, the distance would be 2 km, and for 3 mm hr^{-1}, 75 km. All distances need to be multiplied by 5 to reduce the rainfall error to 21%.
- *Constraint of constant N_w along the path.*
  This is quite restrictive, but the path may be divided into segments provided that the total differential phase change per segment is at least 6°.
- *Interference from hail.*
  A means of recognising hail is needed, otherwise hail will increase Z but not affect Φ_{DP} and so lead to errors in inferred values of N_w.

The ZPHI technique is clearly a very powerful constraint for providing accurate rainfall rates in heavy rain. It is particularly applicable to the tropics where hail is most unlikely. Based on the results of Matrosov et al. (1999), there is scope for extending the ZPHI technique to X-band for moderate rain rates.

5.7 Improved Rainfall Rates When Ice May Be Present

In this section, we will consider two situations when the radar beam may be scanning a region containing icy hydrometeors. Firstly, in the summertime, hail in convective clouds can occur at lower levels where the temperature is above freezing. Secondly, in the wintertime and at longer ranges in all seasons, the beam will be dwelling in regions above the freezing level and can expect to encounter ice. In many papers, only the first situation is addressed (e.g., Brandes et al., 1999), but at a 200 km range, even in the summer, the lowest elevation beam of a NEXRAD radar will be sampling ice.

5.7.1 Estimating Rainfall When Hail May Be Present

Apart from the use of K_{DP} alone, the techniques described in Sect. 5.6 for estimating rainfall rely upon an independent means of identifying the presence of hail. Hail cannot be identified unambiguously from the observed values of Z and Z_{DR}, but its presence will raise Z and depress Z_{DR}. The calibration technique of Goddard et al. (1994) relies on the consistency of the observed values of Z, Z_{DR} at each gate and the total observed phase shift Φ_{DP} along each ray within rain. Smyth et al. (1999) took this argument one stage further, suggesting that the best technique to identify hail is to continually monitor these three variables along each ray and use the failure of consistency as an indication of the presence of hail. When hail is present, then the only approach is to use one of the $R(K_{DP})$ equations to estimate the rainfall.

5.7.2 Estimating Rainfall When the Beam is at or above the Freezing Level

Essentially, we are dealing with the problem of correcting for the vertical profile of reflectivity. This is a major operational problem and is dealt with in this book by Germann and Joss for mountainous regions and by Koistinen et al. for a cold climate. Can polarisation help in this problem? We must distinguish between the stratiform case, when there is an enhanced reflectivity, ΔZ, of up to 10 dB to 13 dB associated with the melting of low density snowflakes and a rapid fall-off of reflectivity in the ice above the bright band, and the convective case, when the higher density graupel particles are not associated with a bright band and the fall of reflectivity with height is less pronounced.

The bright band can easily be identified by its very obvious polarimetric signature; the values of LDR are generally in the range -14 to -18 dB and far higher than that of other precipitation particles. The value of Z is enhanced in such regions, ΔZ is about 10 dB higher than in the rain below. Melting graupel is associated with lower values of LDR of about -22 to -26 dB and values of Z similar to that in the rain below ($\Delta Z = 0$ dB). This suggests that the value of LDR measured at the height of the $0°$ isotherm (obtained with an accuracy of $1°$ or so from an operational model) can be used as the basis of a bright band correction system. Before this is done, the following aspects must be addressed:

- *Is there a reliable relationship between the LDR value and ΔZ?*
  How reliable is it, and what is the quantitative performance of a Z correction scheme based on it?
- *Effect of range on the observed bright band.*
  The bright band is typically about 600 m deep and so at larger ranges will not fill the beam of a 1° radar. This means that the observed ΔZ will be less. LDR should also be lower, but how well related are the two variables?
- *Absence of a bright band in convection.*
  How reliable is the value of LDR which indicates the presence of graupel and the absence of a bright band?
- *Propagation effects at C-band.*
  As the beam propagates through regions of finite LDR, the targets depolarise the incident beam, and the apparent value of LDR rises with range. The effect is negligible at S-band but needs to be quantified at C-band. Is a correction scheme needed at C-band and can one be developed?

Deriving rainfall at the ground, once the radar beam is dwelling only in the dry ice above the melting layer, is a challenging problem which is very important for operational radars at most ranges in the cold season and for cold climates, as extensively discussed in this book by Koistinen et al. The snow occurring at these heights does not have any very marked polarisation characteristics which can be exploited.

5.8 Correction for Attenuation

Gate by gate correction schemes are notoriously unstable (Hitschfeld and Bordan, 1954) and very sensitive to small calibration errors (Hildebrand, 1978). Differential phase shift provides a very powerful technique to correct for attenuation. In heavy rain, attenuation at C-band is a severe problem and affects both Z and Z_{DR}. For example, at 40 mm hr^{-1}, the attenuation, A_{H}, is about 0.5 dB km^{-1} and the differential attenuation, A_{DP}, is about 0.1 dB km^{-1}. In the literature, there are many expressions of the form

$$A_{\mathrm{H}} = \alpha \, K_{\mathrm{DP}}^{b}, \tag{5.27}$$

and

$$A_{\mathrm{DP}} = \beta K_{\mathrm{DP}}^{b}. \tag{5.28}$$

In the Rayleigh limit, A_{H} is proportional to liquid water content (LWC) and for drop shapes which vary linearly with size, K_{DP} is proportional to the product of LWC and D_0, so at S-Band, the value of b in (5.27) is about 0.84. At C-band, the larger drops Mie scatter and absorb more, so that (5.27) is very nearly linear. This suggests that correction for attenuation using K_{DP} or Φ_{DP} at C-band should be reasonably straightforward. In practice, the following factors need to be considered.

- *Temperature effects.*
  Attenuation depends upon the imaginary part of the dielectric constant which changes with temperature, but K_{DP} is a function of the real part which is constant. Attenuation is about twice as large at 0°C as at 20°C. Operational models provide a reliable temperature structure, so, in principle, this effect can be corrected for.
- *The precise value of α and β is a matter of some dispute.*
  Bringi et al. (1990) quote 0.054 and 0.0157 dB °C^{-1}, respectively, at 15°C, but Carey et al. (2000) find that reported values range over a factor of two.
- *Variation with D_0.*
  Changes in the value of D_0 and hence the degree of Mie scattering are probably the cause of this spread (Smyth and Illingworth, 1998; Carey et al., 2000). One method of resolving this difficulty would be an iterative approach, whereby the attenuation-corrected value of Z_{DR} is used to estimate D_0 and hence the correct value of α and β.

Our aim is to correct attenuations of Z to 1 dB and Z_{DR} to 0.2 dB (or better), so clearly, a quite complex attenuation correction is needed in the heaviest rain. There are two alleviating factors. In heavy attenuating rain, the value of Φ_{DP} is large, and it may be possible to use the ZPHI method to correct for Z and avoid the more stringent requirement for correcting Z_{DR}. Smyth and Illingworth (1998) suggested using an additional constraint; the value of Z_{DR} on the far side of a storm where Z is low should be close to 0 dB. Any negative values of Z_{DR} provide a path integrated constraint for the total path integrated value of A_{DP}. A_H (the total path attenuation) can then be derived from A_{DP} through a simple multiplicative factor because both have the same temperature dependence. Bringi et al. (2001b) have combined these two constraints, Φ_{DP} and Z_{DR} on the far side of the rain cell, so that the uncertain values of α and β in (5.27) and (5.28) are no longer prescribed but are derived along with values of D_0; the results are remarkably good, and the technique is confirmed by the agreement with disdrometer values of D_0.

5.9 Identification of Hydrometeors

Thus far, we have discussed identification of hydrometeors in terms of hail and melting snow impacting upon rainfall estimation. Hydrometeor identification itself is important for identifying severe weather (hail), for evaluation of the representation of different hydrometeors both in NWP models and in cloud resolving models (CRMs) and ultimately for assimilating hydrometeor information into such models. Particle classification also leads to a deeper understanding of the processes involved in precipitation formation and growth, so that conceptual and numerical models can be improved. This is especially important in understanding deep convective systems, as discussed in this book by Meischner et al. Straka et al. (2000) provide a very thorough review of the range of values of Z, Z_{DR}, K_{DP}, LDR and ρ_{HV} expected for hail, graupel, rain, rain and wet hail mixtures, and snow crystals and aggregates. For example, hail can be identified by its high value of Z which is accompanied by lower values of Z_{DR} and K_{DP} than would be expected for rain. Liu and Chandrasekar (2000) have extended this classification using a fuzzy logic scheme; the in-situ validation data are very encouraging.

One difficulty that remains is that of identifying different forms of ice crystals. Matrosov et al. (2001) analysed the variation of the depolarisation ratio with elevation angle in terms of the ice particle aspect ratio and found that quasi-circular polarisation provided the best discrimination. In theory, pristine crystals have high density and so should give distinctive polarisation signatures. However, from an extensive analysis of insitu data in stratiform clouds, Korolev et al. (2000) concluded that pristine crystals were rare and that 85% of ice particles were of irregular form. The problem is that the occasional large, low density aggregates, which look spherical to the radar, dominate the radar return and mask the returns from the smaller crystals. This is confirmed by Wolde and Vali (2001) who found that in theory, different crystal habits could be distinguished using a 94 GHz polarimetric cloud radar, but that for real clouds polarimetric signatures were found for only a very few per cent of observations. Ryzhkov et al. (1998) suggested that the IWC should be related to a function of K_{DP}/Z_{DR} which should be insensitive to particle shape. Results were encouraging for regions containing pristine crystals, but the difficulty is that in the more common regions of aggregates, the values of the parameters are low and noisy and the IWC function is ill defined. There is some scope for identifying ice particles by their signatures when they melt; for example, pristine plates will give very high Z_{DR} when they get wet, but after falling 100 m or so will melt completely. We conclude that the polarimetric radar properties can be used to identify the larger hydrometeors but are usually unable to differentiate between the smaller types of ice particles.

5.10 Conclusion

This review has identified some powerful techniques based upon polarimetric observations. In the following discussion, we shall assume that 'new' raindrop shapes of (5.5) are accurate and that natural raindrop size spectra can be represented by a normalised gamma function with a mean shape factor of 5 and a range of 1–10, and an average normalised concentration N_w of 8000 m^{-3} mm^{-1} with a range of up to a factor of ten both larger and smaller. This range leads to a relationship of the form $Z = aR^{1.5}$ with a proportional to $1/\sqrt{N_w}$ and errors of up to a factor of two in rain rate. If we are to estimate rainfall rates to better than 25%, then it may be necessary to account for the increased terminal velocities ($\approx$ 10% at 2 km) aloft (Matrosov et al., 2002).

- *Ground clutter and anomalous propagation*
  They can be recognised at each gate using polarimetric parameters. Particularly powerful is the use of the copolar correlation coefficient as a data quality check.

- *Rain rates from Z and Z_{DR}*
  Z should scale with R as for a given Z_{DR} (or D_0) as N_w changes; this leads to a formula of the form $Z/R = f(Z_{DR})$ given by (5.14) at S-band and (5.15) at C-band. For rain rates to be accurate to 25%, Z must be calibrated to better than 25% (1 dB). Z_{DR} must be accurate to 0.2 dB for $R > 10$ mm hr^{-1} and 0.1 dB for R > 3 mm hr^{-1}. Such Z_{DR} accuracies are difficult to achieve in practice on a scale of 2 km.
- *Rain rate from K_{DP}*
  This technique is attractive because of its immunity to both calibration errors and to attenuation and its insensitivity to hail. However, the exponent in $K_{DP} = aR^b$ is about 1.4 and not very different from those in Z–R relations, so the R derived from K_{DP} has almost the same sensitivity to changes in raindrop spectra as R derived from Z, and the scope for using a path integrated phase shift to infer rainfall over a catchment is limited. In addition, the differential phase measurement is inherently noisy and in practice is also extremely sensitive to small amounts of ground clutter or mismatched sidelobe signals. As a result the R estimates from K_{DP} are not accurate enough for operational work at 2 km resolution unless hail is present.
- *Use of integrated polarisation parameters*
  We conclude that at the 2 km scale needed for an operational system, the additional information from Z_{DR} and K_{DP} is insufficiently accurate to improve rainfall estimates but that the use of integrated polarisation parameters can provide a valuable constraint.
- *Autocalibration of Z and identification of hail*
  Z can be accurately calibrated to 0.5 dB (10%) by exploiting the fact that Z_{DR} and K_{DP} are not independent in rain but that K_{DP}/Z is a unique function of Z_{DR}. The technique is to compute the predicted total theoretical differential phase shift along a ray (Φ_{DP}), by integrating the value of K_{DP} predicted from the observed values of Z and Z_{DR} at each gate and then scaling the Z values so that the predicted value of Φ_{DP} agrees with the observed one. This technique avoids using the noisy observed K_{DP}. The presence of hail can be inferred from the failure of the redundancy exploited in the calibration technique.
- *Drops shape constraint from polarisation redundancy?*
  An alternative approach is to exploit this redundancy by choosing the value of β in the drop shape (5.4) so that the observed Z, Z_{DR} and K_{DP} are consistent, but the accuracy is limited because this approach uses the very noisy observed values of K_{DP}. It seems more sensible to accept the extensive measurements of drop shape and use the redundancy to fix the Z calibration.
- *The ZPHI technique to fix a in $Z = aR^{1.5}$ in heavy rain*
  The observed path integrated differential phase shift, Φ_{DP}, together with the integrated value of Z along the ray can be used in the ZPHI technique to provide a constraint to N_w and hence an estimate of a in $Z = aR^{1.5}$, provided there is no hail present. The technique is limited to rather heavy rainfall because Φ_{DP} at C-band must be at least 32° if the error in R is to be reduced to 21%. Increased phase shifts at X-band may be counteracted by

the extra noise introduced by Mie-scattering effects, such as differential phase shift on backscatter and a lowering of the copolar correlation coefficient.

- *Use of ϕ_{DP} to correct Z and Z_{DR} for attenuation*
  ϕ_{DP} is nearly linearly related to attenuation at C-band but the constant changes with T and D_0. Correction of attenuation using the ZPHI technique is discussed by Testud et al. (2003), this book. Further improvements to the attenuation correction may be possible when the ZPHI Φ_{DP} constraint is combined with the requirement that the total differential attenuation is consistent with the values of Z_{DR} expected on the far side of a heavy rain echo. There is scope for using the ZPHI method at X-band (3.2 cm) provided the rainfall is not too heavy so there is total loss of signal.

- *Integrated Z/Z_{DR} technique in regions of low rainfall*
  In regions of moderate or light rain where the phase shifts are too low for the ZPHI technique, the average value of N_w over a domain can be inferred from the many estimates of N_w from Z and Z_{DR} made at each gate. The random errors associated with noisy Z_{DR} should average zero, and providing Z is accurately calibrated, the mean value of N_w can be estimated, and hence accurate (25%) rainfall estimates obtained down to perhaps $R = 3$ mm hr^{-1}, but this still remains to be demonstrated.

- *Hydrometeor Identification*
  Combined polarisation parameters can be used to distinguish the different types of the larger hydrometeors and this should be useful in evaluating parameterisation schemes in Numerical Weather Prediction (NWP) models and the performance of Cloud Resolving Models (CRMs), and may ultimately be used for assimilation into such models (see Macpherson et al., 2003, this book).

To summarise, we find that polarisation radar has the potential to remove most of the ambiguities which arise when Z alone is measured with a conventional radar. However, the difficulty of estimating rainfall rates at the ground at long ranges when the beam is dwelling in ice above the melting layer remains. In all these applications, it is important to use the latest 'new' raindrop shapes. It may be that the use of the 'linear' drop shapes is responsible for some of the disappointing rainfall estimates reported in the literature. The requirement for pulse to pulse switching of polarisation can be avoided by the use of 'hybrid' (45°) transmission, but this approach will probably not reduce the noise of the differential phase measurement. Ground clutter and 'anaprop' can be unambiguously recognised, but generally, the parameters cannot be estimated accurately enough to contribute independent information to improve rainfall estimates on the operational scale of 2 km. However, when the parameters are integrated over a domain, they can correct for attenuation in C-band systems, provide absolute calibration of Z to 10% and reduce errors in inferred rainfall rates from a factor of two down to about 25%, effectively, by fixing the value of a to be used in $Z = aR^{1.5}$ over a domain. Finally, to accompany these improved rainfall estimates, the polarisation parameters can also provide an estimate of the error which is essential for use in data assimilation.

References

1. Andsager, K., K.V. Beard and N.F. Laird, 1999: Laboratory measurements of axisratios for large raindrops. J. Atmos. Sci., **56**, 2673–2683.
2. Battan, L.J., 1973: *Radar Observations of the Atmosphere*. University of Chicago Press.
3. Blackman, T.M and A.J. Illingworth, 1995: Improved measurements of rainfall using differential phase techniques. *COST 75 Int. Seminar on Weather Radar Systems*, ed C.G.Collier, EUR 16013 EN.
4. Blackman, T.M. and A.J. Illingworth, 1997: Examining the lower limit of K_{DP} rainrate estimation including a case study at S-band. *Proc. 28th Conf. Radar Meteorol.*, AMS.
5. Brandes, E.A., J. Vivekanandan and J.W. Wilson, 1999: A comparison of radar reflectivity estimates from collocated radars. J. Atmos. Oceanic Technol., **16**, 1264–1272.
6. Brandes, E.A., A.V. Ryzhkov and D.S. Zrnic, 2001: An evaluation of radar rainfall estimates from specific differential phase. J. Atmos. Oceanic Technol., **18**, 363–375.
7. Bringi, V.N., T.A. Seliga and S.M. Cherry, 1983: Statistical properties of the dual-polarization differential reflectivity (Z_{DR}) signal. IEEE Trans. Geosci. Remote Sensing, **21**, 215–220.
8. Bringi,V.N., V. Chandrasekar, N. Balakrishnan, and D.S. Zrnic, 1990: An examination of propagation effects in rainfall on radar measurements at microwave frequencies. J. Atmos. Oceanic Technol., **7**, 829–840.
9. Bringi, V.N., Gwo-Jong Huang, V. Chandrasekar and T.D. Keenan, 2001a: An areal rainfall estimator using differential propagation phase; Evaluation using a C-band radar and a dense gauge network in the tropics. J. Atmos. Oceanic Technol., **18**, 1810–1818.
10. Bringi, V.N., T.D. Keenan and V. Chandrasekar, 2001b: Correcting C-band radar reflectivity and differential reflectivity data from rain attenuation: A self consistent method with contraints. IEEE Trans. Geosci. Remote Sensing, **39**, 1906–1915.
11. Bringi, V.N. and V. Chandrasekar, 2001: Polarimetric Doppler weather radar. CUP.
12. Carey, L.D., S.A. Rutledge, D.A. Ahijevych and T.D. Keenan, 2000: Correcting propagation effects in C-band polarimetric radar observations of tropical convection using differential propagation phase. J. Appl. Meteor., **39**, 1405–1433.
13. Caylor, I.J. and A.J. Illingworth, 1992: Polarisation radar estimates of rainfall; correction of errors due to the bright band and to anomalous propagation: *International Weather Radar Networking*. Kluwer, Dordrecht.
14. Chandrasekar, V. and V.N. Bringi, 1988a: Error structure of multiparameter radar and surface measurement of rainfall. Part 1: Differential reflectivity. J. Atmos. Oceanic Technol., **5**, 783–795.
15. Chandrasekar, V. and V.N. Bringi, 1998b: Error structure of multiparameter radar and surface measurement of rainfall. Part II: X-band attenuation. J. Atmos. Oceanic Technol., **5**, 796–802.
16. Chandrasekar, V., V.N. Bringi, N. Balakrishnan and D.S. Zrnic, 1990: Error structure of multiparameter radar and surface measurement of rainfall. Part III: Specific differential phase. J. Atmos. Oceanic Technol., **7**, 621–629.
17. Frost, I.R., J.W.F. Goddard and A.J. Illingworth, 1991: Hydrometeor identification using cross polar radar measurement and aircraft veritication. *Proc. 25th Conf. Radar Meteorol.*, AMS.

18. Goddard, J.W.F., S.M. Cherry and V.N. Bringi, 1982: Comparison of dual-polarization radar measurements of rain with ground based disdrometer measurements. J. Appl. Meteorol., **21**, 252–256.
19. Goddard, J.W.F., J. Tan and M. Thurai, 1994: Technique for calibration of meteorological radars using differential phase. Electron. Lett., **30**, 166–167.
20. Goddard, J.W.F, K.L. Morgan, A.J. Illingworth and H. Sauvageot, 1995: Dual-wavelength polarisation measurements in precipitation using the CAMRA and Rabelais radar. *Proc. 27th Conf. Radar Meteorol.*, AMS.
21. Gorgucci, E., G. Scarchilli and V. Chandrasekar, 1994: A robust estimator of rainfall rate using differential reflectivity. J. Atmos. Oceanic Technol., **11**, 586–592.
22. Gorgucci, E., G. Scarchilli and V. Chandrasekar, 1999a: Specific differential phase estimation in the presence of nonuniform rainfall medium along the path. J. Atmos. Oceanic Technol., **16**, 1690–1697.
23. Gorgucci, E., G. Scarchilli and V. Chandrasekar, 1999b: A procedure to calibrate multiparameter weather radar using properties of the rain medium. IEEE Trans. Geosci. Remote Sensing, **17**, 269–276.
24. Gorgucci, E., G. Scarchilli, V. Chandrasekar and V.N. Bringi, 2000: Measurement of mean raindrop shape from polarimetric radar observations. J. Atmos. Sci., **57**, 3406–3413.
25. Gorgucci, E., G. Scarchilli, V. Chandrasekar,and V.N. Bringi, 2001: Rainfall estimation from polarimetric radar measurements: Composite algorithms immune to variability in raindrop shape-size relation. J. Atmos. Oceanic Technol., **18**, 1773–1786.
26. Hagen, M., 1997: Identification of ground clutter by polarimetric radar, *Proc. 28th Conf. Radar Meteorol.*, AMS.
27. Hall, M.P.M., J.W.F. Goddard and S.M. Cherry, 1984; Identification of hydrometeors and other targets by dual-polarization radar. Radio Sci., **19**, 132–140.
28. Herzegh, P.H. and R.E. Carbone, 1984: The influence of antenna illumination function characteristics on differential reflectivity measurements. *Proc. 22nd Conf. Radar Meteorol.*, AMS.
29. Hildebrand, P.H., 1978: Iteritive correction for attenuation of 5 cm radar in rain. J. Appl. Meteorol., **17**, 508–514.
30. Hitschfeld, W., and J. Bordan, 1954: Errors inherent in the radar measurement of rainfall at attenuating wavelengths. J. Meteorol., **11**, 58–67.
31. Hubbert, J., V. Chandrasekar, V.N. Bringi, and P.F. Meischner, 1993: Processing and interpretation of coherent dual-polarized radar measurements,. J. Atmos. Oceanic Technol., **10**, 155–164.
32. Hubbert, J., and V.N. Bringi, 2000: The effects of three-body scattering on differential reflectivity signatures. J. Atmos. Oceanic Technol., **17**, 51–61.
33. Illingworth, A.J., J.W.F. Goddard and S.M. Cherry, 1987: Polarisation radar studies of precipitation development in convective storms. Q. J. R. Meteorol. Soc., **113**, 469–489.
34. Illingworth, A.J. and I.J. Caylor, 1988: Identification of precipitation particles using dual polarization radar. *10th Int. Conf. Cloud Physics*, Bad Hamburg, Germany.
35. Illingworth, A.J. and I.J. Caylor, 1991: Co-polar correlation measurements of precipitation. *Proc. 25th Conf. Radar Meteorol.*, AMS.
36. Illlingworth, A.J. and T.M. Blackman, 1999: The need to normalise RSDs based on the gamma RSD formulation and implications for interpreting polarimetric radar data. *Proc. 29th Conf. Radar Meteorol.*, AMS.

37. Illingworth, A.J. and M.P. Johnson, 1999: The role of raindrop shape and size spectra in deriving rainfall rates using polarisation radar. *Proc. 29th Conf. Radar Meteorol.*, AMS.
38. Illingworth, A.J., and T.M. Blackman, 2002: The need to represent raindrop size spectra as normalized gamma distributions for the interpretation of polarization radar observations. J. Appl. Meteorol., **41**, 1578–1583.
39. Keenan, T., K. Glasson, F. Cummings, T.S. Bird, J. Keeler and J. Lutz, 1998: The BMRC/NCAR C-band polarimetric (C-POL) radar system. J. Atmos. Oceanic Technol., **15**, 871–886.
40. Korolev, A., G.A. Isaac and J. Hallett, 2000: Ice particle habits in stratiform clouds. Q. J. R. Meteorol. Soc., **126**, 2873–2902.
41. Kozu.,T. and K. Nakamura, 1991: Rainfall parameter estimation from dual-radar measurements combining reflectivity profile and path-integrated attenuation. J. Atmos. Oceanic Technol., **8**, 259–271.
42. Le Bouar, E., J. Testud, T.D. Keenen, 2001: Validation of the rain profiling algorithm 'ZPHI' from the C-band polarimetric weather radar in Darwin. J. Atmos. Oceanic Technol., **18**, 1819–1837.
43. Liu, H.P. and V. Chandrasekar, 2000: Classification of hydrometeors based on polarimetric radar measurements: Development of fuzzy logic and neuro-fuzzy systems, and in situ verification. J. Atmos. Oceanic Technol., **17**, 140–164.
44. Marshall, J.S. and W.M.K.Palmer, 1948: The distribution of raindrops with size. J. Meteorol., **5**, 165–166.
45. Matrosov, S.Y., R.A. Kropfli, R.F. Reinking and B.E. Martner, 1999: Prospects for measuring rainfall using propagation differential phase in X- and Ka-radar bands. J. Appl. Meteorol., **38**, 766–776.
46. Matrosov,S.Y., R.F. Reinking, R.A. Kropfli, B.E. Martner and B.W. Bartram, 2001: On the use of radar depolarization ratio for estimating shape of ice hydrometeors in winter clouds. J. Appl. Meteorol., **40**, 479–490.
47. Matrosov, S.Y., K.A. Clark, B.E. Martner and A. Tokay, 2002: X-band polarimetric radar measurements of rainfall. J. Appl. Meteorol., **41**, 941–952.
48. May,P.T., T.D. Keenan, D.S. Zrnic, L.D. Carey and S.A. Rutledge, 1999: Polarimetric radar measurements of tropical rain at 5-cm wavelength. J. Appl. Meteorol., **38**, 750–765.
49. Pamment, J.A. and B.J. Conway, 1994: Objective identification of echoes due to anomalous propagation in weather radar. J. Atmos. Oceanic Technol., **15**, 98–113.
50. Petersen, W.A., L.D. Carey, S.A. Rutledge, J.C. Knievel, N.J. Doesken, R.H. Johnson, T.B. McKee, T. Vonder Haar, and J.F. Weaver, 1999: Mesoscale and radar observations of the Fort Collins flash flood of 28 July 1997. Bull. Am. Meteorol. Soc., **80**, 191–216.
51. Pruppacher, H.R. and R.L. Pitter, 1971: A semi-empirical determination of the shape of cloud and raindrops. J. Atmos. Sci, **28**, 86–94.
52. Ryzhkov, A.V. and D.S. Zrnic, 1995: Comparison of dual-polarisation radar estimators of rain. J. Atmos. Oceanic Technol., **12**, 249–256.
53. Ryzhkov, A.V. and D.S. Zrnic, 1996: Assessment of rainfall measurement that uses specific differential phase. J. Appl. Meteorol., **35**, 2080–2090.
54. Ryzhkov, A.V. and D.S. Zrnic, 1998a: Beamwidth effects on the differential phase measurement of rain. J. Atmos. Oceanic Technol., **15**, 624–634.
55. Ryzhkov, A.V. and D.S. Zrnic, 1998b: Polarimetric rainfall estimation in the presence of anomalous propagation. J. Atmos. Oceanic Technol., **15**, 1320–1330.
56. Ryzhkov, A.V., D.S. Zrnic and B.A. Gordon, 1998: Polarimetric method for ice water content determination. J. Appl. Meteorol., **37**, 125–134.

57. Ryzhkov, A.V., D.S. Zrnic and R. Fulton, 2000: Areal rainfall estimates using differential phase. J. Appl. Meteorol., **39**, 263–268.
58. Sachidananda, M. and D.S. Zrnic, 1986: Differential propagation phase shift and rainfall estimation, Radio Sci., **21**, 235–247.
59. Sachidananda, M. and D.S. Zrnic, 1987: Rain rate estimates from differential polarisation measurements. J. Atmos. Oceanic Technol., **4**, 588–598.
60. Seliga, T., and V.N. Bringi, 1976; Potential use of radar differential reflectivity measurements at orthogonal polarisations for measuring precipitation. J. Appl. Meteorol., **15**, 69–75.
61. Smyth, T.J. and A.J. Illingworth, 1998: Correction for attenuation of radar reflectivity using polarisation data. Q. J. R. Meteorol. Soc., **124**, 2393–2415.
62. Smyth, T.J., T.M. Blackman and A.J. Illingworth, 1999: Observations of oblate hail using dual polarisation radar and implications for hail-detection schemes. Q. J. R. Meteorol. Soc., **125**, 993–1016.
63. Straka, J.M., D.S. Zrnic and A.V. Ryzhkov, 2000: Bulk hydrometeor classification and quantification using polarimetric radar data: Synthesis of relations, J. Appl. Meteorol., **39**, 1341–1372.
64. Testud, J., E.L. Bouar, E. Obligis, M. Ali-Mehenni, 2000: The rain profiling algorithm applied to polarimetric weather radar. J. Atmos. Oceanic Technol., **17**, 332–356.
65. Testud, J., S. Oury, R.A. Black, P. Amayenc, and D. Xiankang, 2001: The concept of 'normalized' distribution to describe raindrop spectra: A tool for cloud physics and cloud remote sensing. J. Appl. Meteorol., **40**, 1118–1140.
66. Ulbrich, C.W., 1983: Natural variations in the analytical form of the raindrop size distribution. J. Climate and Appl. Meteorol., **22**, 1764–1775.
67. Ulbrich, C.W. and D. Atlas, 1998: Rainfall microphysics and radar properties: analysis methods for drop size spectra. J. Appl. Meteorol., **37**, 912–923.
68. Upton, G. and J.-J. Fernandez-Duran, 1999; Statistical techniques for clutter removal and attenuation detection in radar reflectivity *Cost 75 Int. Seminar, Advanced Weather Radar Systems*, Locarno, EUR 18567 EN.
69. Wilson, D.R., A.J. Illingworth and T.M. Blackman, 1995: The use of Doppler and polarisation data to identify ground clutter and anaprop. *Weather Radar Systems*, Ed. C.G. Collier, Cost-75 EUR 16013 EN, pp. 527–538.
70. Wilson, D.R., A.J. Illingworth and T.M. Blackman, 1997: Differential Doppler velocity: A radar parameter for characterising hydrometeor size distributions. J. Appl. Meteorol., **36**, 649–663.
71. Wolde, M., and G. Vali, 2001: Polarimetric signatures from ice crystals observed at 95 GHz in winter cloud. Part II: Frequency of occurrence. J. Atmos. Sci., **58**, 842–849.

6 Understanding Severe Weather Systems Using Doppler and Polarisation Radar

Peter Meischner, Nikolai Dotzek, Martin Hagen, and Hartmut Höller

DLR, Institut für Physik der Atmosphäre, D-82234 Oberpfaffenhofen

6.1 Introduction

With increasing population, the impact of severe weather on socioeconomic systems is increasing worldwide. This is underlined by spectacular incidences of flash floods, hurricanes, and hail events with their impact on air and ground based transportation systems or big events like the Olympic games (Parker, 2000; Pielke and Pielke, 1999). Further, the interaction with the global climate system, with vertical exchange of trace gases by deep convective systems, on one side, and the increased variability of severe weather as a result of global warming, on the other side, is becoming more and more a focus of research. All these topics call for a better understanding of severe weather systems with the goal of improving forecasts.

An intense interaction of modelling efforts with observations is needed to improve the understanding of the life cycle of weather systems, which are characterised by the development of cloud and precipitation particles in a huge variety of dynamical structures. Frontal systems and mesoscale convective complexes develop a number of characteristic features. Polarimetric Doppler weather radars provide an excellent tool for investigating these features because of their potential to follow the coupled microphysical processes and dynamical developments simultaneously with suitable resolution in space and time.

Although versatile polarimetric and Doppler radar measurements, precisely adapted to the weather system of interest, give the most insight into 3-D structures and processes, complementary measurements such as those of satellites, lightning detection systems and aircraft have been combined in recent field experiments. The most impressive progress of all these efforts is in the availability of comprehensive data sets for comparison with numerical atmospheric modelling results and, increasingly, for data assimilation in numerical weather prediction models, as discussed by Macpherson et al. (2003), this book. The respective resolutions of radar observations and model calculations are converging in space and time.

In this chapter, we will describe the formation and life cycle of selected mesoscale weather systems, including fronts and a variety of deep convective systems investigated with the polarimetric Doppler radar POLDIRAD (Schroth et al., 1988), and the Karlsruhe Doppler radar (Gysi, 1995) during the last decade. These case studies have been selected from a great number of observations as representative of typical midlatitude weather systems producing heavy precipitation and high wind speeds. They demonstrate the improved insight gained by the full potential of Doppler and polarimetric radar measurements – as described by Gekat et al. (2003), this book – and they lead to conceptual models which will be presented. These models not only characterise local observations, but they are consistent with radar measurements reported from across Europe, the United States and other parts of the world.

We present a frontal system, a squall line, a hybrid type storm, a supercell and a tornado. We describe principles known today, underline them with conceptual models, and then show and discuss radar observations.

6.2 Frontal Systems

Frontal systems are among the most recognised and most frequent weather phenomena in mid and high latitudes that are connected with intense precipitation. Numerous case studies have been reported. Radar observations, including Doppler measurements and the use of composites from several radars covering a mesoscale area, have contributed to understanding their complexity and the development of associated precipitation patterns. Excellent overviews are given by Hobbs (1987) and Browning (1990). The state, as summarised by Hobbs (1987) was.

"The regions of relatively heavy precipitation are organised into large mesoscale ($\sim 10^3$–10^4 km^4) rain-bands which are classified into five types: warm-frontal, warm-sector, cold-frontal (wide and narrow), prefrontal cold-surge, and postfrontal bands. The rain-bands themselves are composed of smaller mesoscale areas (~ 10–10^2 km^2) or "cores" of precipitation. In some of the rain-bands the precipitation cores often originate in higher-level generating cells, probably produced by the lifting of potentially unstable air. The generating cells provide "seed" ice crystals, which grow by aggregation and riming as they fall through lower cloud layers produced by large mesoscale or synoptic scale lifting of the air. In these cases the rain-bands move with the velocity of the upper-level winds at the levels of the generating cells and therefore tend to move through the cyclonic storm. Other rain-bands have their origins in lower-level convection. For example, narrow cold-frontal bands are fed by moisture from a low-level, southerly jet of air which the cold front progressively overtakes. The narrow cold-frontal band is therefore anchored to the cold front. Appreciable precipitation on the mesoscale in cyclonic storms is invariably associated with high concentrations (~ 1–$100\ l^{-1}$) of ice particles. Small hills may play an important role in triggering or enhancing mesoscale rain-bands in cyclonic storms in regions where the air is unstable. The precipitation over and downwind of large mountain barriers, on the other hand, may be disrupted because the orography interferes with the low-level, southerly flow of air which provides the front its principal source of moisture."

Both reviews by Hobbs (1987) and Browning (1990) show detailed conceptual models for the most typical systems investigated so far.

Many more details of such air flow patterns both generally and in connectioin with the small scale precipitation cores, however, can be investigated by the VAD and ECUW techniques mentioned in Gekat et al. (2003), this book. Recently

6 Severe Weather Systems   169

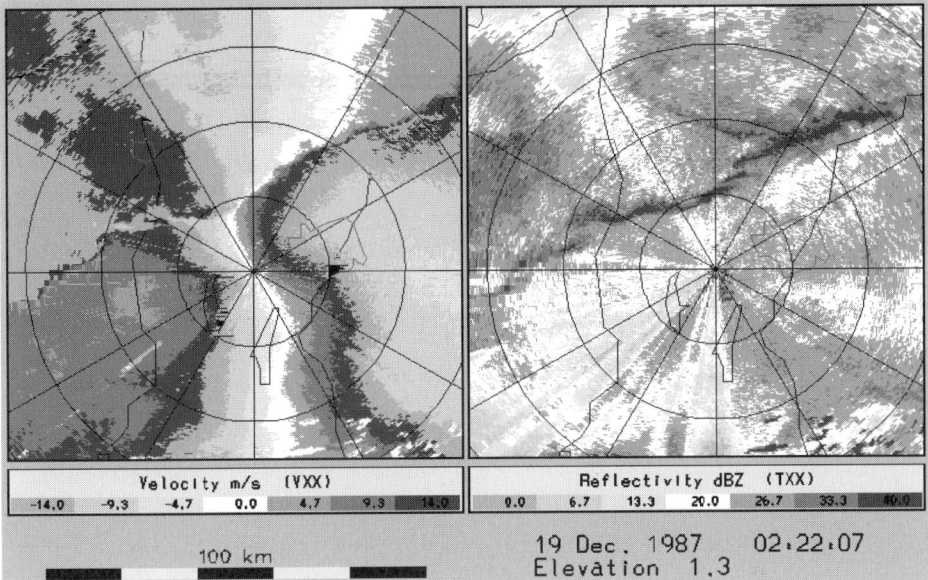

Fig. 6.1. Radar picture (PPI) of the front at 0121 UTC on 19 December 1987. Doppler velocity on the left, reflectivity on the right. Range circles are every 20 km. Areas surrounded by black lines are (from left to right) Augsburg, Ammersee, Starnberger See and Munich. The color version is given in color Fig. 14. on page 323

available polarimetric measurements also allow the formation of precipitation to be described in much more detail.

A cold front in southern Germany, connected with pronounced narrow rainbands, has been analysed in such detail and the orographic impact of the Alps has been documented by Hagen (1992).

Figure 6.1 shows the radar reflectivity and the Doppler velocity of that cold front from 19 December 1987, as it approached the radar site at Oberpfaffenhofen from the northwest.

The narrow rain-band structure as well as the velocity jump are easily recognisable. A section showing the measured Doppler velocities across the front (Fig. 6.2), gives the wind flow with a strong updraft resulting in the pronounced narrow rain-band. The lowering of the zero degree isotherm from the prefrontal to postfrontal situation is clearly recognisable by the bright band, representing the melting layer.

The narrow rain-band is composed of individual rain cores, arranged with the local wind field structure, as analysed in detail by VAD and ECUW techniques Hagen (1992). Developments in the rain-bands when approaching the Alps are documented by Fig. 6.3. The radar-observed rain-band together with the wind shear are confirmed by the analysed synoptic surface observations, where available. The intense precipitation of more than 20 mm/h lasted for a few minutes

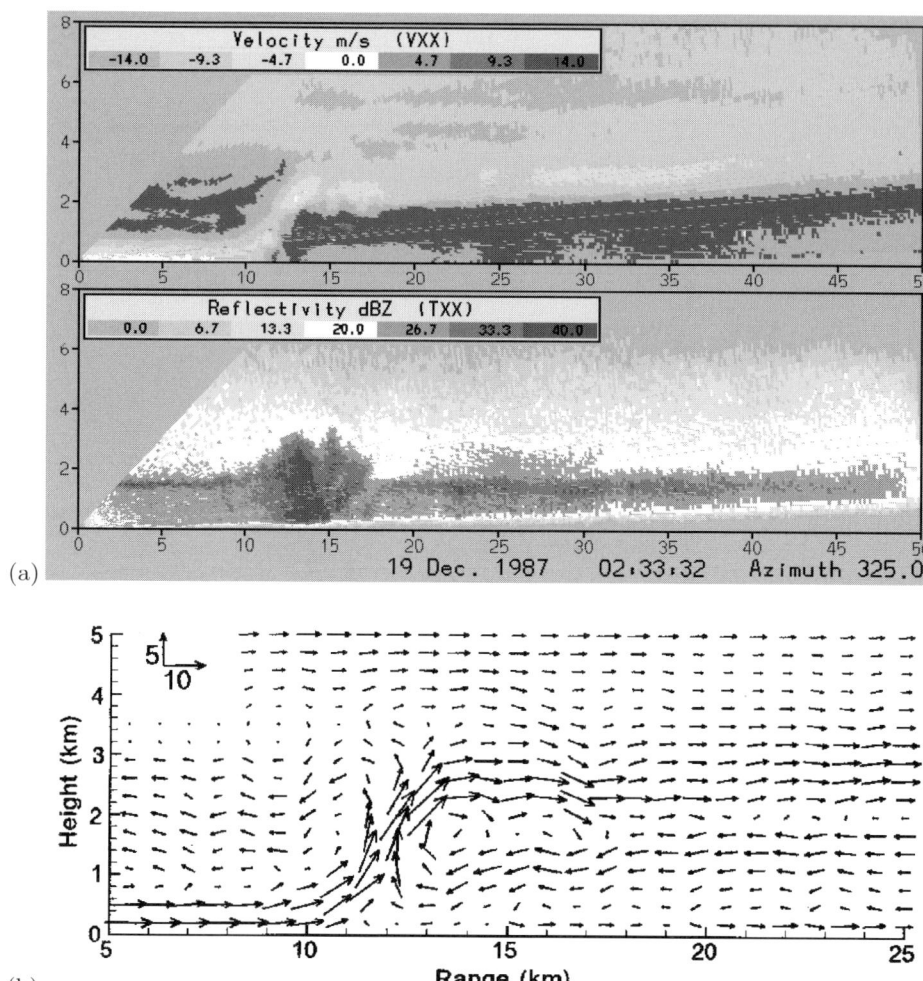

Fig. 6.2. (a) Radar picture (RHI) showing Doppler velocity (*top*) and reflectivity (*bottom*) perpendicular to the front at 0133 UTC (*along cross section A B in Fig. 6.3*). (b) Wind vectors as derived from the Doppler velocity. In this figure, the front is moving to the left. More details are to be recognised in color Fig. 15 on page 324

only, but the rain-bands maintained their general structure for more than 6 hours travelling a distance of about 300 km.

One main finding of this study is that this ordered local flow pattern at the rain-band cores finally is destabilised, together with a breakdown of the southwesterly prefrontal low-level jet or conveyor belt, when the front approaches the Alps. Therewith, the front movement is decelerated when the first mountain ridge is reached and retarded by some hours at the main ridge of the Alps.

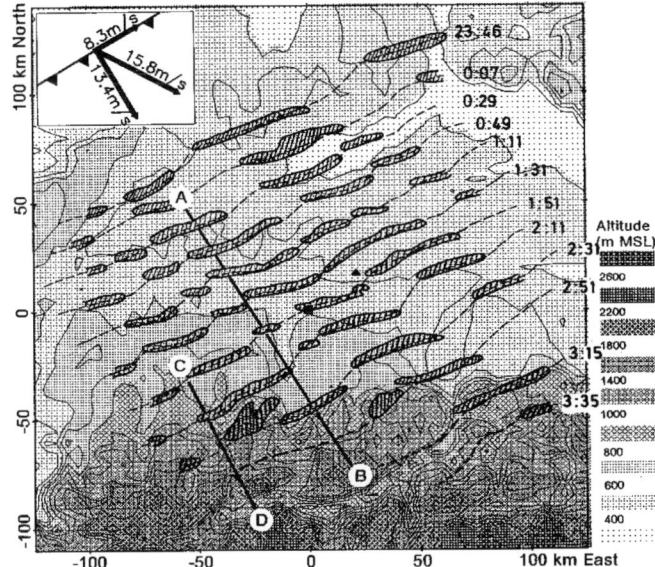

Fig. 6.3. Position of the cold front at the indicated times (UTC) on 18 and 19 December 1987. Areas with high reflectivity along the front are hatched, — is the wind shear zone. The data are based on PPI-scans with 1.5° elevation. ■ Radar Oberpfaffenhofen, (0,0 km). The orography (m MSL) is shaded. The inset shows the propagation of the precipitation cores. A cross section along A B is shown in Fig. 6.2

Some details of the precipitation process can be followed by polarimetric measurements (Fig. 6.4). This figure shows the combination of reflectivity and differential reflectivity of a RHI scan across the rain-band. The temperature profile at München-Oberschleissheim at 0144 UTC, about 5 minutes after the frontal passage, is also shown. At a height of about 1.5 km and to the right and left of the precipitation core, Z_H and Z_{DR} are enhanced due to melting ice particles, representing the bright band. In the updraft region, Z_{DR} is greater than 0.4, suggesting big drops which are suspended in the updraft. Bigger drops are also indicated by an extended region of large Z_H in the updraft. Above the freezing level, such big drops start to freeze and rime, forming graupel. Also snow aggregates mixed in from above grow by riming. The occurrence of graupel and large ice aggregates is indicated by the fact that in the upper part of the updraft, a considerable part of the high reflectivity region (above 34 dBZ) shows low Z_{DR}. With the airflow, the graupel and aggregates are transported to the left which is outside of the updraft. There they fall and melt giving the narrow shaft of precipitation core to the rear of the updraft. In this shaft, with $Z_H \approx 45$ dBZ and $Z_{DR} \approx 2$ dB, a median drop diameter of 1.7 mm is estimated, following Seliga and Bringi (1976).

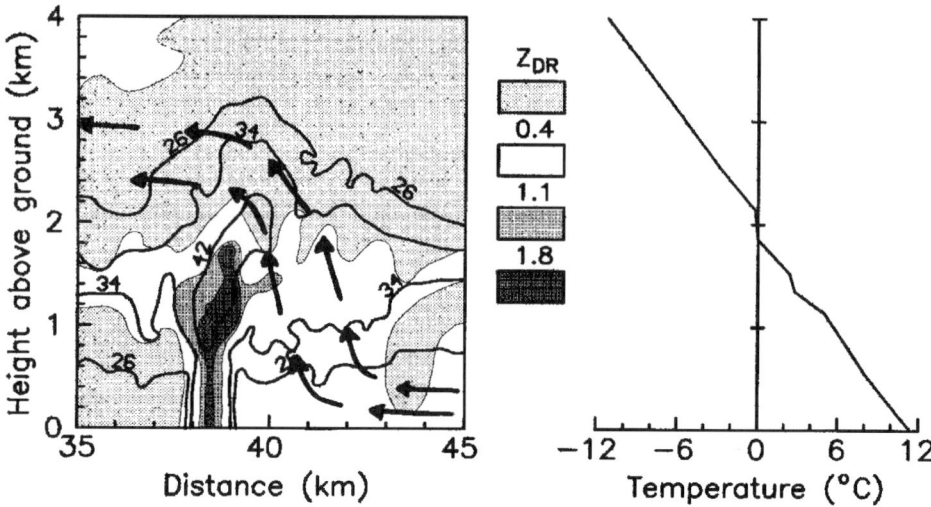

Fig. 6.4. Reflectivity (Z_H) contour lines and differential reflectivity Z_{DR} shaded of a RHI scan at 0228 UTC with azimuth 170°. The air flow is estimated from Doppler measurements. The right side shows the temperature profile at München-Oberschleißheim at 0144 UTC

6.3 Deep Convective Systems

6.3.1 Storm Classification Scheme

The manifestation of thunderstorm cells is closely related to the microphysical processes responsible for the formation of precipitation. A cell is defined by closed contours of radar reflectivity which, in turn, is a function of the number concentration, size distribution and thermodynamic phase of the particles involved. In their developing stages, the cells coincide with individual updraft turrets. Later on, reflectivity maxima may also be caused by particle growth in favoured regions of a developed storm circulation, for example, in the main updraft of a mature thunderstorm.

According to Foote (1985), the storms may be classified as

- ordinary or single cells
- multicellular storms
- supercell storms.

These are more or less archetypal structures representing fixed points on an essentially continuous scale of storms with the single cell as the basic unit. Nevertheless, these types can be used to represent classes of storms having the same basic features in common.

The characteristics of a **single cell storm** are shown schematically in Fig. 6.5. The complete cycle of precipitation development and fallout takes place in one cell. Particles are not transferred into another cell. At time t_1, the cell is in its cumulus stage. It consists exclusively of cloud droplets or small ice crystals not yet detectable by S- or C-band radar. At time t_2, a first radar echo has developed due to the presence of rimed ice particles, graupel, snow, or raindrops. The cell now is in its growing stage. Intense riming of these particles resulting in centimeter sized hailstones occurs at t_3 when the cell enters its mature stage. While the upper part of the cloud is still growing, a first downdraft has formed associated with the fallout of precipitation. At t_4 (decaying stage), the cloud releases its precipitation to the ground in connection with an intensified downdraft often called a downburst. In the upper part of the storm, the updraft air has spread horizontally on reaching an equilibrium level forming the typical thunderstorm anvil. If the hailstones are large enough that they do not melt completely during fallout below the melting level, they hit the ground and may cause damage. But due to the transient nature of the cell and its relatively low propagation speed, the resulting hailswath at the ground is usually small. The duration of the hailfall is only a few minutes. The typical time interval between each consecutive snapshot in Fig. 6.5 is 5 minutes. Single cell storms may occur either in isolation or organized in large scale complexes or lines. Here, they can coexist and interact with other single cells or even form multicell or supercell storms.

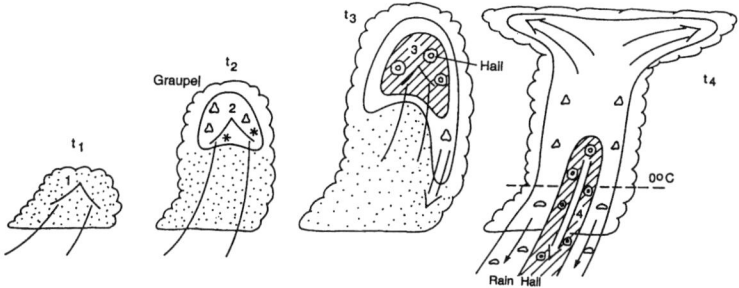

Fig. 6.5. Schematic diagram of a single cell thunderstorm. Four different stages of cloud development are shown (t_1 through t_4). Graupel forms at t_2, hail at t_3 and precipitation falls out of the cloud at t_4. The *solid lines* indicate contours of radar reflectivity with increasing values toward the cloud's interior

In **multicell storms** (Fig. 6.6), cells in all stages of development coexist, each one forming a part of the same storm system. One large updraft provides the root for the individual towers which become clearly visible in the higher parts of the storm. The hail formation processes are active simultaneously, the different stages acting in different cells. In contrast to that, they are active consecutively in time within the single cells. Because of the multicellular structure of the storm, an exchange of particles among the different cells may occur.

Graupel falling in the growing cells (feeder cells) above the low level inflow region can be swept toward the storm's center where they enter the main updraft region. Following such trajectories, these particles are recirculated into the storm

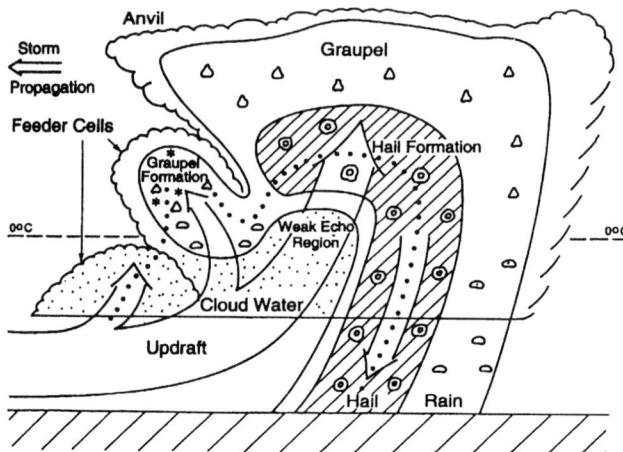

Fig. 6.6. Schematic diagram of a multicell thunderstorm. A succession of cells in different stages of development can be seen contributing to a common cloud mass. Graupel particles are forming in the newly grown cells in the left part of the figure. Later, in the mature stage of the cell, these graupel can act as hailstone embryos. Hail formation largely takes place right at the top of the weak echo region (WER). The *solid lines* indicate contours of radar reflectivity with increasing values toward the cloud's interior. The trajectory of an ice nucleus growing into a hailstone is indicated by the *dotted line*

and start to rise again. In the main updraft, they may encounter very high liquid water content areas and can grow into large hailstones while crossing the main updraft horizontally. On the downwind side, hail particles enter the storm's main downdraft and rapidly fall down to the ground. The processes described here give rise to the typical radar structure of these cells: a forward flank (vault-like) overhang of particles growing into hail on top of a weak echo region (WER) of high liquid water content (cloud water, usually not detected by radar). Updraft speeds in the WER are too high for graupel or hail particles to penetrate these regions in appreciable amounts. If the storm is less intense or if the contributing cells have a wider spacing, the graupel originating in the feeder clouds may fall out of the cloud before being able to recirculate. A WER does not exist in this case. Hailfalls from such systems are less intense than those resulting from vaulted multicells. Multicell storms are more persistent than single cells, typically lasting 10 to 60 minutes. The associated hailswaths are in the range 10 to 100 kilometers in length (Höller et al., 1994).

Supercell storms are formed by a single large updraft (Fig. 6.7) which does not break up into multiple branches.

The graupel particles which can later grow into hail are formed at the fringes of the updraft zone on the upshear side (with respect to midlevel shear) of the

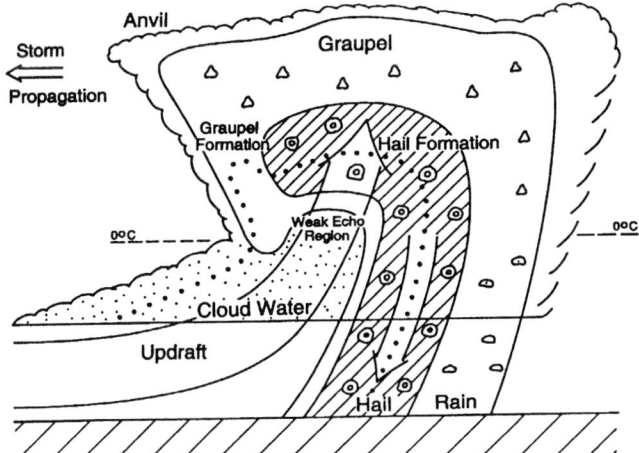

Fig. 6.7. Schematic diagram of a supercell thunderstorm. The storm is formed by a large, quasi-stationary cell. All the different stages of hail formation (particle trajectory indicated by the *dotted line*) take place within the same cell but at different locations. The *solid lines* contour the radar reflectivity with increasing values toward the cloud's interior

storms. These regions play the part of the feeder cells in the multicellular case. The subsequent phase of hail growth is basically similar in multicells and supercells, except for the larger updraft size of supercell storms. Supercells have a much higher persistence than the other types of storms. Their quasi-steady structure is maintained for more than half an hour (by definition, see Browning, 1977), but these cells can last for several hours – extreme cases of up to 12 hours duration have been reported by Paul (1973) – and the corresponding hail-swath can extend some hundred kilometers in length. The "Munich hailstorm" of 12 July 1984 showed such characteristics (Höller and Reinhardt, 1986).

Between the discrete multicell development and the steady supercells, transitional forms of storms can be observed. These storms have been termed weak evolutionary or hydrid storms (Nelson and Knight, 1987; Nelson, 1987). One example will be described below.

In the scale above cloud scale, the mesoscale, the guideline for classifying storm organisation is the shape of the radar echo. Four basic configurations have been identified (Höller, 1994):

I: Isolated storms. Thunderstorms are more or less isolated from each other and are not embedded in common reflectivity contours. Such isolated storms may appear as singular events or be irregularly scattered.
C: Clusters or complexes of storms. Thunderstorms are closely grouped together, embedded in a circumscribed reflectivity contour. A variety of clusters might coexist at a given instant at different locations in the observational

area. A cluster does not have a preferred direction of alignment (orientation) for the contributing storms.

L: Line oriented storms. Thunderstorms have a preferred direction of alignment. They may appear as a line of isolated storms or as a linear complex circumscribed by a common reflectivity contour. The typical maximum dimension should be more than twice as large as the typical minimum dimension; otherwise the system is classified as a cluster.

SL: Squall lines. This is a special form of line oriented sytems. The typical maximum dimension is very large compared to the typical minimum dimension. The length to width ratio is at least five to one, and the length of the line is at least 50 km, persisting for at least 15 min. In many cases the system develops a quasi two dimensional structure. In contrast to the line oriented storms, squall lines are often characterised by a leading line of heavy convection followed by a region of widespread stratiform precipitation.

Figure 6.8 gives statistics of the mesoscale organisation of the thunderstorms observed during a period of 6 years in southern Germany (Höller et al., 1994; Hagen et al., 1999).

The typical size or length of a convective complex has been used as an ordering criterion. Order increases from small scale individual cells (I) to clusters or complexes (C), lines (L) and squall lines (SL). Two different classification schemes are shown: (MAX) where only the highest order convective sytem is counted per day and (ALL) where multiple structures observed on the same day also considered. The results clearly indicate the dominance of line structures. This especially holds for the MAX case, while in the ALL case, the number of clusters is considerably increased but still smaller than the number of lines. We also note a remarkable number of squall lines (SL). The severity of weather phenomena like heavy gusts or large hail clearly increases with the degree of organisation from single cell storms toward squall lines and supercell storms.

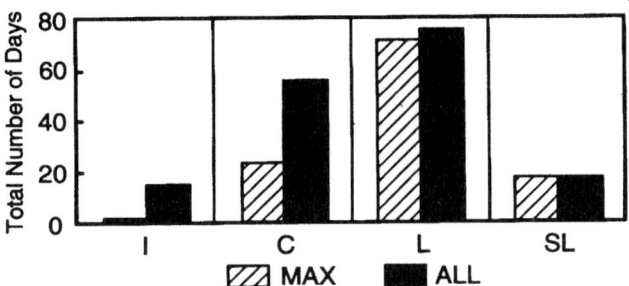

Fig. 6.8. Large scale organisation of convection in the 6-year period 1987–1992. Total number of days as a function of convective type: (I) individual cells, (C) cloud cluster, (L) line structures, and (SL) squall lines. MAX: only maximum daily development is counted, ALL: all different types of convection are added

6.3.2 Squall Lines

In summertime situations, lines of intense thunderstorms, so-called squall lines, some 100 km in length, often develop at convergence lines connected with or preceding cold fronts approaching from the west. The special prefrontal situation must be characterised by high potential instability at lower levels up to about 5 km ASL and high convective available potential energy. A large temperature difference between the prefrontal and postfrontal air masses is essential.

Such squall lines in southern Germany have been analysed and described in detail by Meischner et al. (1991) and Haase-Straub et al. (1997).

Figure 6.9 shows a typical cross section, as deduced from such observations. The system is moving at a speed of some 30 km/h to the right, in Europe normally to the east. The moving wedge-shaped cold airmass near ground forces lifting of the warm humid air, thus initiating or increasing further convection. This moving cold surface layer may be induced by the melting and evaporation of the precipitation from more or less individual thunderstorms which merge to lines moving ahead of a cold front. Along these lines, enhanced local lifting is forced again, initialising deeper convection.

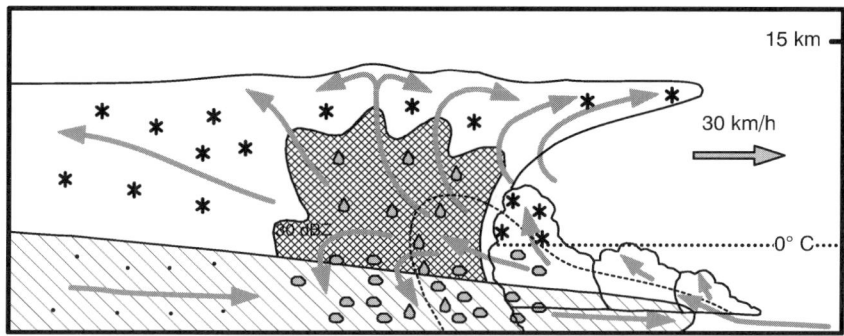

Fig. 6.9. Conceptual model cross section of a squall line observed on 1 July 1987 in southern Bavaria. The cross section shows a deep cell in its mature state with a well-developed anvil cloud. It is displayed in accordance with the polarimetric and Doppler radar measurements

Once formed, the overall structure of a squall line may be maintained for several hours by mainly internal processes. The principal characteristics are the front inflow layer at low and midlevels, the rear inflow extending underneath the front inflow and culminating in the low level gust front circulation, and high level anvil outflow extending to the front and rear.

Newly growing cells embedded within the inflow area ahead of the main precipitation cores represent smaller updrafts. Heavy precipitation is formed at the leading edge of the line (in the eastern part), while the western part shows more or less a stratiform character with widespread precipitation. The heavy precipitation – a mixture of melting hail and rain – hits the ground in a forward direction, thus contributing to the forward moving, cool, air pool or even controlling its strength.

The greatest precipitation is formed in areas where feeder cells merge with the main system. These areas are collocated with the main updraft, and the merging feeder cells explosively reach a mature stage of deep convection, thus forming a new leading edge.

Ice particles, after having reached their final sizes, which depend upon the updraft speed and the LWC available, fall first within the slanting frontal inflow and then within the wedge-shaped rear inflow surface layer. Cooling by melting and evaporation contributes to the surface cold pool. The forward forcing is indicated by the gusts, which are clearly connected with the precipitation hitting the ground.

Situations have been observed where the gust front has already proceeded about 20 km ahead of the main precipitation cores (Meischner et al., 1991). In such situations, new isolated convective cells grow well ahead of the older ones, often not merging with them.

The precipitation in the primary core in such a situation is much less intense than when the feeder cells are coupled more closely with the main system. So, if the feeder cells are decoupled from the main system by the forward accelerated cold pool, the process of forming heavy precipitation is decreased. A further consequence of the decrease in precipitation intensity would be a decreased contribution to the cold air pool and a deceleration of the gust front. This feedback mechanism might be of importance in maintaining the squall line system in an optimal state and explaining its longevity. Further, an oscillation in system strength indicated by the vertical velocity and discussed by Meischner et al. (1991), Rotunno et al. (1988) and Weisman et al. (1988), can be explained in this way.

Figure 6.10 now shows a radar sequence of the position of a squall line on 21 July 1992, oriented south–north, together with the preceding gust fronts and an additional cloud line approaching from the northwest. The velocity pattern across a squall line typically shows a 2-D structure, as indicated by the measured Doppler velocity (Fig. 6.11). We see strong convergence in the updraft region and divergence at the outflow at cloud top. The measurement further shows the mid-level rear inflow as well as the gust front preceding the squall line itself that contributes to the lifting ahead.

The precipitation formation process, especially hail production via snow aggregates, graupel and recirculated raindops that freeze on reaching low temperature regions, is monitored by polarimetric measurements (Fig. 6.12).

6 Severe Weather Systems   179

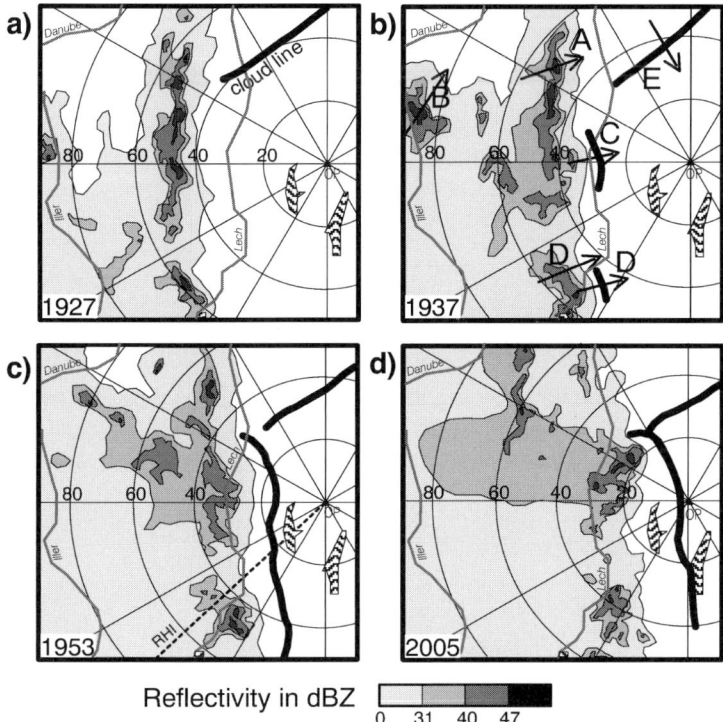

Fig. 6.10. PPI reflectivity patterns of a squall line from the DLR radar; distances are given in kilometres. Gust fronts and the cloud line are indicated by *bold lines* at times (a) 1927 UTC, (b) 1937 UTC, (c) 1953 UTC and (d) 2005 UTC, marked by capital letters. *Arrows* point in the corresponding directions of propagation. The *dotted line* "RHI" gives the orientation of the RHI of Fig. 6.11

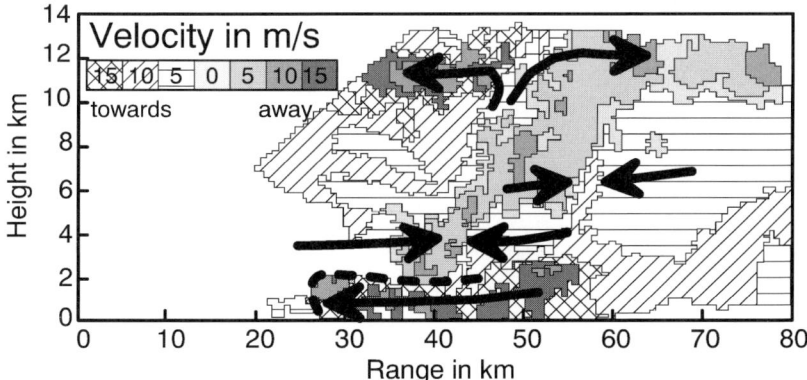

Fig. 6.11. Range height cross section of a squall line showing Doppler velocity. The *arrows* indicate the relative flow; the *dashed line* marks the gust front flow. Note that because of folding, an observed speed of 10 ms^{-1} away from the radar must be interpreted as 22 ms^{-1} toward the radar

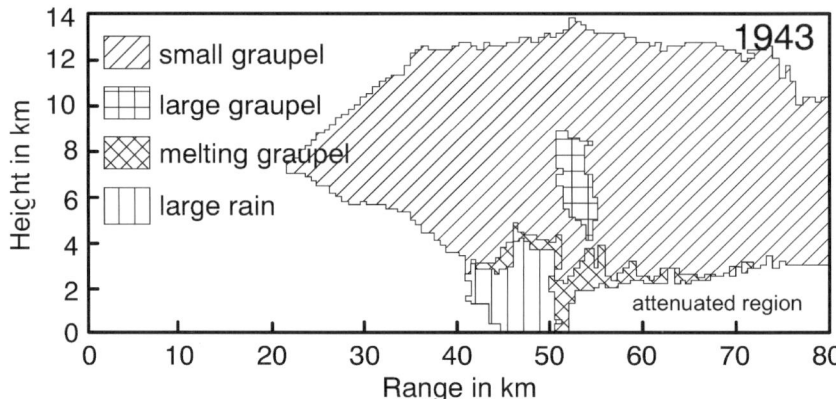

Fig. 6.12. Cross section of the squall line with precipitation classified according to Höller et al. (1994). Number at upper right corner indicates time of observation

6.3.3 A Hybrid Type Storm

Figure 6.13 summarises findings for a V-shaped hybrid type thunderstorm showing features of both a multicell storm and a supercell storm (Höller et al., 1994). The precipitation formation during the life cycle is revealed by polarimetric and Doppler radar measurements.

Rain is found in region R_1, originating from the melting of graupel from the newly grown cell (G). Some of these raindrops recirculate into the weak echo region (W), where they are carried up to heights well above the freezing level. Rain in region R_2 can have its origin either from drops passing through R_1 at relatively low trajectories or from melting or shedding hailstones in the main downdraft. The horizontal component of the velocity field causes a size sorting of rain, mixed particles, and wet hailstones (HW) on their way to the ground. The position of the gust front is marked by GF.

The general evolution of this storm from 30 June 1990 during its lifetime of about two and a half hours is given in Fig. 6.14.

We can recognise the initialisation phase which lasts roughly until 1315 UTC, the mature phase from 1315 to 1455, and then the decaying phase. The direction of propagation is basically from WSW, which corresponds to the direction of the midlevel winds. A typical feature of severe storms is the weak echo region, WER or W in this figure, indicating the main updraft area in front of the moving storm where the upward inflow velocity is greatest and precipitation particles are suspended aloft.

The overall flow pattern can be estimated by using the measured Doppler velocity and subtracting the ground relative system speed. Figure 6.15 shows these storm relative Doppler velocities for three different elevation angles and for a vertical cut at constant azimuth.

Selected reflectivity contours are superimposed. The arrows indicate possible streamlines. The low-level flow (Fig. 6.15a) is characterised by an anticyclonic

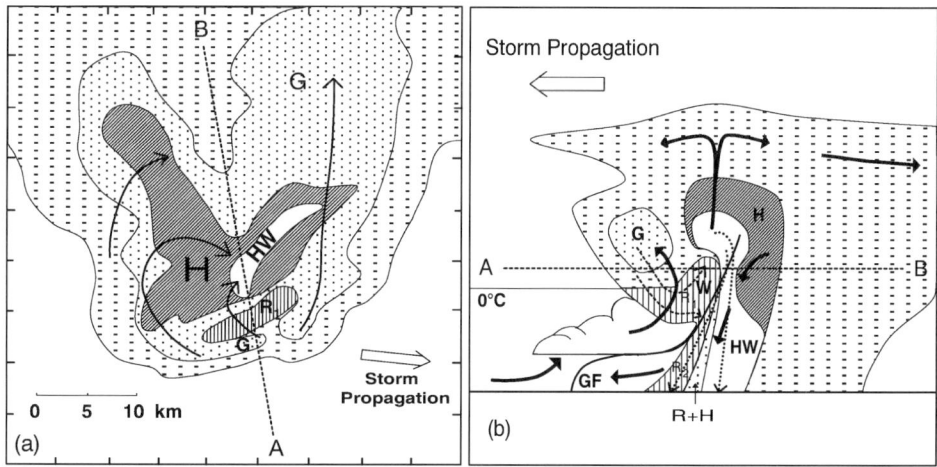

Fig. 6.13. Schematic diagram of the main dynamical and microphysical features of the hailstorm of 21 July 1998 in sourthern Germany. Panel (a) shows a typical midlevel (about 5 km height) cross section. The *arrows* indicate horizontal streamlines. Panel (b) shows a typical vertical cross section along A–B. *Solid arrows* indicate streamlines; *dashed* and *dotted* arrows represent particle trajectories

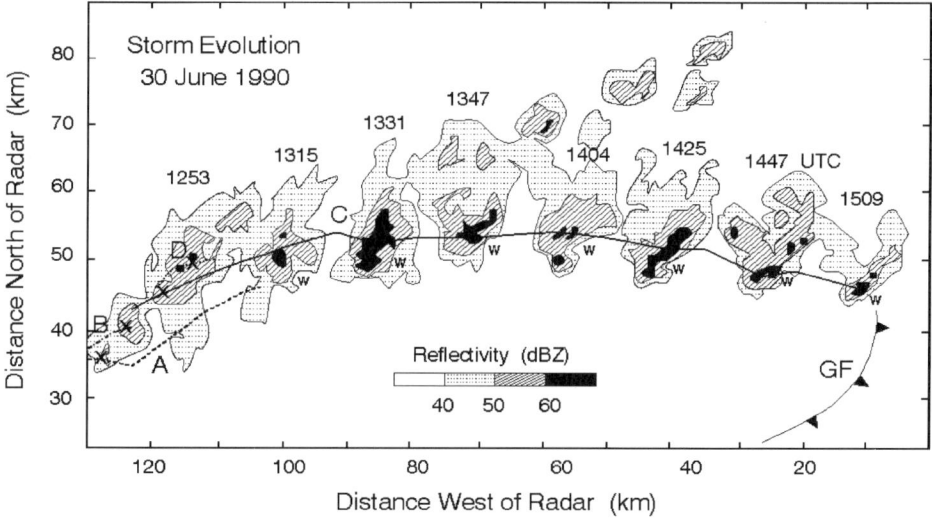

Fig. 6.14. Evolution of the storm's reflectivity pattern: 40, 50, and 60 dBZ contours are shown for different times at those elevations demonstrating most clearly the weak echo structure (W). Tracks of four different reflectivity centers (A to D) at 2.2° elevation as well as their position at 1253 UTC (x) are indicated. The position of the gust front G, marked by a thin line echo at 1509 UTC, is also indicated

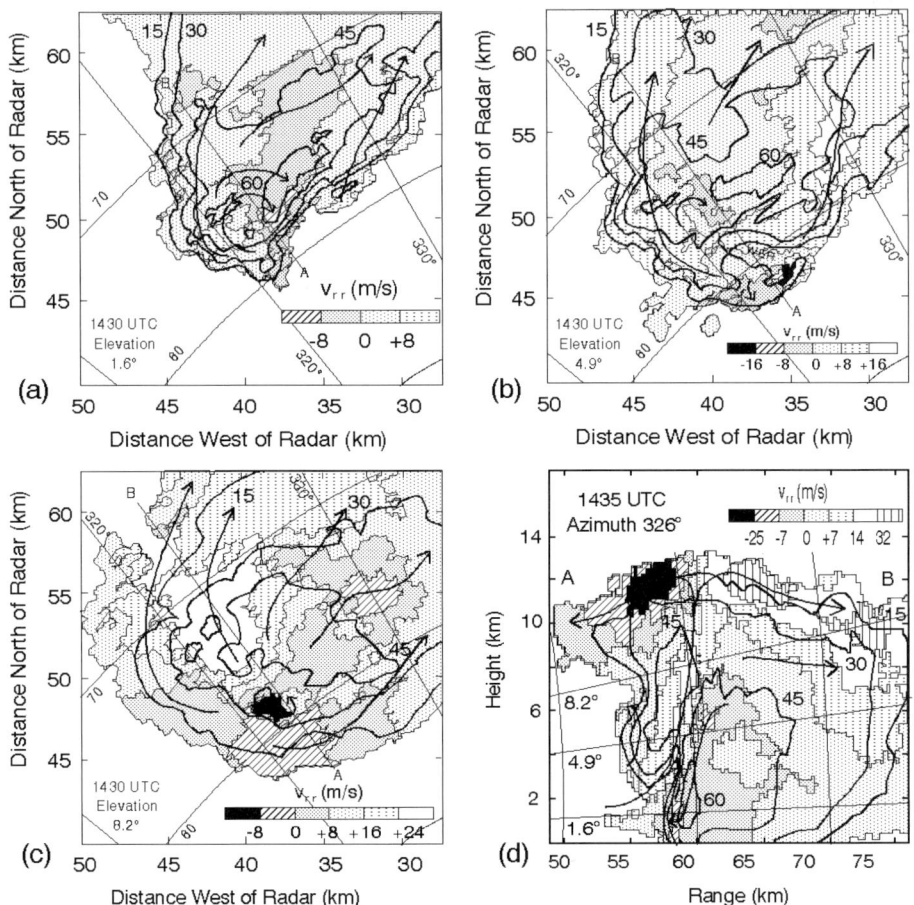

Fig. 6.15. Storm-relative Doppler (radial) velocity v_{rr} (*shaded areas*), reflectivity contours (*solid lines*) labeled in dB, and possible streamlines (*solid arrows*) as suggested by the v_r field. A radar centered spherical coordinate system is shown as an overlay. Radial distances are indicated in km. Storm scans at 1430 UTC are shown at different elevation angles of (a) 1.6°, (b) 4.9°, and (c) 8.2°. In (d), a vertical section taken at 1435 UTC is presented. Taking into account the storm advection, the vertical scan corresponds to a section A–B in (a)–(c)

rotation in the southern parts. At midlevel (Fig. 6.15b), the anticyclonic rotation has weakened in the high reflectivity zone, and a cyclonic rotation can be seen at the southeastern edge of the WER. At upper levels (Fig. 6.15c), the diverging outflow on top of the WER marks the location of the main updraft. By following the high reflectivity cores with time, the growth of precipitation in the updraft around the southern part of the storm and its fall close to the updraft core can be followed. The flow structure shown in Fig. 6.15d, of course, is only a snapshot,

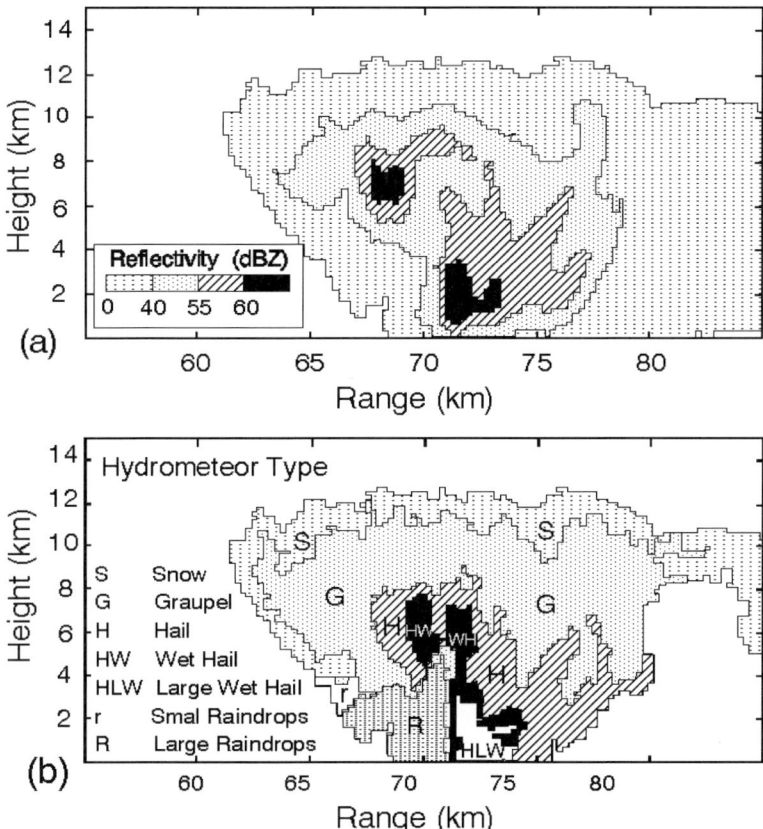

Fig. 6.16. Vertical section through the storm's core during its vigorous phase at 1414 UTC. (a) Reflectivity Z_{HH}, (b) Hydrometeor type

but many features remained essentially unchanged for more than a half hour, so the flow structure may be regarded as typical.

Polarimetric measurements, performed simultaneously, provide insight into the microphysical developments of precipitation. As discussed in more detail by Illingworth (2003), this book, and in a number of other publications, for example, Straka and Zrnić (1993), Höller et al. (1994), and Balakrishnan and Zrnić (1990), the combined use of measured polarimetric parameters allows the classification and discrimination of different types of hydrometeors and the following of microphysical processes. Among these are the initiation of precipitation as liquid or ice processes, the subsequent growth of graupel and hail, shedding of drops from hail, melting of ice particles, and recirculation of raindrops. Figure 6.16 gives a RHI of the storm in its mature stage. Figure 6.16a shows the reflectivity, indicating the growth of precipitation above the updraft, and a precipitation package reaching the ground behind the updraft. Figure 6.16b gives the particle classification for this situation.

6.3.4 A Supercell Storm

An example demonstrating the improved insight into storm development and storm internal processes by a multisensor approach during well-coordinated field experiments is the supercell storm of 21 July 1998, observed during the European Lightning Nitrogen Oxides Project, EULINOX (Höller and Schumann, 2000). The aim of this project was to contribute to the understanding, modelling and parameterisation of lightning-induced production of nitrogen oxides in thunderstorms in order to assess its contribution to ozone production in the upper troposphere. Storm structure and dynamics have been observed by multiple Doppler radars, the microphysical developments by polarimetric radar, the lightning by 2-D and 3-D detection systems, and the nitrogen oxides outflow and turbulence structures in the anvil region by the Falcon research aircraft (Höller and Schumann, 2000; Meischner et al., 2001). This comprehensive data set is, among other things, suitable for comparison with modelling results. Is has been compared with detailed simulations from the mesoscale MM5 model.

The storm developed in the Allgäu region WSW of the DLR radar site at Oberpfaffenhofen at about 100 km distance and 240° azimuth; see Fig. 6.17a.

Following the intensification phase, the cell began to split into a northern and a southern part with lightning more pronounced in the southern cell. The northern cell soon began to decay while the southern one developed into a supercell structure with a hook echo and an arc-shaped gust front signature extending south of the main cell.

Such characteristics of cell development are due to the prevalent instability and shear conditions, as have been reported from observations as well as numerical simulations (Wilhelmson and Klemp, 1981). The characteristics found for this case give good indications for the actual simulation by the mesoscale models. The splitting and cell development characteristics have been picked up by the MM5 model simulations. Figure 6.17 shows impressive agreement between radar observations and model results.

The southern, right moving storm was not an archetypal supercell storm. It also developed multicell-like features which were superimposed on the general supercell-like pattern. For reasons of simplicity, these details will be neglected in the following discussion, and the general term "supercell" will be used for it.

Shortly after splitting, the cells were located in a region suitable for dual-Doppler observations. The two radars at Oberpfaffenhofen OP and Hohenpeißenberg HP performed volume scans which allowed for a reconstruction of the 3-D wind field. For deriving the vertical velocity components in the cloud region, zero vertical wind was assumed at cloud top as an upper boundary condition. Due to deficiencies caused by this assumption and because of imperfect synchronisation of the measurements, the 3-D wind fields derived and shown in Fig. 6.18 have to be interpreted carefully.

The dominant wind field structures are the strong horizontal divergence at midlevels (Fig. 6.18a) and a distinct cyclonic rotation, as seen at the southern edge. A vertical section (Fig. 6.18b) shows the main updraft coinciding with

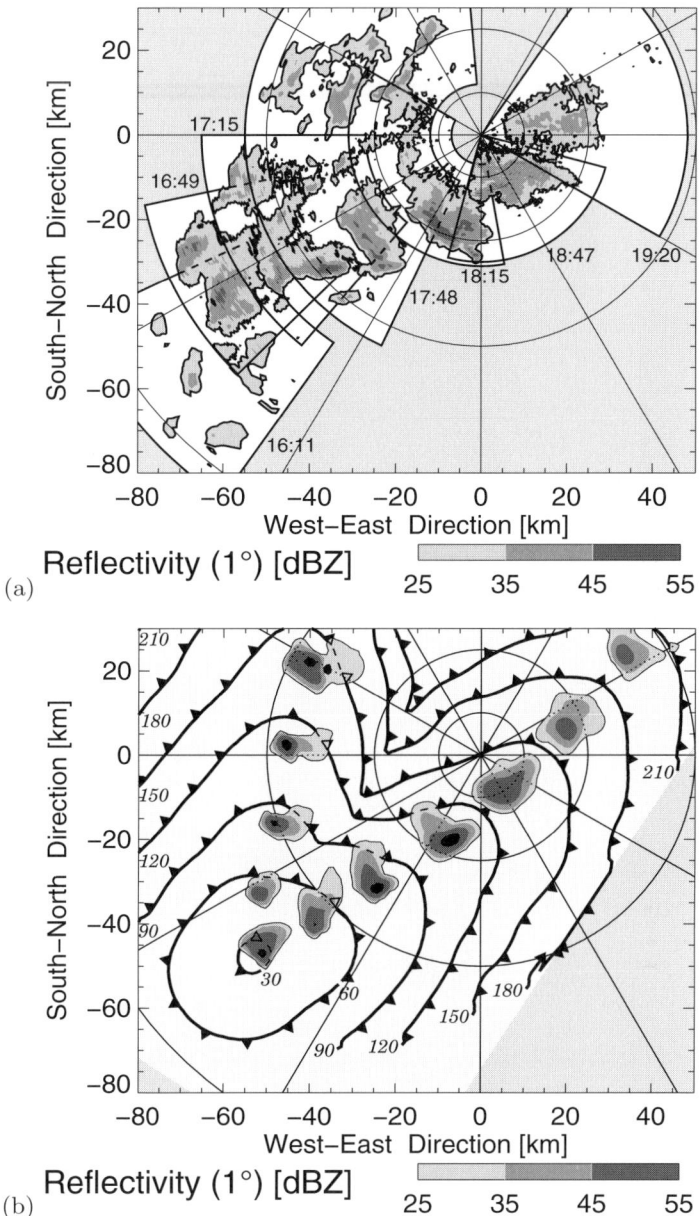

Fig. 6.17. (a) POLDIRAD radar reflectivity measurements at 1° elevation for the 21 July 1998 supercell thunderstorm; data from seven different scans are overlaid to illustrate the development. (b) Simulated reflectivity every 30 min at 1° elevation. Radar position with respect to the modelled thunderstorm corresponds to POLDIRAD location; surface gust front (−0.5° perturbation isotherm) is denoted by thick barbed lines. Figure by courtesy of Th. Fehr

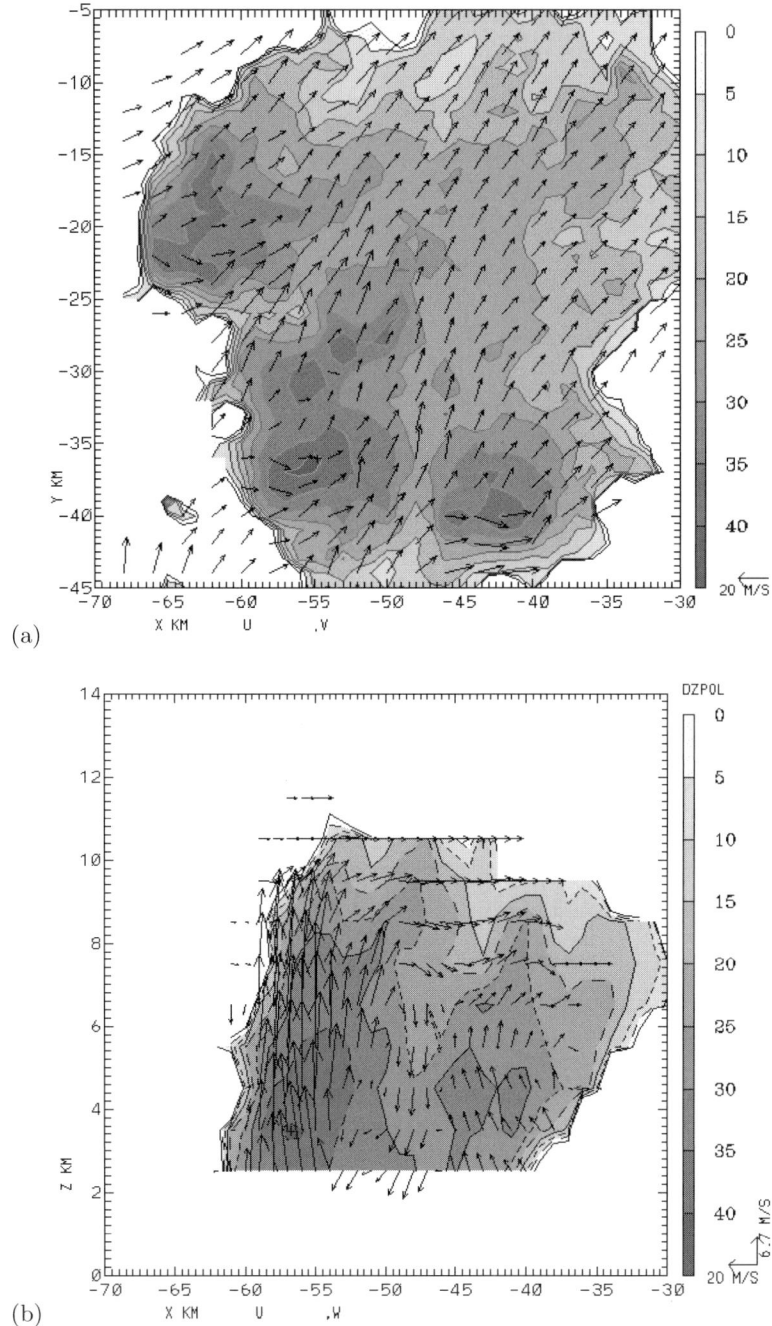

Fig. 6.18. Dual-Doppler wind field analysis for 21 July 1998 at 1657 UTC. (a) Radar reflectivity and horizontal wind at 5.5 km height; (b) radar reflectivity and vertical wind in an W–E section through the most severe cell at 35 km south of POLDIRAD

6 Severe Weather Systems   187

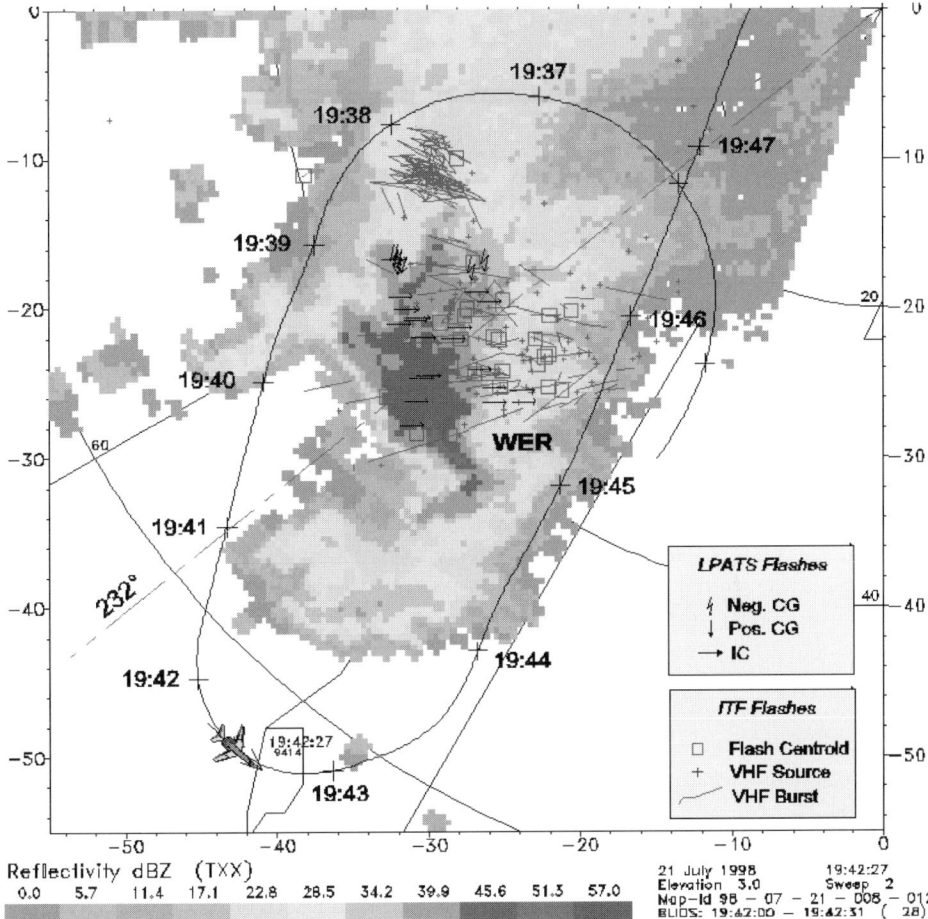

Fig. 6.19. The supercell of 21 July 1998. POLDIRAD radar reflectivity at 1742 UTC taken at 3° elevation. FALCON flight track with the position of the FALCON aircraft indicated at the time the radar scan was taken. Flashes from a 45-s period are shown. Negative, positive and intracloud flashes as measured with the Lightning Position And Tracking Systems LPATS are indicated as well as flashes indicated by VHF interferometric measurements (ITF); see Höller and Schumann (2000). Color Fig. 16 on page 325 shows structures more clearly

the reflectivity core. The updraft speed is 24 m/s, whereas downdraft strength reaches about 8 m/s.

As an example of the typical radar and lightning structures, Figs. 6.19 and 6.20 show a comparison of polarimetric radar parameters and the flash characteristics as inferred from 2-D and 3-D lightning detection systems; see Höller and Schumann (2000). At 40 km distance from POLDIRAD; the radar scan height is about 2 km above ground, thus Fig. 6.19 represents low-level features. The weak echo region, WER, and the hook echo extension toward the south can clearly be identified. The WER is most likely connected to the position of the main updraft of the storm.

One consistent feature is that the flashes occur preferentially in the regions located downwind (NE) of the main storm updraft. Most of the flashes detected were identified as intracloud flashes. This finding applies to longer periods of the storm development and thus has important implications for nitrogen oxides assessment.

The vertical sections shown in Fig. 6.20 demonstrate a close connection between cloud microphysical developments and lightning. From the reflectivity scan, it can be noted that discharges dominate in regions of moderate reflectivity values. The core of the storm (high reflectivity) is relatively free of flashes, as is the large anvil and the stratiform-like precipitation area at closer range. The hydrometeor classification scheme, described by Höller et al. (1994), uses differential reflectivity (Z_{DR}) and linear depolarisation ratio (LDR) for inferring hydrometeor type. From this analysis, it can be concluded that the VHF sources are concentrated in the graupel region extending in the upper parts of the convective cell (Dotzek et al., 2001).

Some interesting features to be noted from Figure 6.20c are (i) at around 40 km range, a mixture of hail and rain reaches the ground at the southwestern edge of the storm and (ii) the radar echo of the Falcon aircraft can be seen at 55 km range and 9 km height, which is in agreement with the aircraft flight track data shown in Figure 6.19 at 1941 UTC.

Figure 6.21 shows the linear depolarisation ratio (LDR) and radar reflectivity at 1739 UTC when the storm was in an intense phase of development. The feature to be noted here is a structure of very high LDR values between 7 and 12 km height at the top of the convective cell structures, the updraft branches. LDR increases progressively with range up to −13 dB. Such a structure is indicative of ice particles being tilted with their major axis into the vertical direction. The maximum LDR effect is expected with tilting angles close to 45°. Under the influence of a strong vertical electric field, the alignment of pristine ice crystals can deviate from the normal horizontal direction (Caylor and Chandrasekar, 1996). Electric fields of 100 kV m^{-1} can align plates of less than 0.6 mm diameter, while for the alignment of larger plates (1.7 mm), fields of 200 kV m^{-1} are necessary.

6.3.5 Tornadic Storms

During recent years, it has been realised that in Europe several hundred tornadoes occur each year. Tornadoes cause considerable damage (Dotzek 2002b, http://www.tordach.org) but are not always recognised because of their small scale. Recent research by radar observation has increased the understanding of these weather events in Europe.

The Upper Rhine valley in southwest Germany (Fig. 6.22) is a region of frequent thunderstorm development (Finke and Hauf, 1996; Hagen, 1999), connected with severe weather, such as hail, downbursts, and tornadoes (Dotzek et al., 2000; Dotzek, 2001). Since the early 1990s, an operational C-band Doppler

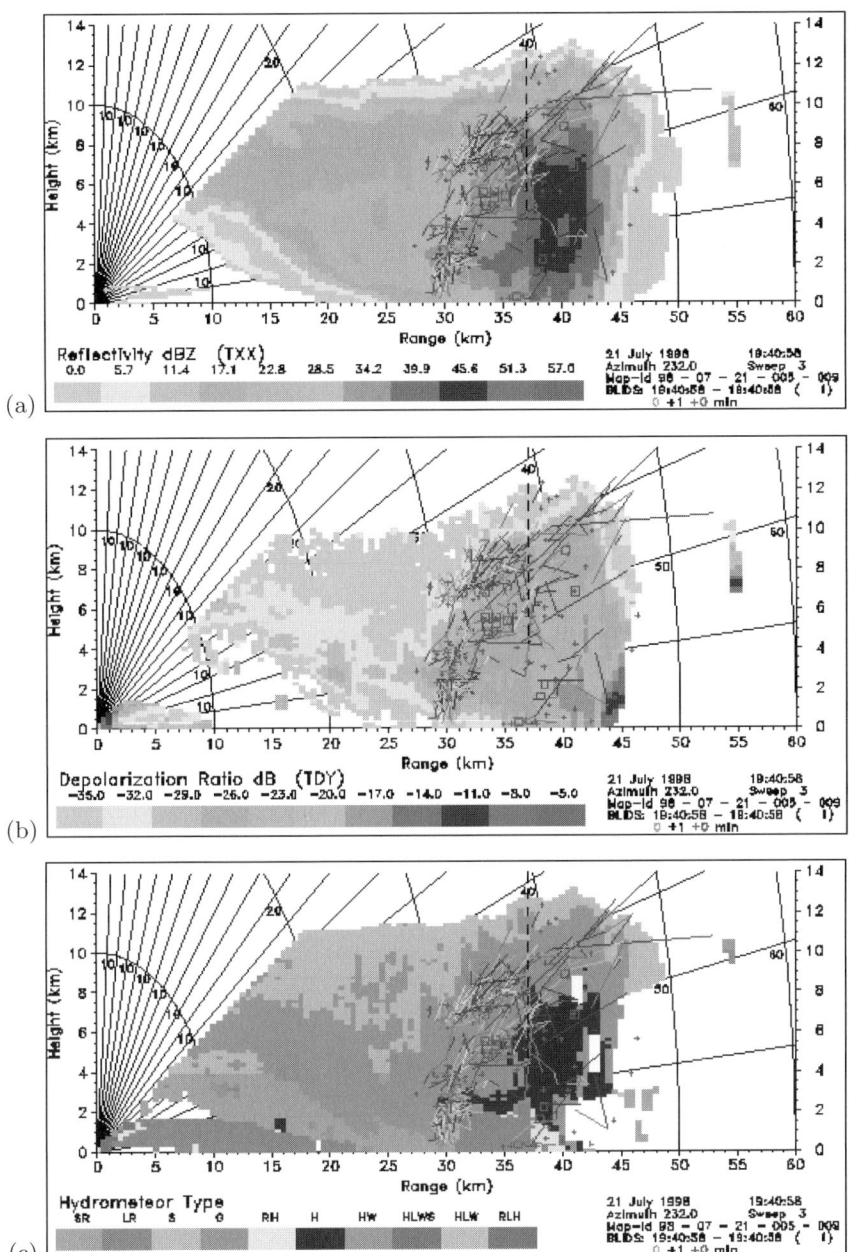

Fig. 6.20. Vertical section of the supercell storm of 21 July 1998 at 1741 UTC along the 232° azimuth from POLDIRAD, as shown in Fig. 6.19. ITF flashes are plotted for a 30-s interval around the nominal scanning time from an azimuth interval of ±10°. The Falcon aircraft is visible in all parameters shown at a range of 55 km and 9 km height SW of the storm. (a) reflectivity factor, (b) linear depolarisation ratio, (c) particle type, to be recognised more clearly in color Fig. 17 on page 326

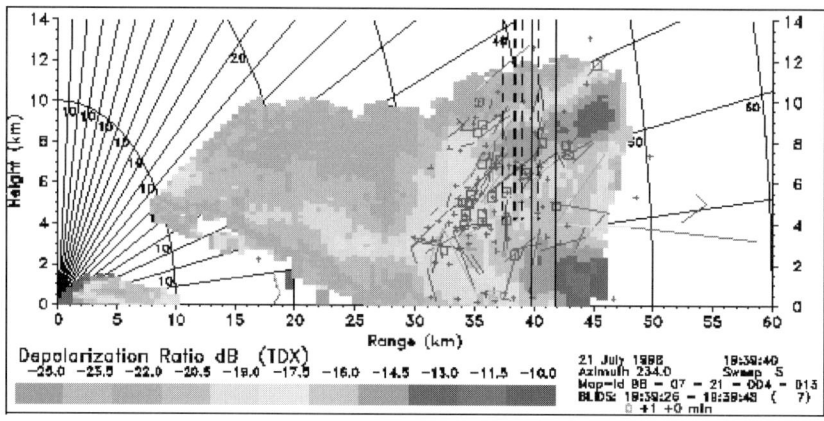

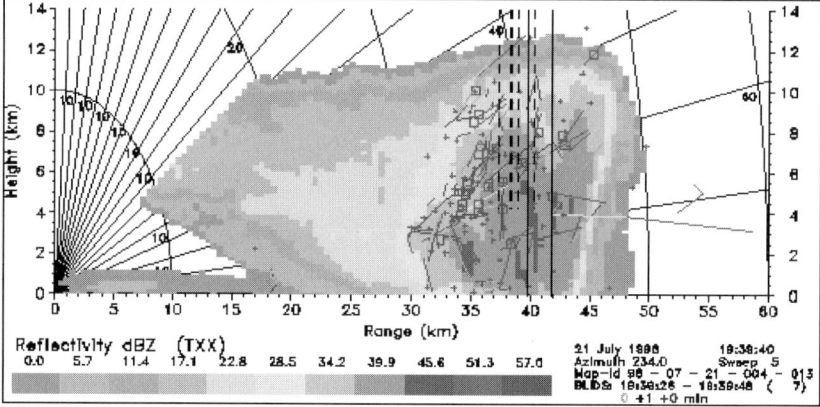

Fig. 6.21. Vertical section of the supercell storm of 21 July 1998 at 1739 UTC along the 234° azimuth from POLDIRAD. ITF flashes are plotted for a 30-s interval around the nominal scanning time from an azimuth interval of ±10°. Top: Linear depolarisation ratio, bottom: reflectivity factor. See also color Fig. 18 on page 327

radar operated by the Institut for Meteorology and Climate Research IMK at Research Center Karlsruhe has been used to study deep convection over this highly structured terrain (Gysi, 1995).

Figure 6.23 shows five tornadic storm tracks observed since 1992. The 1995 and 1996 cases were described by Hannesen et al. (1998, 2000). Both storms were so called minisupercells (Kennedy et al., 1993). The 1995 tornado was confirmed by eyewitnesses, newspaper reports and damage analyses from the German Weather Service. For the 1996 tornado, newspaper reports and damage maps from forest authorities are available.

Analysis of further tornadoes, as shown in Fig. 6.23, underline that during the summer months, tornadic storms in this region are most probable if warm and moist air from the Mediterranean is advected across southern France (Morris,

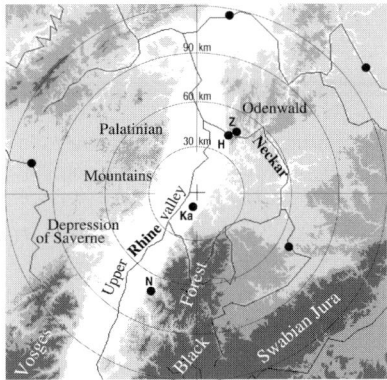

Fig. 6.22. The Karlsruhe area. Rivers are given as *solid lines*, orography in 200-m steps (*white* = below 200 m, *darkest grey* = above 600 m ASL). **H** = Heidelberg, **Ka** = Karlsruhe

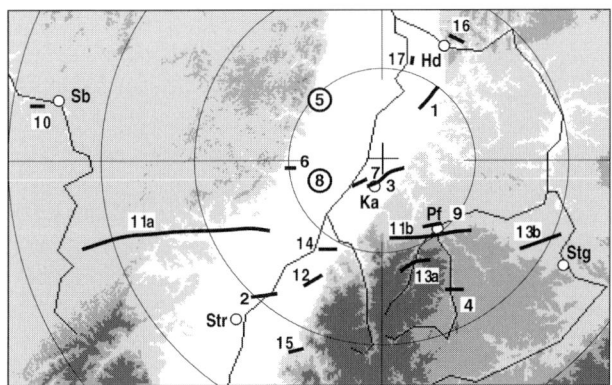

Fig. 6.23. Tornadoes in the Upper Rhine valley region: (1) 29 Jul 1845, (2) 24 May 1878, (3) 4 Jul 1885, (4) 1 Jul 1895, (5) 11 May 1910, (6) End Sep 1913, (7) 7 Jun 1952, (8) 10 Aug 1959, (9) 13 Aug 1952, (10) 27 Apr 1960, (11 a,b) 10 Jul 1968, (12) 8 May 1985, (13 a,b) 23 Jul 1986, (14) 21 Jul 1992, (15) 9 Sep 1995, (16) 23 Jul 1996, (17) 6 Oct 2002. Ka = Karlsruhe, Hd = Heidelberg, Sb = Saarbrücken, Str = Strasbourg, Stg = Stuttgart, Pf = Pforzheim. Orography shaded in 200 m steps, *white* = below 200-m asl, *dark grey* = above 600 m ASL. The + denotes the location of the Karlsruhe radar, range rings every 30 km

1986), along the Rhône valley, the Swiss Jura mountains and the Belfort Gap into the Rhine valley. If this synoptic condition is terminated by a cold front, typically approaching from W–NW, or by positive vorticity advection with the passage of a shortwave upper level trough, the inversion layer above the near surface high θ_e air may be removed, and deep moist convection can be induced. This also results in a large low-level shear situation, which may continue upward. The hodograph then displays veering winds becoming stronger with height, a necessary condition for supercell formation. The strong tornado of 10 July 1968 is an impressive example of such a situation (Nestle, 1969).

Two cases will be discussed now.

On 9 September 1995, the synoptic setting was similar. Warm and moist air moved into the southwestern parts of Germany at the southern flank of a deep low. A midtropospheric trough passed the area around noon, increasing the atmospheric instability.

Several rain showers and a few thunderstorms were observed. At 1055 UTC, one of these storms spawned a tornado near Oberkirch–Nußbach at the eastern flank of the Upper Rhine valley. The tornado swath was about 1 km long and 50 m wide. From the trees' fall pattern, a counterclockwise rotating vortex could be reconstructed. According to the TORRO and Fujita intensity scales (Meaden, 1976; Fujita, 1981), the reported damage indicates a T2/F1 tornado intensity, a weak tornado.

The situation was characterised by a CAPE of 440 $J\,kg^{-1}$, a level of free convection, LFC = 1.3 km and level of neutral buoyancy, LNB = 7.3 km ASL. The wind profile leads to a storm relative helicity SRH of 105 $J\,kg^{-1}$ and an energy helicity index EHI $\approx$ 0.3, values typically found in the environment of weak tornadoes (Kerr and Darkow, 1996). The bulk Richardson number Ri_b of 25 further gives evidence for possible supercell thunderstorm formation (Weisman and Klemp, 1984).

The left panel in Fig. 6.24 shows the fast ascent of the 40 dBZ contour to more than 4 km ASL, indicating strong updrafts within the cloud. Radar detected cloud tops reached heights of about 8 km. This is in good agreement with the estimated LNB of 7.3 km.

The right panel shows the Doppler velocity at an elevation of 1.0° at 1252 LST, just before the tornado was observed at Nußbach. A vortex signature is present, but superimposed on a convergence signature: The area of positive Doppler velocity is located a bit closer to the radar than the area of negative velocity. A horizontal shrinking and vertical stretching of the vortex associated with this convergence may finally have initiated the tornado.

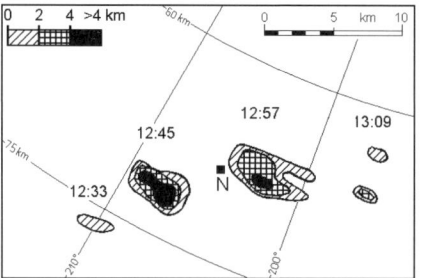 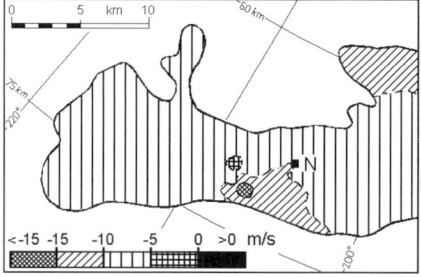

Fig. 6.24. Time evolution of the 40 dBZ reflectivity echotop (*left*, in km ASL). PPI of Doppler velocity, elevation 1.0°, at 1252 LST (*right*). N = Nußbach

The intensification of the mesocyclone and the tornado touchdown are in phase with the maximum updraft. It is, therefore, suggested that vortex stretching below the overshooting cloud top is the main source for the increase of vorticity and helicity prior to formation of the tornado vortex. A similar case of a fast developing mesocyclone with high vertical extension of high reflectivity has also been found by Linder and Schmid (1996).

The radar observed cloud tops at ≈ 8 km ASL fit well with the atmospheric stratification with the special heat and moisture conditions in the Rhine valley. Typical supercell features were observed with a well-defined mesocyclone at lower levels and divergence at higher levels. The small overall size, however, would make it classified as a "mini" supercell (Kennedy et al., 1993).

Mostly undisturbed by the flat orography, the storm developed to the tornadic phase, while the downward generation of the funnel was facilitated by a low-level vorticity increase above the upsloping terrain at the eastern flank of the valley. However, the small supercell was too weak to maintain the tornado vortex there, and it lasted for only a few minutes.

The rather small value of CAPE supports the analysis of Kerr and Darkow (1996), which limits the main role of CAPE to the strength of convective initiation while storm relative winds and helicity favour evolution to a tornadic supercell.

On 23 July 1996, a strong convergence line extended from northern Germany toward the Alps. In Switzerland and Germany, the CAPE values were about 700 J kg^{-1}, the tropopause was at about 10 km ASL, and intense thunderstorms developed. At 1700 UTC, a multicell complex had formed over the Palatinian Mountains and entered the flat Upper Rhine valley. About one hour later, a tornado touched down. The site **T** in Fig. 6.25 is indicated by an arrow.

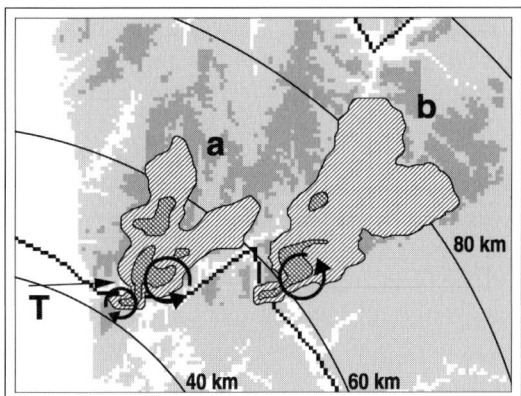

Fig. 6.25. Reflectivity and Doppler velocity at 1808 UTC (a) and 1828 UTC (b). Grey-scale shading gives terrain height, as in Fig. 6.22. The *hatched region* shows the 20 dBZ area, *cross hatching* for Z $\geq$ 43 dBZ. *Large open ring arrows* denote location of the mesocyclone, the *small ring arrow* indicates the low-level anticyclonic rotation at the hook echo tip. Tornado position is marked by the *thin straight arrow* (**T**)

The funnel touched down shortly after 1800 UTC, uprooting mature beech and spruce trees and injuring several people driving in cars. The mesovortex of this strong tornado developed to the left side of the cell's ground track and then crossed over the N–S oriented lateral valley of the Neckar river, causing a 5-km forest damage swath within 5 min. During and after this stage, the storm produced very heavy precipitation and moved E, only slowly decaying and becoming multicellular again.

Figure 6.25 shows reflectivity and Doppler velocity at 3° elevation for 1808 UTC (a) and 1828 UTC (b). The tornado position is indicated by the arrow. The storm area is indicated by 20 dBZ, hatched, and 43 dBZ, cross-hatched reflectivity thresholds. Vortex signatures derived from Doppler velocity are given by circular arrows.

In Fig. 6.25a, about the time of tornado touchdown, the radar echo is dominated by a core of high reflectivity factor just E of the tornado position. Attenuation is indicated by the V-shaped storm contours, with a reflectivity gap at larger distance from the radar. The reflectivity factor reached 55 and 60 dBZ, indicating heavy rain or hail.

The Doppler data show a mesocyclonic vortex signature (MVS) on the eastern, forward flank of the storm core. Significant mesocyclonic rotation is clearly detectable from 1728 UTC on, intensifying between 1748 and 1808 UTC over the hilly terrain at the time of the most intense growth. At this time, the MVS diameter was about 8 km and its vertical dimension 3 km, extending from roughly 2 to 5 km above ground level. The mesocyclone had a velocity difference of 23 to 27 m s^{-1} corresponding to a horizontal shear of 0.003 s^{-1}.

At the southern edge of the storm, a small but distinct hook echo extends outward from the Cb core. Its curvature would imply the presence of an anticyclonic tornado. This rare type requires an environmental wind vector backing with height instead of veering, as in the typical right moving supercell situation.

The Doppler velocity data indeed showed a vortex signature. Its rotational sense was anticyclonic and consistent with the anticyclonic hook echo. The tornado damage swath, however, is closer to the track of the mesocyclone, but as the 3° radar beam is about 2 km above the terrain, it cannot be completely ruled out that the funnel extended from the anticyclonic hook echo northward to the observed damage path.

Concerning the tornado's sense of rotation, we argue that this small supercell was of the usual cyclonic, right moving type and the J-shaped hook echo was induced by interaction of the rear flank downdraft with local topography. Figure 6.26 shows relevant orographic structures together with the observed mesocyclone's ground track and a probable low-level flow field. Orographic anticyclonic vorticity most probably is generated by the isolated Königstuhl mountain rising up to 600 m ASL. With the mountain ranges north of the Neckar river, the Königstuhl likely served as a nozzle for the westerly airflow, inducing convergence, lifting, updraft intensification and cloud growth, as the storm entered the Neckar valley and became tornadic. Due to the orographically enhanced horizontal mass flux and the sudden broadening of the Neckar valley east of the

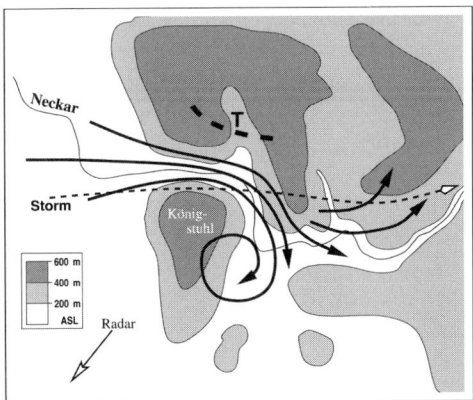

Fig. 6.26. Small scale sketch map of the probable low-level flow field, *bold solid arrows*, around Königstuhl mountain, the storm track, *open dashed arrow*, and tornado damage swath, *bold dashed line*. Direction to Karlsruhe radar is also indicated

Königstuhl, the "nozzle outlet," it is likely that in the wake of this mountain, a transient lee vortex developed during the passage of the tornadic storm.

The tornadoes presented fit with a German tornado climatology (Wegener, 1917; German TorDACH data in http://www.tordach.org). It shows that July is the most likely month for tornado outbreaks, with a mean of 10 each year (Dotzek, 2001; Dotzek, 2002a).

Both tornadic thunderstorms presented were of the minisupercell type (Kennedy et al., 1993). Nowcasting this type of storm with operational C-band Doppler radars is difficult because of the small structures and short lifetimes. Advanced weather radar measurements, however, promise to increase our knowledge and understanding of the specific genesis mechanisms, leading to improved forecasts.

6.4 The Future

Three lines of investigation are apparent today for improving the forecast of severe weather events with the help of radar observations:

Nowcasting by following detailed radar measurements and extrapolating them in time and space with the knowledge of typical developments of the different weather systems under the actual meteorological conditions. This, of course, needs detailed radar measurements preferably with the most advanced techniques such as Doppler, polarisation and bistatic Doppler, which in Europe are so far available only at a few places. Experienced observers, well trained in meteorology and advanced radar meteorology, are also needed. But, when affordable, this will be the most reliable method in the near future.

More economical, and following the successful trend in general weather forecasting, local forecasts with steadily increasing space and time resolution will be done by numerical weather prediction models. Some shortcomings af current cloud resolving models are in the descriptions of the various growth mechanisms of ice particles and their interaction with the other microphysics and 3-D kinematics. Progress is expected through verification of model results with appropriate observations, thereby improving the reliability and performance toward levels required by operational applications.

The final step then will be directly to assimilate radar observations in high resolution numerical weather prediction models. The current state of knowledge has been described by Macpherson (2003), this book. Essential for that is a profound knowledge and control of instrumental and methodical errors in weather radar measurements under all weather conditions, as well as fast exchange and processing of data across networks with a common and high quality standard. An exciting challenge!

References

1. Balakrishnan, N. and D.S. Zrnić, 1990: Use of polarization to characterise precipitation and discriminate large hail. J. Atmos. Sci., **47**, 1525–1540.
2. Browning K.A., 1990: Organisation and internal structure of synoptic and mesoscale precipitation systems in midlatitudes. *Radar in Meteorology*. Am. Meteorol. Soc., Boston, pp. 433–460.
3. Browning K.A., 1977: The structure and mechanisms of hailstorms. In Foote G.B. and C.A. Knight (Eds.): *Hail: A Review of Hail Science and Hail Suppression*. Am. Meteorol. Soc. Boston, pp. 1–43.
4. Caylor, I.J. and V. Chandrasekar, 1996: Time-varying ice crystal orientation in thunderstorms observed with multiparameter radar. IEEE Trans. Geosci. Remote Sensing, **34**, 847–858.
5. Dotzek, N., 2002a: Severe local storms and the insurance industry. J. Meteorol., **27**, 3–12, 189.
6. Dotzek, N., 2002b: An updated estimate of tornado occurrence in Europe. Submitted to Atmos. Res., **56**, 233–252.
7. Dotzek, N., 2001: Tornadoes in Germany. Atmos. Res. **56**, 233–252.
8. Dotzek, N., H. Höller, C. Théry, and T. Fehr, 2001: Lightning evolution related to radar–derived microphysics in the 21 July 1998 EULINOX supercell storm. Atmos. Res. **56**, 335–354.
9. Dotzek, N., G. Berz, E. Rauch, and R. E. Peterson, 2000: Die Bedeutung von Johannes P. Letzmanns "Richtlinien zur Erforschung von Tromben, Tornados, Wasserhosen und Kleintromben" für die heutige Tornadoforschung. Meteorol. Zeitschr., **9**, 165–174.
10. Finke, U., and T. Hauf, 1996: The characteristics of lightning occurrence in southern Germany. Atmos. Phys. **69**, 361–374.
11. Foote, G.B., 1985: Aspects of cumulonimbus classification relevant to the hail problem. J. Rech. Atmos., **19**, 61-74.
12. Fujita, T.T., 1981: Tornadoes and downbursts in the context of generalized planetary scales. J. Atmos. Sci., **38**, 1511–1534.
13. Gysi, H., 1995: Niederschlagsmessung mit Radar in orographisch gegliedertem Gelände. Dissertation, Univ. Karlsruhe.
14. Haase-Straub S.P., M. Hagen, T. Hauf, D. Heimann, M. Peristeri and R.K. Smith, 1997: The squall line of 21 July 1992 in Southern Germany: An observational case study. Beitr. Phys., **48**, 231–381.

15. Hagen, M., B. Bartenschlager, and U. Finke, 1999: Motion characteristics of thunderstorms in southern Germany. Meteorol. Appl., **6**, 227–239.
16. Hagen, M. 1992: On the appearance of a cold front with a narrow rain-band in the vicinity of the Alps. Meteorol. Atmos. Phys. **48**, 231–248.
17. Hannesen, R., N. Dotzek, and J. Handwerker, 2000: Radar analysis of a tornado over hilly terrain on 23 July 1996. Phys. and Chem. of the Earth, Part B, **25**, 1079–1084.
18. Hannesen, R., N. Dotzek, H. Gysi, and K. D. Beheng, 1998: Case study of a tornado in the Upper Rhine valley. Meteorol. Zeitschr., **7**, 163–170.
19. Hobbs, P.W., 1987: Organisation and structure of clouds and precipitation on the mesoscale and microscale in cyclonic storms. Rev. Geophys. Space Phys., **16**(4), 741–755.
20. Höller H. and U. Schumann, eds., 2000: *EULINOX – The European Lightning Nitrogen Oxides Project.* DLR-FB 2000-28.
21. Höller H., V.N. Bringi, J. Hubbert, M. Hagen and P.F. Meischner, 1994: Life cycle and precipitation formation in a hybrid-type hailstorm revealed by polarimetric and Doppler radar measurements. J. Atmos. Sci., **51**, 2500–2522.
22. Höller, H. and M.E. Reinhardt, 1986: The Munich hailstorm of July 12, 1984 - Convective development and preliminary hailstone analysis. Beitr. Phys. Atmos. **59**, 1–12.
23. Kennedy, P.C., N.E. Westcott and R.W. Scott, 1993: Single–Doppler radar observations of a mini–supercell tornadic thunderstorm. Mon. Weather Rev., **121**, 1860–1870.
24. Kerr, B.W. and G.L. Darkow, 1996: Storm-relative winds and helicity in the tornadic thunderstorm environment. Wea. Forecasting, **11**, 489–505.
25. Linder, W. and W. Schmid, 1996: A tornadic thunderstorm in Switzerland exhibiting a radar-detectable low-level vortex. *Proc. 12th Int. Conf. Clouds Precip.*, Zürich, pp. 577–580.
26. Meaden, G.T., 1976: Tornadoes in Britain: their intensities and distribution in space and time. J. Meteorol., **1**, 242–251.
27. Meischner, P.F., R. Baumann, H. Höller and T. Jank, 2001: Eddy dissipation rates in thunderstorms estimated by Doppler radar in relation to aircraft in situ measurements. J. Atmos. Oceanic Technol., 1609–1627.
28. Meischner, P.F., V.N. Bringi, D. Heimann and H. Höller. 1991: A squall line in Southern Germany: Kinematics and precipitation formation as deduced by advanced polarimetric and Doppler radar measurements. Mon. Weather Rev. **119**, 678–701.
29. Morris, R.M., 1986: The Spanish plume – testing the forecaster's nerve. Meteorol. Mag., **115**, 349–357.
30. Nelson, S.P. and N.C. Knight, 1987: The hybrid multicellular-supercellular storm – an efficient hail producer. Part I: An archetypal example. J. Atmos. Sci. **44**, 2042–2059.
31. Nelson, S.P., 1987: The hybrid multicellular-supercellular storm – an efficient hail producer. Part II: General characteristics and implications for hail growth. J. Atmos. Sci. **44**, 2060–2073.
32. Nestle, R., 1969: Der Tornado vom 10.07.1968 im Raum Pforzheim. Meteorol. Rdsch., **22**, 1–3.
33. Parker, D.J. (Ed.), 2000: Floods. Routledge, London and New York.
34. Paul, A.H., 1973: The heavy hail of 23–24 July 1971 on the western prairies of Canada. Weather **28**, 463–471.

35. Pielke Jr., R., and R. Pielke Sr. (Eds.), 1999: *Storms*, Vol. II, Routledge Hazards and Disasters Ser. 2, pp. 103–132. Routledge, London and New York.
36. Rotunno, R., J.B. Klemp and W.L. Weisman, 1988: A theory for strong, long-lived squall lines. J. Atmos. Sci., **45**, 406–426.
37. Schroth A.C., M.S. Chandra and P.F. Meischner, 1988: A C-band coherent polarimetric radar for propagation and cloud physics research. J. Atmos. Oceanic Technol., **5**, 803–822.
38. Seliga, T.A. and V.N. Bringi, 1976: Potential use of radar differential reflectivity measurements at orthogonal polarizations for measuring precipitation. J. Appl. Meteorol., **15**, 69–76.
39. Straka, J.M. and D.S. Zrnić, 1993: An algorithm to deduce hydrometeor types and contents from multiparameter radar data. *Proc. 26th Conf. Radar Meteorol.*, AMS, pp. 513–515.
40. Weisman, U.M., J.B. Klemp and R. Rotunno, 1988: Structure and evolution of numerically simulated squall lines. J. Atmos. Sci., **45**, 1990–2013.
41. Weisman, M.L. and J.B. Klemp, 1984: The structure and classification of numerically simulated convective storms in directionally varying wind shears. Mon. Weather Rev., **112**, 2479–2498.
42. Wegener, A., 1917: *Wind- und Wasserhosen in Europa*. Friedrich Vieweg & Sohn, Braunschweig.
43. Wilhelmson, R.B. and J.B. Klemp, 1981: A three-dimensional numerical simulation of splitting severe storms on 3 April 1964. J. Atmos. Sci., **38**(8), 1581–1600.

7 Precipitation Measurements from Space

Jacques Testud

CETP, F-78140 Velizy, France

7.1 Correcting for Attenuation: A Major Problem with a Space Borne or an Airborne Radar

Weather radars on airborne or spaceborne platforms have been conceived to overcome the limitations of ground based systems. While spaceborne systems aim to sample the precipitation on a global scale, airborne radars are used for two purposes:

1. to serve as a *demonstrator* for future spaceborne systems and as such, help development and test of retrieval algorithms;
2. to be used in *field experiments* where the flexibility of the mobile platform is exploited for specific purposes.

The first and only example up to now of weather radar in space is the TRMM Precipitation Radar (PR) launched by NASA and NASDA in November 1997. TRMM was designed to monitor the precipitation systems in the tropics (its inclination of orbit is 35°). The TRMM PR operates at 13.8 GHz and scans across track at ± 17° about the nadir. With an altitude of orbit of 360 km this scanning capability provides a swath of 220 km, with a footprint at nadir of 4×4 km^2 (obtained with a 2×2 m^2 antenna). The TRMM PR collects 8.8×10^6 samples in a day covering about 1.4×10^8 km^2, i.e. almost twice the surface of the tropical belt. However, these samples are not uniformly distributed: sampling is less dense near the equator, where the revisit time is about 1 sample per 3 days on average. To overcome this insufficient revisit time, TRMM associates with the PR a 5-channel passive microwave imager (TRMM Microwave Imager or TMI), whose swath is about 800 km. Combining the three-dimensional picture of the precipitation field delivered by the PR within the 220 km central swath with the two-dimensional picture within the 800 km swath provided by TMI allows deriving an improved precipitation product. The procedures to achieve this combined product is beyond the scope of this chapter. Documentation about these procedures may be found in Kummerow et al. (1998), Bauer (2001) and Bauer et al. (2001).

From its launching in November 1997, the scientific exploitation of the TRMM data has covered various research fields, the most spectacular being the documentation of the El Niño/La Niña episode of January 1998 – December 2000, the analysis of numerous hurricane observations. The experiments of assimilation, of the TRMM data in operational numerical weather prediction (NWP) models relate to the general topic of radar data assimilation as presented by Macpherson et al. in this book. The success of TRMM drives plans for the future "Global Precipitation Mission" that anticipates a constellation of satellites covering the full globe with a three-hour revisit time. On the "mother ship" of the constellation, GPM plans to deploy an advanced precipitation radar, *implementing the dual frequency technique.*

Table 7.1 summarises the main airborne precipitation radar systems that are developed today. ARMAR (Jet Propulsion Laboratory, USA) and CAMPR (Communication Research Laboratory, Japan) are Ku-band systems designed to serve as demonstrators of the TRMM Precipitation Radar. However, their additional capabilities (Doppler for both systems and fore and nadir viewings for CAMPR) allow them to document the microphysics and the dynamics of the observed weather systems. Four X-band systems have been built: EDOP (NASA-GSFC) with Doppler capability and fore and nadir viewings, the tail radars of the two NOAA-ERL P3 aircrafts, and the ELDORA-ASTRAIA radar developed by NCAR and CETP. The last three systems are operated according to the dual helical scanning strategy described in Fig. 7.1: the antenna system rotates about an horizontal axis collinear with the aircraft fuselage and scans (alternately or simultaneously) two cones oriented forward (fore) and backward (aft). Associated with the aircraft motion, this scanning strategy allowing with collecting at each point of the three-dimensional space, the two independent components of the air motion required to perform a three-dimensional wind field synthesis.

In *field experiments*, airborne radar systems allow investigating storm events difficult or impossible to document from the ground. Oceanic storms like hurricanes, tropical oceanic mesocale convective complexes (MCCs), or midlatitude frontal systems are out of reach of ground based systems. For many years, the Hurricane Research Division of NOAA-ERL has been sending its P3 surveillance aircraft equipped with the above mentioned X-band tail radar to document the dynamics and microphysics of hurricanes in the Gulf of Mexico. The observation of oceanic storms over the warm pool of the West Pacific was an important component of TOGA-COARE (1992–1993) in which three airborne Doppler weather radars were used: the two systems on board the NOAA P3s aircraft and ELDORA-ASTRAIA on the Electra aircraft of NCAR. The same airborne Doppler weather radar systems were used in FASTEX (1997) to better understand frontal dynamics and rapidly deepening depressions over the North Atlantic.

In other applications, the *flexibility* of the airborne system is used to better sample particular phenomena. During VORTEX, ELDORA-ASTRAIA delivered unprecedented observations of Oklahoma tornadic storms thanks to its particular sampling strategy and to the ability of the aircraft to manoeuvre close to the storm. In the Mesoscale Alpine Program (MAP), the storm flow over the complex alpine orography could be documented by the airborne radars on board the NCAR-Electra and NOAA-P3 aircraft, providing viewing angles less subject to beam blocking than with the ground based systems.

a

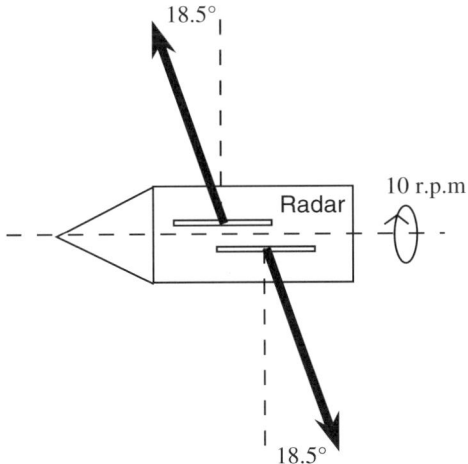

b

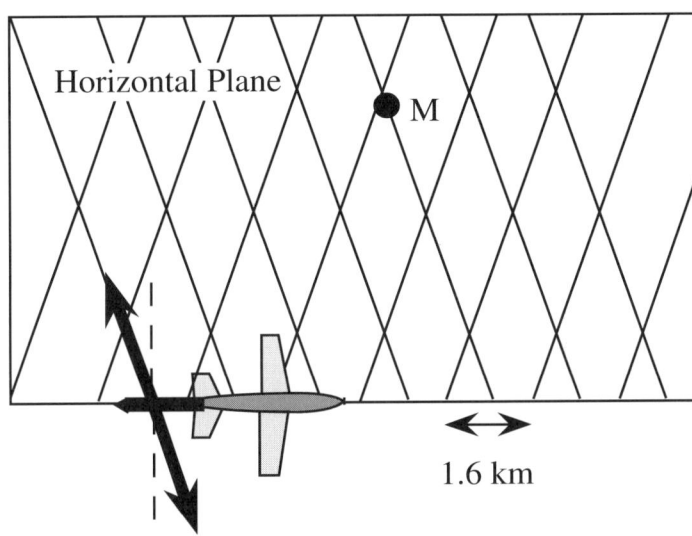

Fig. 7.1. Principle of the dual beam airborne Doppler radar. (a) The antenna system is rotating about an horizontal axis collinear with the fuselage; one antenna is looking forward (fore) and the other backward (aft). (b) Owing to the aircraft motion, a dual helical scan is obtained. Each point M of the three-dimensional space is sampled successively by the fore and the aft antenna with a time shift of about 2 minutes at 20 km range (for an aircraft velocity of 135 m/s). The "helical resolution" is related to the angular velocity of the antenna system: it is 800 m for a rotation at 10 revolutions per minute (rpm). With ELDORA-ASTRAIA , the use of coded pulses allows collecting more independent samples and speeding up the antenna to 40 rpm, thus achieving a helical resolution of 200 m

Table 7.1. Spaceborne and airborne weather radar developed today

Name Platform (Organism)	Frequency	Sampling Stragegy	Antenna technology	Transmitter technology
Precip.Radar TRMM (NASDA)	Ku-band 13.0 GHz	± 17° across track (about nadir)	Slot guide antenna electrical Scan	Distributed SSPA frequency coding (2)
ARMAR NASA-DC8 (NASA-JPL)	Ku-band 14 GHz	± 17° across track (about nadir)	Offset parabola	Pulse compression
CAMPR G4 (CRL)	Ku-band 14 GHz	180° helical scanning Dual beam (0° and 45°) Doppler Polarimetric	Offset parabola	Simple pulse
EDOP ER2 (NASA-GSFC)	X-band (9.3 GHz)	Nadir and 45° fore Across track scanning	Offset parabola	Simple pulse
Tail radar P3-43 (NOAA)	X-band (9.3 GHz)	360° Helical scanning pseudo-dual (± and −20°) Doppler	Parabola central feed	Simple pulse
Tail radar (NOAA)	X-band	360° Helical scanning (± and −18°) Doppler	2 slot guide antennae	Simple pulse
Eldora-Astraia (NCAR-CETP)	X-band (9.3 GHz)	360° Helical scanning Dual beam (± and −18°) Doppler	2 slot guide antennae (back to back)	Frequency coding (4) (240°/s)

When operating from space, the distance of operation is much farther than from the ground. For instance, the altitude of the TRMM mission is 350 km. To perform accurate measurement, the antenna beam should capture the small scale heterogeneity of the rain field, otherwise "nonuniform beam filling" effects will bias the rain rate estimate. Various studies of the rain cell statistics have shown that the typical rain cell size is about 4 km. Thus the footprint resolution of a radar operating from space should not exceed this value. With a radar orbiting at 350 km altitude as the TRMM PR, this condition imposes an angular resolution (or beam width) ϵ for the antenna around 0.7°. With the TRMM PR, the angular resolution is 0.71° (at 3 dB one way), obtained with a 2×2 m^2 slot guide array at 13.8 GHz. This choice of the radar frequency is dictated by the maximum antenna size compatible with the space vehicle and launcher: given the size of the antenna L, its angular resolution ϵ is inversely proportional to the radar frequency.

With an airborne radar, the distance of operation may be adjusted by the aircraft manoeuvre, but the radar should operate over a range similar to the extension scale of the observed weather system, i.e. typically 15–20 km for a radar looking downward near to nadir or 30–50 km for a radar performing helical scanning (at horizontal or slantwise incidence). Again, a high resolution antenna is required (between 1.5 and 2° at 3 dB one way), which imposes the use of a frequency prone to attenuation. With ELDORA-ASTRAIA, the dual helical scanning is obtained from a system of two 1.4×1.4 m^2 slot guide arrays operating at 9.3 GHz (X-band). The two arrays are mounted back to back (see Fig. 7.1 and look, respectively, at 18.5° backward (aft) and 18.5° forward (fore). The choice of X-band in ELDORA-ASTRAIA is a trade-off to satisfy, on the one hand, an acceptable angular resolution and, on the other hand, a sufficient penetration capability through the most intense weather systems.

In operational ground based radars where the limitation of antenna size is less stringent, the choice of C- band or S- band is generally made since it leads to moderate or negligible attenuation by rain. With classical radars, the standard estimate for rain is based on an empirical Z–R relationship like

$$R = aZ^b, \tag{7.1}$$

where Z is the radar reflectivity (mm^6/m^3) sixth moment of the raindrop size distribution if Rayleigh scattering is assumed; equivalent reflectivity, otherwise, R is the rain rate (in mm/h), and a and b are empirically determined coefficients.

In ground based radars (7.1) is often applied without consideration of any along path attenuation, i.e. the "apparent" reflectivity Z_a measured by the radar is used as Z. With airborne and spaceborne radars, this simple approach does not hold since the along path attenuation may reach factors of 1000 or 10000 (i.e. several tens of dB). Thus the retrieval algorithms should necessarily take this major effect into account.

The problem of correcting the measured reflectivity Z_a for along path attenuation was addressed already by Hitschfeld and Bordan (1954). Z_a is related to the "true" (nonattenuated) reflectivity Z as:

$$Z_a(r) = Z(r) \exp\left[-0.46 \int_0^r A(s) \mathrm{d}s\right], \tag{7.2}$$

where r is the range along the radar beam and A is the specific attenuation in dB/km. Under the hypothesis that the specific attenuation A (in dB/km) is related to the radar reflectivity Z through a power-law relationship as $A = \alpha Z^\beta$, they derived the exact mathematical solution that may be expressed as

$$Z(r) = \frac{Z_a(r)}{[1 - \alpha I(r)]^{1\beta}}, \tag{7.3}$$

with

$$I(r) = 0.46\beta \int_0^{r_m} Z_a^\beta \mathrm{d}s \tag{7.4}$$

(r is the radial distance along the beam).

Hitschfeld and Bordan (1954) recognized that though mathematically exact, the solution given by (7.3) was numerically unstable. When the along path attenuation is large, the denominator in (7.3) is close to zero, thus a small error in α or a small calibration error in Z_a induces a large error in Z. To overcome this stability problem, Hitschfeld and Bordan recommended performing a backward integration from a far bound r_m, where a boundary condition $Z(r_\mathrm{m})$ has to be defined. The solution then is expressed as

$$Z(r) = \frac{Z_\mathrm{a}(r)}{[Z_\mathrm{a}(r_\mathrm{m}/Z(r_\mathrm{m}) + \alpha I(r)]^{1/\beta}} \quad (7.5)$$

with

$$I(r) = 0.46\beta \int_r^{r_\mathrm{m}} Z_\mathrm{a}^\beta \mathrm{d}s. \quad (7.6)$$

Hitschfeld and Bordan (1954) suggested estimating $Z(r_\mathrm{m})$ from a local raingauge. The difficulty to practically implement this approach explains why it was not used with great success in operational application with ground based radar. Nevertheless, the pioneering work of these authors, forgotten for many years, constitutes the basis of the "rain profiling algorithm" developed for the TRMM precipitation radar. It also inspired the most efficient techniques to correct for attenuation of airborne weather radar data, and it allowed the recent development of promising polarimetric techniques for operational ground based radars.

The definition of the "inverse model" is fundamental to any algorithm development. By "inverse model" is presently understood the set of relationships between integral parameters of the rain drop size distribution (DSD) that should be introduced in any retrieval algorithms. The above cited Z–R and A–Z relationships are essential for what we call "inverse model". It is a standard approach to represent these relations by power laws, though it is recognized that the coefficients in these power laws are subject to the natural DSD variability. The next section investigates the theoretical foundation of this representation and presents a parametric dependence of the coefficients on the DSD variability.

7.2 Statistical Properties of the Drop-Size Distribution and Parameterisation of the Rain Relations

The topic of the correct DSD is central in radar meteorology, i.e. in the problem of the relationships between radar observables (like equivalent radar reflectivity Z) and physical parameters (such as rainfall rate R or precipitation liquid water content LWC). With spaceborne or airborne radars, an additional difficulty is related to the indispensable correction for attenuation that implies the A–Z relationship, itself largely dependent on the variability of the DSD. Driven by the problem of interpreting classical weather radar data, the parameterisation of the DSD by a single parameter (generally R) has been proposed by several authors. Since the earlier parameterisation by Marshall and Palmer, the most synthetic work is probably that by Ulbrich (1983), who presented a systematic approach for calculating the relationships between any couple of integral parameters of the distribution, assuming a gamma shape for the DSD. In the framework of the present chapter, it is also relevant to cite Sempere Torres et al. (1994) which describes a general approach to parameterising the DSD from a single parameter (e.g., R). While Marshall and Palmer's parameterisation implies that all the moments of the distribution are related through powerlaw relationships, Sempere Torres (1994) postulated the existence of such powerlaws to establish a normalisation of the DSD as $N(D, R) = R^\alpha g(R^{-\beta}D)$ where R^α and R^β are the normalisation factors in concentration and diameter, respectively, and $g(X)$ is the "intrinsic shape" of the normalised DSD. The objective of this section is twofold:

1. to propose an approach for the normalisation of the DSD *independent* of *any* assumption about its mathematical shape; or *any* postulate about the relationship between its moments;
2. to propose a new parameterisation of the relationships between integral parameters of the DSD.

7.2.1 The Concept of Normalised DSD

a) Theory.
The physical characterisation of any observed raindrop spectrum raises three questions:

1. What "rain intensity" corresponds to this spectrum?
2. What is the "mean" drop diameter?
3. What is the "intrinsic" shape of the raindrop size distribution?

In order to characterise "rain intensity," two parameters may be considered: the liquid water content (LWC) (in g/m^3) or the rainfall rate R (in mm/h). Traditionally, R is more often used than LWC. However, it may be argued that LWC is a better parameter than R since it has a clearer physical meaning (at altitude, R is subject to vertical air motion and change in terminal fall velocity related to air density). There is no ambiguity in the definition of LWC. It is simply proportional to the third moment of the drop-size distribution $N(D)$ (N: number of particles per unit volume and per interval of diameter, in m^{-4}; and D: drop diameter in m). More specifically,

$$\text{LWC} = \frac{\pi \rho_w}{6} \int_0^\infty N(D)^3 \mathrm{d}D. \tag{7.7}$$

Here, the "mean" diameter is defined as the "volume weighted" mean diameter D_m (generally referred to as the "mean volume diameter") defined as the ratio of the fourth to the third moments of the DSD:

$$D_\mathrm{m} = \frac{\int_0^\infty N(D)D^4 \mathrm{d}D}{\int_0^\infty N(D)D^3 \mathrm{d}D}. \tag{7.8}$$

D_{m} is in fact very close to the median volume diameter D_0 generally used by radar meteorologists (Ulbrich, 1983). The definition of the "intrinsic" shape is intimately related to the concept of normalisation of raindrop spectra. Normalisation is interesting when it is intended to compare the shape of two spectra that not have the same liquid water content LWC and/or mean volume diameter D_{m}. Thus, normalisation should be defined in such a way that the "intrinsic shape" is independent of LWC and/or D_{m}. A general expression of the normalisation of the DSD is

$$N(D) = N_0^* F(D/D_{\mathrm{m}}), \qquad (7.9)$$

where N_0^* is the scaling parameter for concentration, and D_{m} is the scaling for diameter. F(X) in (7.9) denotes the "normalised DSD" describing the "intrinsic" shape of the DSD (noting $X = D/D_{\mathrm{m}}$). Testud et al. (2001) stated that, in order the normalised function F for be independent of LWC and D_{m}, the N_0^* parameter should be proportional to $\mathrm{LWC}/D_{\mathrm{m}}^4$. More specifically, they recommended the following particular definition of N_0^*:

$$N_0^* = \frac{4^4}{\pi \rho_{\mathrm{w}}} \frac{\mathrm{LWC}}{D_{\mathrm{m}}^4} \qquad (7.10)$$

(ρ_{w}: density of liquid water).

The interesting point of (7.10) is that, for an exponential DSD as $N(D) = N_0 \exp(-\Lambda D)$

$$N_0^* = N_0. \qquad (7.11)$$

This equality justifies the notation of the parameter N_0^* and allows us to give a simple physical interpretation of it: *Whatever the shape of an observed DSD, the corresponding N_0^* is the intercept parameter of the exponential DSD with the same LWC and D_{m}*. It is interesting to note that the normalisation defined by (7.8) and (7.9) is very similar to that proposed by Sekhon and Srisvatava (1971), Willis (1984) and Testud et al. (2000) (except for the choice of D_{m} instead of D_0). But while the DSD was assumed to be exponential in Sekhon and Srisvatava (1971), gamma in Willis (1984) and in Testud et al. (2000), in the present approach, *there is absolutely no assumption on the shape of the DSD*. N_0^* is referred to in the following as the *"normalised intercept parameter"* of the DSD.

b) Intrinsic shape of the DSD.
The above defined normalisation has been applied to the following data sets:

1. Airborne microphysical data collected by PMS probes on board the NCAR Electra during 21 flights in TOGA-COARE (West Equatorial Pacific, October 1992 – February 1993).
2. Five-day record of a ground based disdrometer operated by BMRC (Bureau of Meteorology Research Centre) at Darwin.
3. Nine-month record of a ground based disdrometer operated by ETH at Zürich.

4. Three-month record of a ground based disdrometer operated by Meteo-France at Trappes.

Data sets (1) and (2) are representative of tropical oceanic convection, while (3) and (4) are representative of mid-latitude weather systems.

Figure 7.2 illustrates the application of the normalisation to the TOGA-COARE data to identify the "intrinsic" shape of the DSD. Four rain categories are considered: stratiform, convective < 10 mm/h, convective 10 to 30 mm/h, and convective > 30 mm/h.

In Fig. 2a, an average raindrop spectrum is calculated by integration over all spectra of the category; then it is normalised. One can see that the integrated spectra for all categories follow closely the exponential. This is a classical result: when raindrop spectra are integrated for a long time or a large number of samples, one always gets an exponential distribution. Figure 2b demonstrates the value of the normalisation in identifying the "true shape" of the DSD. Here, each individual spectrum (integrated over 6 s, that is, 800 m along the aircraft track) is normalised beforehand; then the average "normalised" spectrum is calculated for each category. It can be seen that the "normalised" shapes of the DSD for the various categories are very close to one another, but depart significantly from the exponential. In particular, the tail of the distribution is definitely subexponential and could be represented by a gamma DSD with $\mu = 2$. However, such a gamma DSD would underestimate the concentration of small drops with normalised diameter < 0.7. We therefore prefer the "modified exponential" illustrated in Fig. 2c, defined as

$$N(D) = N_0^* \exp\left[a - 4\frac{D}{D_\mathrm{m}} - s\sqrt{\left(\frac{D}{D_\mathrm{m}} - X_0\right)^2 + b}\right], \qquad (7.12)$$

with $s = 1.5$, $b = 0.06$, $X_0 = 1.124$, $a = 0.705$.

c) Statistics of N_0^*.

Figure 7.3 displays the average intrinsic shape of the DSD observed during five rainy days at Darwin (Australia) from a ground based disdrometer, together with the corresponding histogram of N_0^*. Each raindrop spectrum here is integrated over one minute. As opposed to the remarkable stability of the intrinsic shape (again close to the "modified exponential"), the N_0^* parameter is characterised by a large variability. The N_0^* histogram spreads over at least two decades and exhibits a bimodal shape that corresponds to a distinct behaviour between stratiform and convective rain. Typical values of N_0^* in this climatic zone are 2.2×10^6 m^{-4} for stratiform rain and 2×10^7 m^{-4} for convective. The same bimodal shape of N_0^* histogram and differential behaviour between stratiform and convective was observed with the airborne microphysical data collected in TOGA-COARE (see Testud et al., 2001).

Figures 7.4 and 7.5 compare the normalised shapes of the DSD and corresponding N_0^* histograms from the disdrometers of Trappes (France) and Zürich (Suisse), respectively. It appears that the stability of the intrinsic shape of the

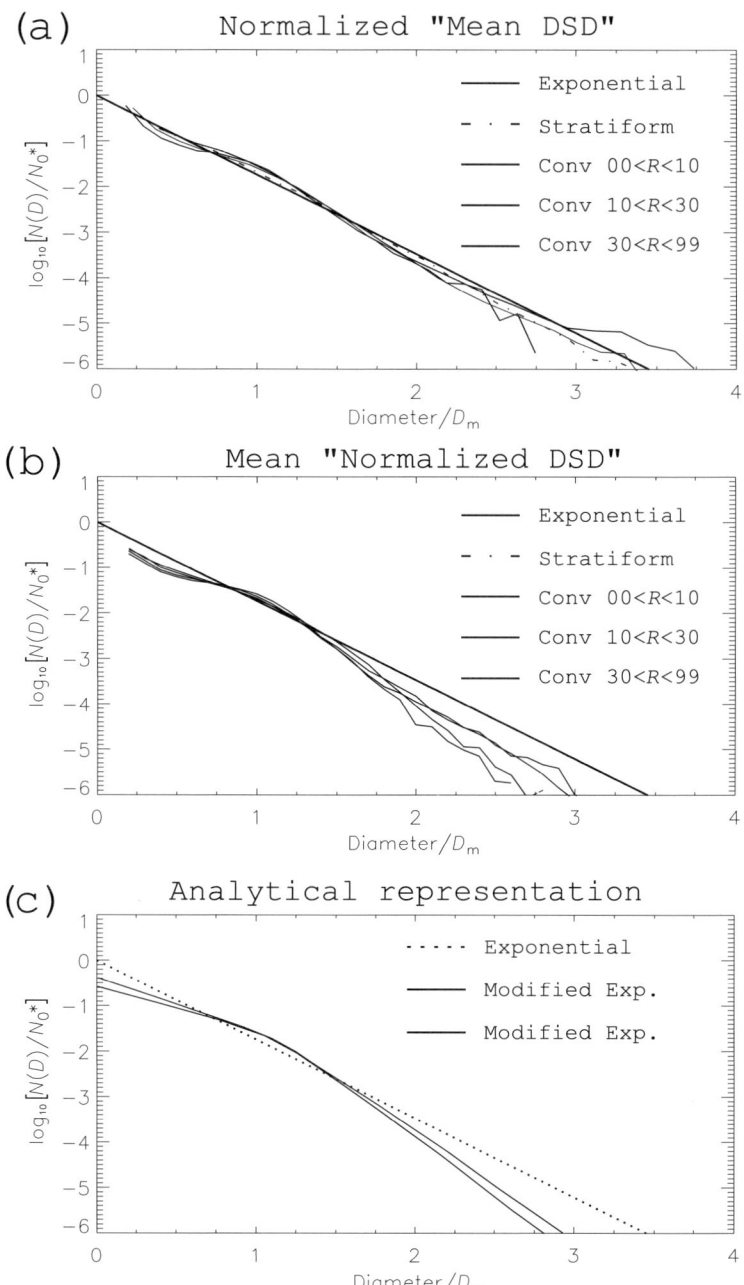

Fig. 7.2. Shape of the raindrop size distribution derived from the TOGA-COARE airborne microphysical data for four rain categories. (a) Normalisation is applied to average spectra. (b) Normalisation is applied before averaging. (c) "Modified exponential" proposed to represent the DSD

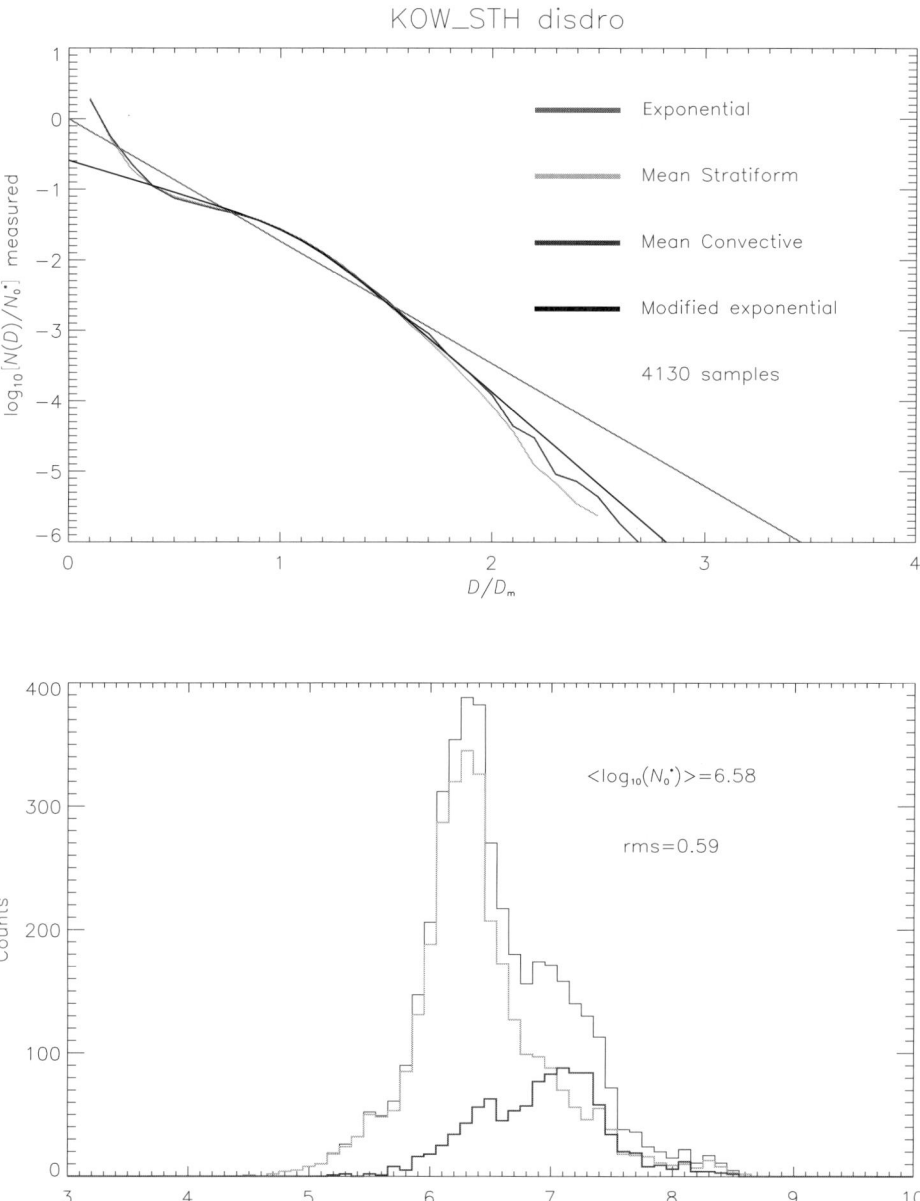

Fig. 7.3. DSD statistics derived from a ground based disdrometer near Darwin (Australia) during 5 rainy days of January 1998. Upper diagram: observed normalised shape of the DSD distinguishing the two rain categories stratiform or convective, compared with two theoretical shapes, exponential and "modified exponential." Lower diagram: corresponding histograms of N_0^* for convective, stratiform and all together

210   J. Testud

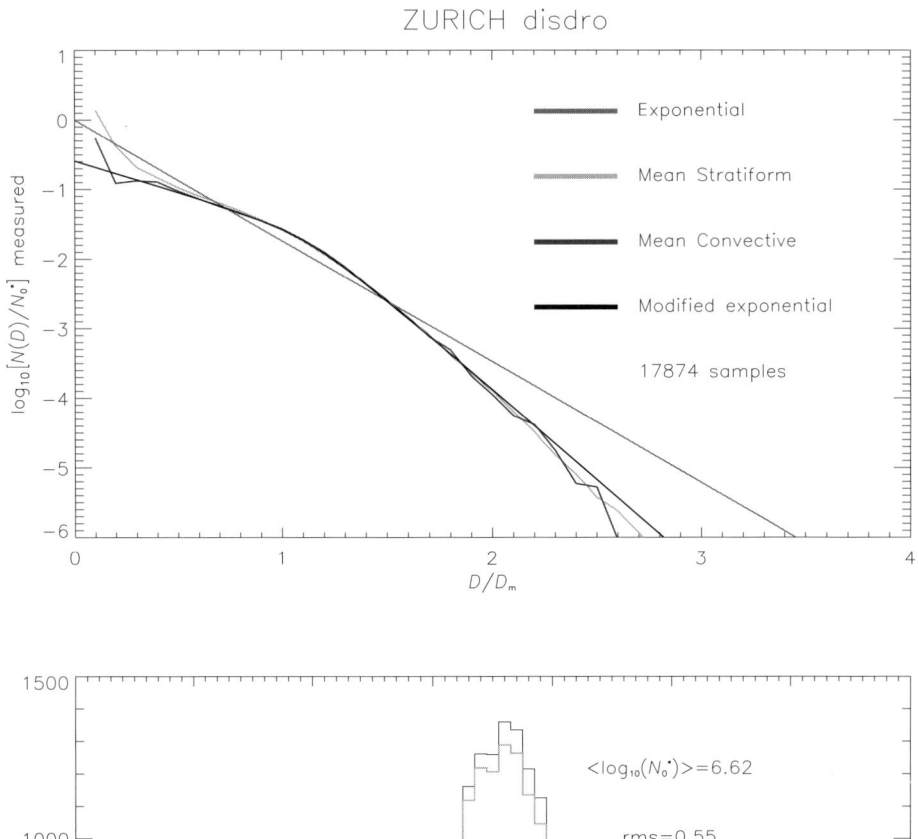

Fig. 7.4. Same as Fig. 2 but for nine months of observation (in 1993) by a disdrometer at Zürich (Switzerland)

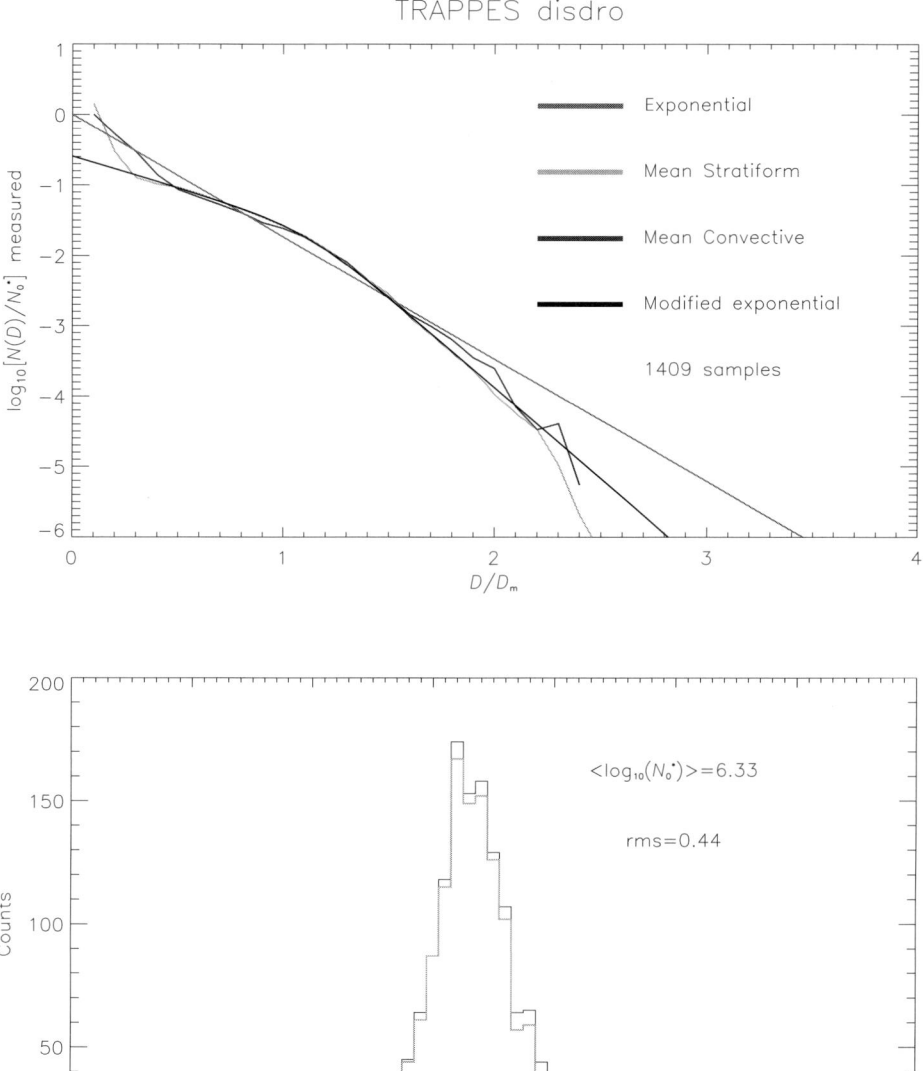

Fig. 7.5. Same as Fig. 2 but for three months of observation (May 12th to July 9th, 1999) by a disdrometer at Trappes (France)

DSD is quite remarkable, independent of the climatic zone and type of precipitation – stratiform or convective, and is well represented by the "modified exponential". Meanwhile, the N_0^* histograms show differences. At Trappes and Zürich, the global histogram is rather unimodal, and the difference between stratiform and convective is less marked. A shift toward higher values of N_0^* with convective rain is nevertheless perceptible.

7.2.2 The Inverse Model for Rain Radar Retrieval

a) Relationships Between Moments of the DSD
Moments of the DSD represent more or less most of the integral parameters of the DSD of interest in the "rain relations." For example, the liquid water content LWC is proportional to M_3, the rainfall rate R to $M_{3.67}$ (assuming that the terminal fall velocity is $V_t \propto D^{0.67}$); the radar reflectivity Z to M_6 and the specific attenuation A to M_3 (the last one, under the Rayleigh approximation). A general expression of the ith order moment of the DSD is

$$M_i = \int N_0^* F(D/D_\mathrm{m}) D^i \mathrm{d}D = N_0^* D_\mathrm{m}^{i+1} \xi_i \tag{7.13}$$

where ξ_i is the ith order moment of the *normalised distribution* $F(X)$:

$$\xi_i = \int F(X) X^i \mathrm{d}X \tag{7.14}$$

Thus, between two moments of order i and j, we have the following relationship:

$$\frac{M_i}{N_0^*} = \xi_i \xi_j^{-\left[\frac{i+1}{j+1}\right]} \left[\frac{M_j}{N_0^*}\right]^{\left[\frac{i+1}{j+1}\right]} \tag{7.15}$$

with N_0^* ranging typically between 10^6 and $10^8 m^{-4}$.

Equation (7.15) shows that, *when normalised by N_0^*, the relationship between two moments of order i and j of the DSD is a powerlaw whose exponent is $(i+1)/(j+1)$, independent of the shape of the DSD*. Taking the particular example of the Z–R relationship, as $Z = M_6$ and $R \propto M_{3.67}$ (with a powerlaw as $V_t \propto D^{0.67}$ for terminal fall velocity), (7.15) sets the exponent of the $Z/N_0^* - R/N_0^*$ relationship to $7/4.67 = 1.499$. Thus it may be written that

$$Z/N_0^* = a \left[R/N_0^*\right]^{1.499} \quad \text{or} \quad Z = aN_0^{*(-0.499)} R^{1.5} \tag{7.16}$$

For R in mm/h, Z in mm$^6 m^{-3}$, N_0^* in m^{-4}, and with the terminal fall velocity law $V_t = 386.6 D^{0.67}$ (V_t in m/s, D in m; Atlas and Ulbrich, 1977), we have $a = 5.2 \times 10^4 \xi_6 \xi_{3.67}^{-1.499}$. For the modified exponential distribution, $\xi_6 = 0.034995$ and $\xi_{3.67} = 0.023441$. Traditionally, empirical power-law Z–R relationships are established from a linear correlation between $log_{10}(Z)$ and $log_{10}(R)$ on a given

Table 7.2. The Z–R relationships parameterised by N_0^*, as determined empirically (from the TOGA-COARE data set) or theoretically (with, as shape function, the Modified Exponential)

Empirically derived (full TOGA-COARE data set)	$Z/N_0^* = 4.73 \times 10^{-5} [R/N_0^*]^{1.494}$ (with $\rho^2 = 0.988$)
Theoretical (with Modified Exponential)	$Z/N_0^* = 4.87 \times 10^{-5} [R/N_0^*]^{1.499}$

data set. Each of them is specific for a particular rain type or climate. As opposed to the traditional Z–R relationship, (7.16) now specifies a relationship *randomised* by N_0^*, and, as long as the normalised shape is stable, it represents a "universal" relationship that applies to *any* type of rain under *any* climate. Table 7.2 presents a test of validity of (7.16) based on a linear correlation analysis between $log_{10}(Z/N_0^*)$ and $log_{10}(R/N_0^*)$ applied to the full TOGA-COARE data set. Quite good agreement is found between the "universal" relationship as well concerning the exponent of the power law as for the coefficient. The correlation coefficient is very close to 1 (ρ^2=0.9888) while it is only 0.838 with a standard Z–R relationship fitted to the same data.

b) General Model of the "Rain Relationships"

Extensive variables (that is, variables proportional to particle concentration) like rainfall rate R, liquid water content LWC, equivalent radar reflectivity Z_e, specific attenuation A, specific differential phase shift K_{DP} are integral parameters of the DSD. Their general expression is:

$$P = \int f_P(D) N(D) \mathrm{d}D, \qquad (7.17)$$

where P denotes the parameter in question and $f_P(D)$ the corresponding weight of an individual raindrop of diameter D. In the general case, $f_P(D)$ cannot be expressed as $\propto D^a$; thus P is not proportional to a moment of the DSD. Nevertheless, introducing the representation of $N(D)$ by (7.8), (7.17) transforms as

$$P = N_0^* D_m \int f_P(X \cdot D_m) F(X) dX \qquad (7.18)$$

Equation (7.18) shows that, if the intrinsic shape is stable, P/N_0^* is a function of D_m *only*. Thus, considering two parameters P_1 and P_2, by eliminating D_m between their two expressions, given (7.18), we may establish a *functional* relationship between P_1/N_0^* and P_2/N_0^*. The same reasoning stands with "intensive variables" (whose definition is independent of particle concentration) like the differential reflectivity Z_{DR} or the backscattering phase shift δ, except that there is no need to normalise by N_0^*.

Figure 7.6 illustrates the useful "rain relationships" pertaining to the TRMM precipitation radar. The $A/N_0^* - Z_e/N_0^*$, $R/N_0^* - Z_e/N_0^*$ and $R/N_0^* - A/N_0^*$

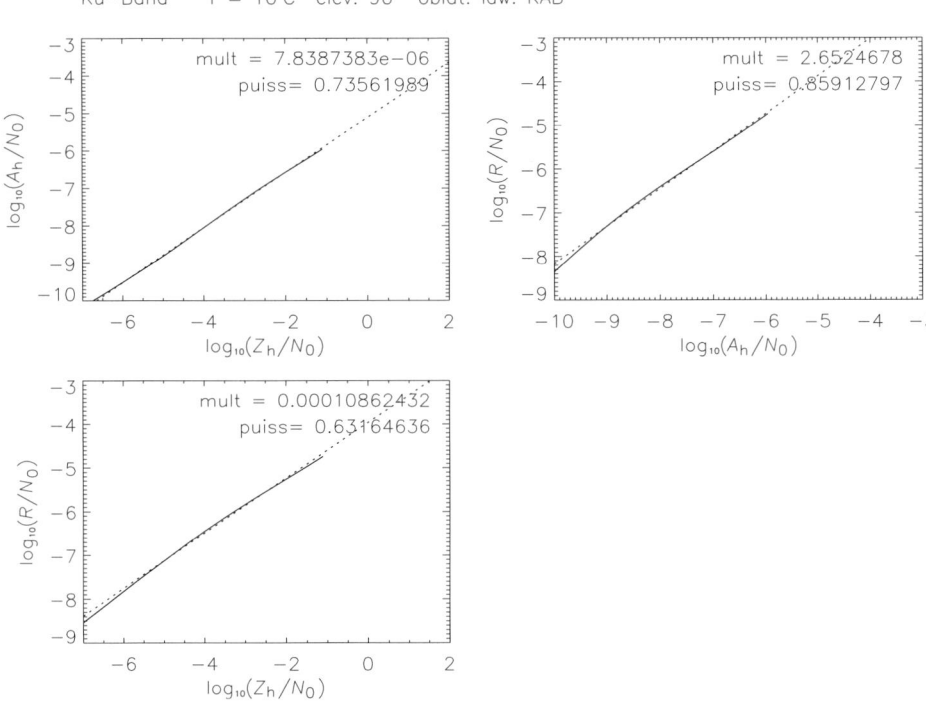

Fig. 7.6. Inverse model for a spaceborne 14 GHz radar (10°C temperature, nadir incidence). Left diagrams, top: normalised specific attenuation A/N_0^* against normalised radar reflectivity Z/N_0^*; bottom: normalised rain rate R/N_0^* against Z/N_0^*. Right diagram: normalised rain rate R/N_0^* against A/N_0^*. In all diagrams, the assumed DSD shape is the "modified exponential." The *dotted line* represents the best power-law fit to the functional relationship

relationships are calculated for rain at 10°C and for the nadir incidence of the radar beam, using a T-matrix scattering model (laws for particle terminal fall velocity and raindrop oblateness, as defined in Le Bouar et al. (2001)). The shape of the DSD is assumed "modified exponential," as defined by (7.12). The three relationships appears well represented by power laws (straight lines in theses log-log diagrams). Such relations constitute the "inverse model" for spaceborne radar.

The same kind of consideration applies as well for the definition of the inverse model for a ground based polarimetric radar. Figure 7.11 investigates various relationships among specific attenuation A, radar reflectivity Z, specific differential phase shift K_{DP}, specific differential attenuation $A_{\mathrm{H}} - A_{\mathrm{V}}$, differential radar reflectivity Z_{DR}, and backscattering phase shift δ. These relations are parameterised by N_0^* when the two variables are extensive (like A and Z or A and R), and are independent of N_0^* when the two variables are intensive (like R/A and Z_{DR} or δ and Z_{DR}). It is interesting to note that the relationships depend only slightly on the shape parameter of the DSD (assumed "modified exponential" or gamma with $\mu = -1$ to 4). This confirms the robustness of the inverse model parameterised by N_0^*, independent of the shape of the DSD. Moreover, the apparent stability of the "intrinsic" shape of the DSD underlines this robustness.

7.3 Algorithms for Rain Retrieval with a Spaceborne Weather Radar

As said before, weather radars on airborne or spaceborne platforms operate at attenuated frequencies. With the TRMM precipitation radar (at 13.8 GHz), despite the near nadir operation that limits the path length through rain, the path integrated attenuation in heavy rain may reach 10 to 20 dB (one way).

7.3.1 Single Frequency Spaceborne Radar

The current concept used for spaceborne radar is the "rain profiling" algorithm. As said before, this concept is derived from Hitschfeld and Bordan (1954) (referred to hereafter as HB54). Equation (7.5) expresses that if the solution $Z(r_\mathrm{m})$ is known at the far range r_m, then a stable solution for the profile $Z(r)$ may be derived for $r < r_\mathrm{m}$. To determine the solution at $r_\mathrm{m} = r_\mathrm{s}$ (the near surface range gate), the spaceborne radar exploits the observation of the sea surface used as a reference target. Assuming that the scattering cross section of the surface σ_0 varies at a scale larger than the precipitation field, one may derive the path integrated attenuation PIA by comparison of the apparent σ_0 observed at pixels outside and inside the precipitation field. PIA provides an estimate of the unknown ratio $Z_\mathrm{a}(r_\mathrm{m})/Z(r_\mathrm{m})$ in (7.5). This is the basic approach used in data analysis of the TRMM precipitation radar. Nevertheless, in the algorithm actually implemented in the TRMM PR data processing, it should be taken into consideration that the uncertainty in the PIA estimate (associated with the natural variability of the ocean cross section) does not allow treating low rainfall rates. Thus, below a certain PIA threshold, the classical estimate of a Z–R relation (tuned for stratiform rain) is preferred. In practice, an hybrid algorithm has been developed for the operational algorithm of the TRMM precipitation radar (Iguchi and Meneghini, 1994; Iguchi et al. 2000).

Remember that the HB54 algorithm has been formulated under the assumption of a functional relationship between the specific attenuation A and the reflectivity Z of the power-law type: $A = \alpha Z^\beta$. To be consistent with the inverse model described in the previous section, where the A/N_0^* versus Z_e/N_0^* relationship is approximated by a power-law relationship as

$$A/N_0^* = \alpha_0 \left[Z_\mathrm{e}/N_0^*\right]^{\beta_0}, \tag{7.19}$$

we should write

$$\alpha = \alpha_0 \left[N_0^*\right]^{1-\beta} \text{ and } \beta = \beta_0, \tag{7.20}$$

216   J. Testud

which shows that an implicit assumption of the A–Z relationship is that N_0^* is *constant* along the path, as recognised by Nakamura (1991) or Kozu et al. (1991). This seems to be contradictory the extreme variability of N_0^* mentioned in the previous section. Indeed a great variability of N_0^* is observed from one precipitation event to the other or between convective and stratiform rain. Nevertheless, for a given rain event and a given rain type, N_0^* appears to be stable (Testud et al., 2001), which justifies the above hypothesis.

An alternate way to formulate the HB54 solution is to express it with respect to the specific attenuation A. Our specific approach to reformulate the solution is the following:

$$A(r) = A(R_\mathrm{s}) \frac{Z_\mathrm{a}^\beta(r)}{Z_\mathrm{a}^\beta(r_\mathrm{s}) + A(r_\mathrm{s}) \cdot I(r, r_\mathrm{s})}, \tag{7.21}$$

where

$$I(r, r_\mathrm{s}) = 0.46b \int_r^{r_\mathrm{s}} Z_\mathrm{a}^\beta \mathrm{d}s. \tag{7.22}$$

In (7.21), the profile of specific attenuation $A(r)$ is expressed as a function of the specific attenuation near the surface $A(r_\mathrm{s})$ (unknown). It is interesting to note that with this formulation, $A(r)$ is independent of the DSD variability (N_0^* is eliminated in this formulation) and independent of the radar calibration. However, $A(r_\mathrm{s})$ is introduced as a new unknown and should be determined using an external constraint. Also, since the primary parameter considered in the retrieval is $A(r)$, the rain rate should be further estimated from A, according to our inverse model, as

$$R = c \left[N_0^*\right]^{1-d} A^d. \tag{7.23}$$

The external constraint, supplied by the "path integrated attenuation" (PIA) estimated from the surface echo used as a reference target, is expressed as

$$\int_0^{r_\mathrm{s}} A(u) \mathrm{d}u = \mathrm{PIA}. \tag{7.24}$$

Substituting (7.20) in (7.23), it is straightforward to derive the expression of $A(r_\mathrm{s})$ as

$$A(r_\mathrm{s}) = \frac{Z_\mathrm{a}^\beta(r_\mathrm{s})}{I(0, r_\mathrm{s})} \left\{\exp\left(0.46\beta \cdot \mathrm{PIA}\right) - 1\right\}, \tag{7.25}$$

with

$$I(0, r_\mathrm{s}) = 0.46\beta \int_0^{r_\mathrm{s}} Z_\mathrm{a}^\beta \mathrm{d}s. \tag{7.26}$$

Moreover, comparing $A(r_s)$ and $Z_e(r_s)$, N_0^* may be estimated as

$$N_0^* = \left[\frac{1}{\alpha}\frac{1-\exp(-0.46\beta \cdot \text{PIA})}{I(0,r_s)}\right]^{\frac{1}{1-\beta}}. \tag{7.27}$$

Profiles $A(r)$ and $R(r)$ may subsequently be determined using (21) and (23). Thus the rain profiling algorithm not only allows us to correct the reflectivity profile for attenuation but also to adjust the N_0^* parameter in the R estimate.

With the TRMM PR, the accuracy of the estimate of $A(r_s)$ and N_0^* derived from (7.25) and (7.27) are subject to the statistical uncertainty and bias in the scattering cross section of the surface σ_0. It is obviously for heavy rain that the algorithm is expected to provide the best results (since the relative uncertainty in PIA will be then minimised). It is also interesting to note that the retrieved $A(r)$ profile [through (7.21) and (7.25)] is not subject to radar calibration C (in dB), while N_0^* derived from (7.24) is strongly dependent on C as

$$\Delta\left(log_{10} N_0^*\right) = \frac{C\beta}{10(1-\beta)}. \tag{7.28}$$

It is possible to exploit the statistical stability of the N_0^* histogram to check the calibration of the radar. Such a formulation was used in an application to a data set of the TRMM precipitation radar data by Ferreira et al. (2001).

7.3.2 Dual Frequency Spaceborne Radar

A dual frequency radar (14 and 35 GHz or 14 and 24 GHz) was considered in several precipitation missions from space. Two arguments support the use of such an instrument:

1. It allows an extension of the dynamics of the rain profiling algorithm toward low rain rates. As mentioned above, an hybrid algorithm is needed to cover the full dynamics of the 14 GHz radar (typically 0.1 to 100 mm/h at 14 GHz). At low rain rate, the attenuation being insufficient for application of the rain profiling algorithm, the operational algorithm of TRMM switches to the classical Z–R relationship, less accurate because it is subject to the N_0^* variability. The combination with a higher frequency, more subject to attenuation, would allow covering the lowest rain rates.
2. A dual frequency algorithm may be applied on the limited dynamic range common to the two frequency channels (0.1 to 15 mm/h typically). Such a dual frequency algorithm would operate *without* the information of a PIA as derived from the surface echo. Thus it can operate in special conditions where the heterogeneity of the σ_0 field of the surface makes the PIA estimate unreliable as, for example, in a hurricane environment or over mountainous areas.

We will restrict ourselves to a brief discussion of the dual frequency algorithm. Two approaches are possible, the "differential" or "integral" technique.

In the differential approach, the derivative of Z_1–Z_2 along the beam (subscript 1 or 2 refers to radar frequency 1 or 2, respectively) is a measure of the differential attenuation A_1–A_2. This is obtained without assuming "$N_0^* = $ constant" along the profile. This approach was investigated by Iguchi and Meneghini (1995). However, it seems that this technique requires an integration time (required to collect enough independent samples) incompatible with an operation from space. The "integral" approach remains in line with the "rain profiling" algorithm. With the dual frequency radar, the "external" constraint (provided by the ocean surface echo used as a reference target) is replaced by a *mutual constraint* that expresses the consistency of the along path attenuations over a segment $[r', r'']$ sampled in common at the two frequencies. This second approach that is numerically much more stable than the differential one (at the cost of maintaining the assumption of "$N_0^* = $ constant" along the profile) has been investigated by Meneghini and Nakamura (1990), Marzoug and Amayenc (1994), and Testud et al. (1992). In line with our present formulation, a possible mutual constraint could be expressed as

$$\int_{r'}^{r''} A_1(u) du = \int_{r'}^{r''} p N_0^{*(1-q)} A_2^q(u) du \tag{7.29}$$

where $[r', r'']$ represent a common interval of data sampled by the two frequency channels and $A_1 = p N_0^{*(1-q)} A_2^q$ is the power-law relationship relating the specific attenuations at the two frequencies.

7.4 Airborne Dual Beam Doppler Radar

The concept of dual beam airborne Doppler radar has been conceived as equivalent to a mobile ground based dual Doppler radar in order to sense remote regions such as the open ocean or mountainous areas. These systems are mostly X-band systems. The principle is illustrated in Fig. 7.1: the antenna system mounted in the aircraft tail, rotates about a horizontal axis collinear with the aircraft fuselage. The radar feeds two antennas mounted back to back, one looking 18.5° forward (fore) and the other 18.5° backward (aft). The antenna rotation associated to the aircraft motion produces a dual helical scan allowing sampling any point of the three dimensional space along *two viewing angles*. The time shift between the samplings along the "fore" and "aft" viewing angles (typically 2 minutes at 20 km range) is small enough when compared with the characteristic evolution time of a storm (typically 20 minutes), in order to apply a "quasi-stationary" hypothesis in the data processing.

Dual beam (or pseudo-dual beam) radars have been the object of joint development by NOAA and CETP, on the one hand, and by NCAR and CETP, on the other hand. The first pseudo-dual beam airborne Doppler radar experiment was accomplished in July 1989 in Florida during CAPE, with the NOAA-P3 tail radar equipped with a dual beam antenna of CETP. It is also worth mentioning the FAST methodology (Fore and Aft Scanning Strategy) used on the

NOAA P3-42 aircraft (Jorgensen et al. 1996). The ELDORA-ASTRAIA radar, jointly developed by NCAR and CETP and mounted on the NCAR Electra (see Hildebrand et al., 1996), flew for the first time in TOGA-COARE (November 1992 – February 1993).

The three-dimensional wind field synthesis from a dual beam airborne Doppler radar uses essentially the same technique as that from a ground based dual Doppler radar. Airborne measurements, however, allow following and analysing weather developments in areas where no ground based radar is available and following a moving storm staying in an optimised measurement distance. Such observations of cloud systems, as presented by Meischner et al. (2003), this book, can be complemented. The retrieval is based on the measurement, at each point of the three-dimensional space, of two independent components of the air motion, and should be constrained with the air mass continuity equation (expressed under its anelastic form), with special attention to the boundary condition. These techniques are now well established and will not be reviewed here. For details, see Testud and Chong (1983), Chong et al. (1983), Chong and Testud (1983), Chong and Testud (1996), Scialom and Lemaitre (1990), and Protat et al. (1998).

The above mentioned airborne radars operate at X-band. The choice of this frequency is a trade-off between antenna size and beam width. However at X-band, the along path attenuation may be quite severe for heavy rain, and the correction of the "apparent" reflectivity for attenuation is a key issue obtaining satisfactory rain rate retrieval. Two algorithms were developed at CETP for that purpose, called "stereoradar" and "dual beam." Both take advantage of the two viewing angles realised by the dual beam scanning strategy. The sampling scheme is presented in Fig. 7.7; α_1 and α_2 are the angles that the forward (fore) pointing and backward (aft) pointing beams, respectively, make with the aircraft track; s_1 and s_2 are the abscissas along the forward (fore) and backward (aft) beam, respectively.

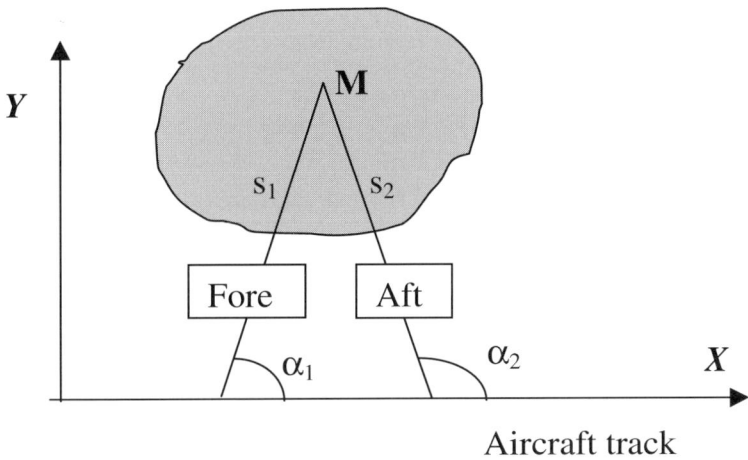

Fig. 7.7. Sampling scheme with the dual beam airborne radar

The "Dual Beam" Algorithm

The "dual beam" algorithm (Testud and Oury, 1997) is an integral algorithm of the same family as those for spaceborne radars described above. The starting point is to assume that N_0^* is uniform within the particular rain cell of interest, that is, along the two paths corresponding to the two viewing angles. Using a A–Z_e relationship at X-band as (7.19), we may apply the HB54 original algorithm (7.3) along each path (aft or fore) reaching point M sampled by the airborne radar. The "corrected" equivalent reflectivity at point M, according to (7.3), may be written as

$$Z_{e1,2}(M) = \frac{Z_{a1,2}(M)}{\left(1 - \alpha \, [N_0^*]^{1-\beta} \, I_{1,2}(M)\right)^{1/\beta}}, \quad (7.30)$$

with

$$I_{1,2}(M) = 0.46b \int_0^{r_{1,2}} Z_{a1,2}^\beta \, ds_{1,2}. \quad (7.31)$$

Subscript 1 or 2 in (7.30) and (7.31) refers to the forward and backward looking antenna, respectively; r_1 (or r_2) is the distance from point M to the radar seen by the fore or the aft antenna, respectively.

Now it may be written that the "corrected" equivalent reflectivity should be the same along the two viewing angles

$$Z_{e1}(M) = Z_{e2}(M). \quad (7.32)$$

It follows that

$$\alpha \, [N_0^*]^{1-\beta} = \frac{Z_{a1}^\beta - Z_{a2}^\beta}{Z_{a1}^\beta I_2 - Z_{a2}^\beta I_1} \quad (7.33)$$

and

$$Z_e = \left(\frac{Z_{a1}^\beta I_2 - Z_{a2}^\beta I_1}{I_2 - I_1} \right)^{1/\beta} \quad (7.34)$$

and

$$A = \frac{Z_{a1}^\beta - Z_{a2}^\beta}{I_2 - I_1}. \quad (7.35)$$

Equations (7.34) and (7.35) express that at any point M of the three-dimensional space, the corrected reflectivity Z_e and the specific attenuation A may be calculated as a function of the apparent reflectivity fields observed from the two antennas. Equation (7.33) shows that in addition, at any point M, an estimate of N_0^* may be derived. However it should be realised that this estimate is not a local "estimate"; it represents an "average" estimate along the two paths reaching point M (since N_0^* has to be assumed constant along these two paths to perform the dual beam algorithm).

As previously mentioned with the TRMM radar, N_0^* may be used in two ways:

1. if an independent N_0^* statistic is available (for example from in situ microphysical data), the radar derived N_0^* histogram allows one to calibrate the radar;
2. once the radar has been calibrated, N_0^* may be used to tune the $R - A$ or $R - Z_e$ relationship of the inverse model used to estimate R.

Stereoradar Algorithm
In contrast to the dual beam algorithm, the "stereoradar algorithm" developed by Kabèche and Testud (1995) is a differential one. Taking the along beam derivative of the apparent (attenuated) reflectivity Z_a (given by [7.2]) for each viewing angle,

$$\frac{1}{Z_{a1,2}} \frac{\partial Z_{a1,2}}{\partial r_{1,2}} = \frac{1}{Z_e} \frac{\partial Z_e}{\partial r_{1,2}} - 0.46A. \tag{7.36}$$

Subtracting the two equations eliminates A as

$$\frac{1}{Z_e} \left/ \left(\frac{\partial}{\partial r_1} - \frac{\partial}{\partial r_2} \right) Z_e \right. = \frac{1}{Z_{a1}} \frac{\partial Z_{a1}}{\partial r_1} - \frac{1}{Z_{a2}} \frac{\partial Z_{a2}}{\partial r_2} \tag{7.37}$$

or, taking account of the geometry (see Fig. 7.7):

$$\frac{1}{Z_e} \left[(cos\alpha_1 - cos\alpha_2) \frac{\partial}{\partial X} + (sin\alpha_1 - sin\alpha_2) \frac{\partial}{\partial Y} \right] Z_e = \frac{1}{Z_{a1}} \frac{\partial Z_{a1}}{\partial r_1} - \frac{1}{Z_{a2}} \frac{\partial Z_{a2}}{\partial r_2} \tag{7.38}$$

where X and Y are standard cartesian coordinates as illustrated in Fig. 7.7.

Equation (7.38) is a differential equation for Z_e that was obtained *without any assumption* about a $Z_e - A$ relationship. Thus, it applies to any type of hydrometeor. In attempting to solve (7.38) for Z_e, two problems are met: (i) not always, can appropriate boundary conditions be defined, and (ii) the calculation of the right-hand side of (7.38) from the data is delicate because it involves differentiation. Examples of successful resolution of (7.38) can be found in Kabèche and Testud (1995) and Oury et al. (1999).

Hybrid Algorithm
The dual beam algorithm is very easy to implement, but it nevertheless has the shortcoming of collapsing when $I_1 = I_2$ [see (7.34) and (7.35)], which generally happens at the centre of the sampled precipitation cell. The stereoradar algorithm is very powerful in the sense that it does not make any assumption about

the type of hydrometeor (our inverse model parameterised by N_0^* is very reliable for rain but is certainly incorrect for hail). But with most data sets, it is very difficult to define the boundary conditions required at the edges of the precipitation cells. The hybrid algorithm developed by Oury et al. (2000) couples the dual beam and the stereoradar algorithms. It uses the dual beam algorithm to generate the boundary conditions needed for the stereoradar algorithm. With this strategy, the hybrid algorithm is much more flexible than the stereoradar or the dual beam alone. Thus, the hybrid algorithm is particularly indicated when an extensive application to a large data set is anticipated.

7.5 Validation of the Algorithm Product with Airborne or Space Borne Radars

The validation of the algorithm product from an airborne or a spaceborne radar is not easy since any comparison with a ground truth (like a rain-gauge network) is extremely difficult, if not impossible, because of the completely different sampling strategies of the two types of instruments. *Cross-validation* approaches should be preferred. They are illustrated in Fig. 7.8 and Fig. 7.9. In Fig. 7.8, the hybrid algorithm for the dual beam airborne radar is evaluated using simultaneous observations of the same squall line by two NOAA-P3 aircraft flying on both sides on the squall line (see Fig. 7.9). With the classical estimate of the reflectivity consisting of selecting the max reflectivity between fore and aft beams, the correlation between the Z estimates by the two aircraft is very poor. After application of the hybrid technique (independently to the two data sets) the correlation is considerably improved. Figure 7.9 goes beyond, showing that between the "max of the four beam" reflectivity (using the two aircraft data set still) is considerably biased with respect to that retrieved by the hybrid technique. The impact on the rain rate estimate (in the same Fig. 7.9) appears quite considerable. This may totally change the interpretation of the squall line dynamics.

Figure 7.10 shows the results of cross-validation between the TRMM-PR and the P3-42 aircraft tail radar in their quasi-simultaneous observation of hurricane Bonnie in the Gulf of Mexico.

The origin of Bonnie was a large and vigorous tropical wave that moved over Dakar, Senegal, on 14 August 1998. Under a favorable upperlevel wind environment, Bonnie gradually strengthened and became a hurricane at 0600 UTC 22 August when it was located about 200 nautical miles north of the eastern tip of Hispaniola. Bonnie moved on a general west–northwest heading and reached maximum winds of 100 knots and a minimum pressure of 954 mb about 150 nautical miles east of San Salvador in the Bahamas. The hurricane then drifted northward for a period of 18 to 24 hours. Thereafter, the subtropical ridge re-intensified, forcing Bonnie to move northwestward and then northward toward the coast of North Carolina while the hurricane maintained winds of 100 knots. After a slight weakening, the eye of Bonnie passed just east of

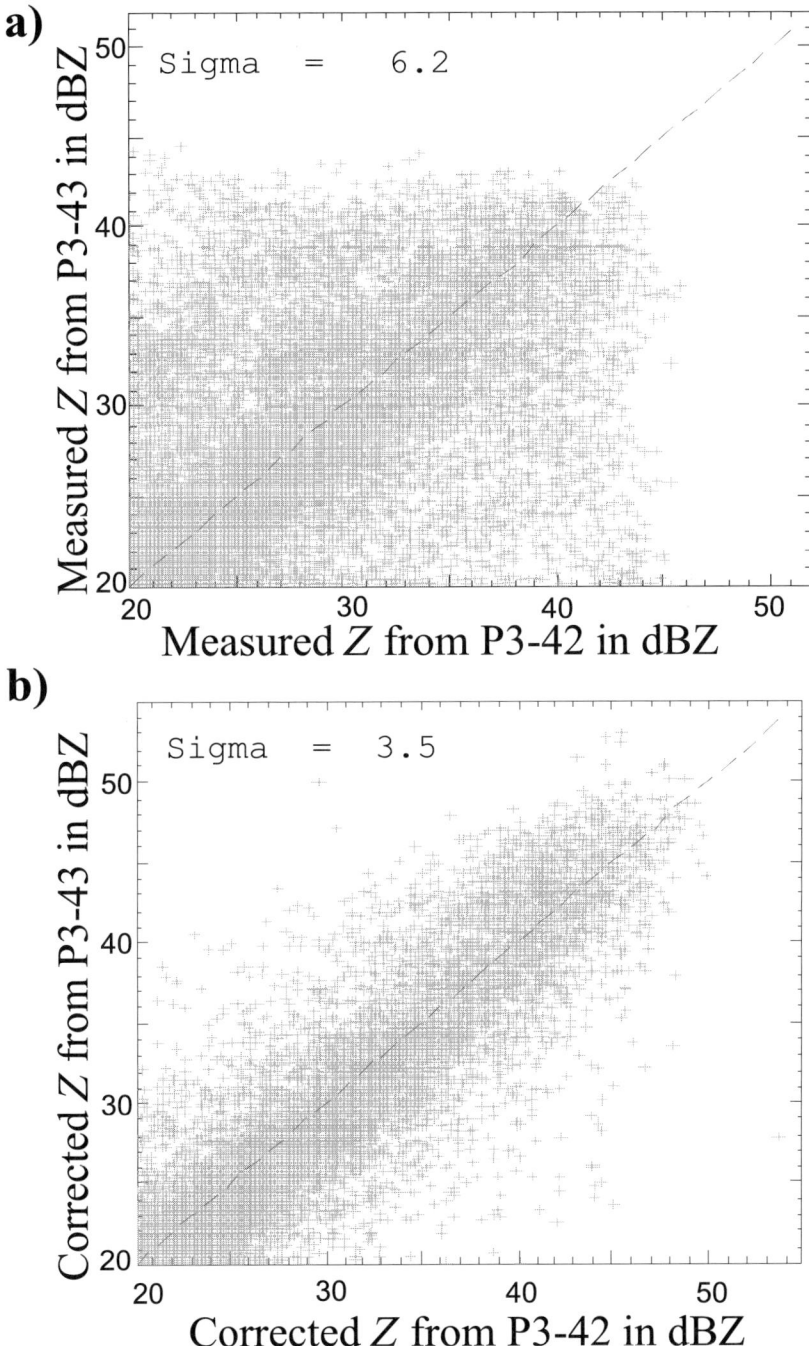

Fig. 7.8. Cross-validation of the hybrid technique using quasi-simultaneous but independent observations by the P3-42 and P3-4" NOAA aircrafts of the 9 February 1993 Toga-Coare squall line

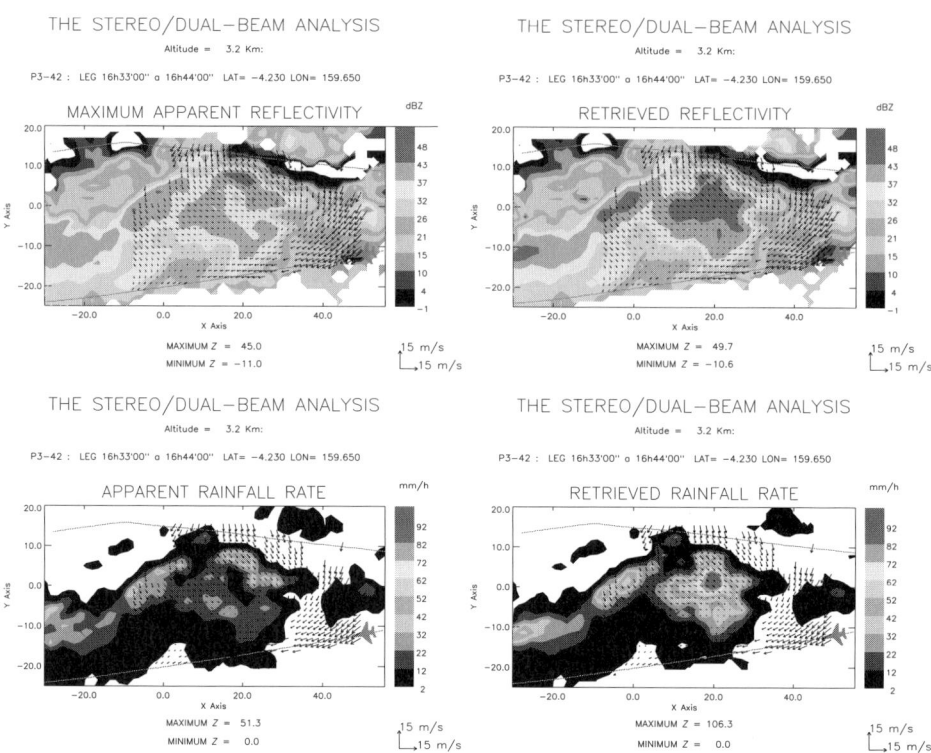

Fig. 7.9. Horizontal sections of the precipitation field at 3.2 km altitude in the 9 February 1993 Toga-Coare Squall line observed by the two NOAA-P3 aircraft each equipped with a dual beam (or pseudo-dual beam) X-band radar (*aircraft trajectories thin lines*). Top left shows the maximum reflectivity field observed from the four available beams and bottom left the corresponding "apparent" rain rate. The top right panel gives the corrected reflectivity determined from the hybrid technique and bottom right the corresponding "corrected" rain rate. The background wind field corresponds to 0.8 km altitude. See color Fig. 19 on page 328

Cape Fear around 2130 UTC 26 August and then made landfall near Wilmington.

At the time of the quasi-simultaneous observations by the TRMM-PR and the P3 tail radar (August 26th, 1998, around 11:37Z), Bonnie was at its very mature stage, just before starting decaying. Figure 7.10 perfectly illustrates the agreement of airborne and spaceborne radar observation when corrected for attenuation.

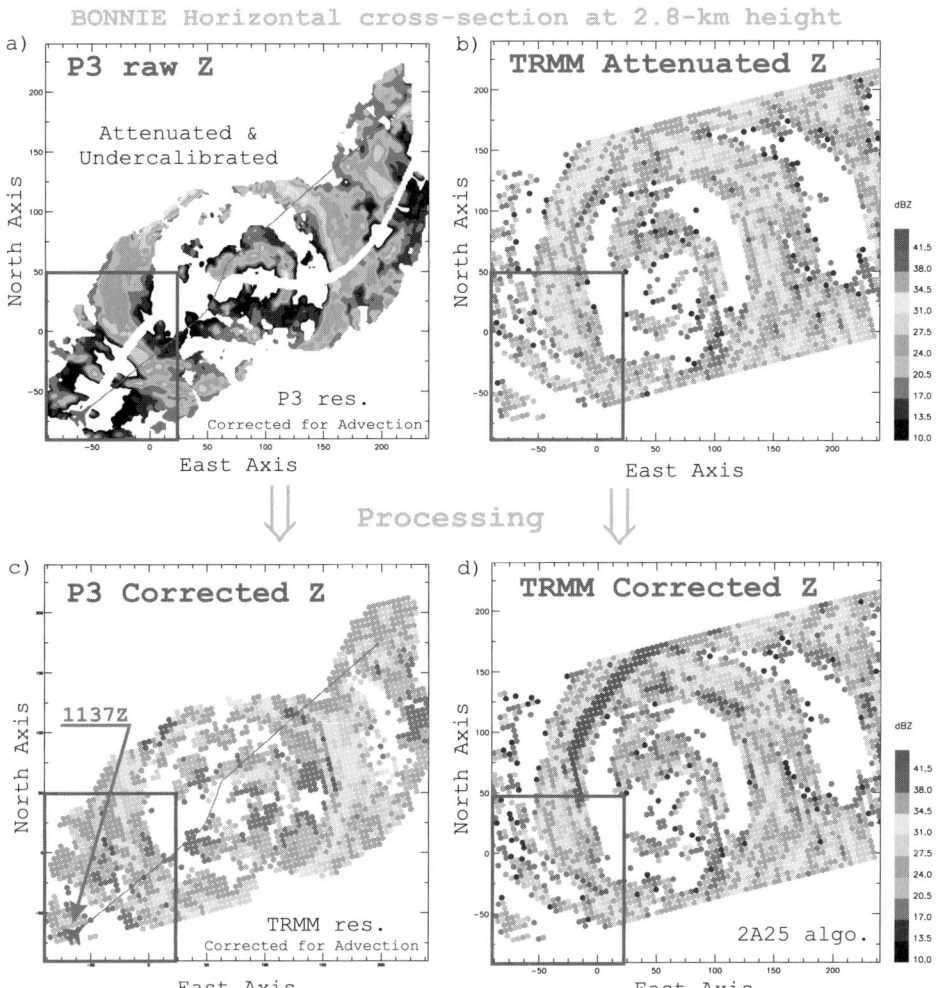

Fig. 7.10. Quasi-simultaneous sampling of hurricane Bonnie by the TRMM precipitation radar and by the NOAA P3-42 aircraft. Top diagrams: raw reflectivity field (max between "fore" and "aft" beams) of the P3 compared with the "attenuated" reflectivity observed by the TRMM PR. Bottom diagrams: Retrieved reflectivity from the aircraft using the hybrid technique compared with the "corrected" reflectivity of the TRMM-PR using the "rain profiling algorithm". Note the particularly good coincidence in the SW corner where the time coincidence of the samplings is within ± 5 min. The color version is given in color Fig. 20 on page 329

7.6 The Spaceborne Technology Applied to Ground Based Polarimetric Radar: Algorithm ZPHI

The limitation of the classical radar (measuring Z only) for estimating the rainfall rate has been recognised for long time. Four major causes of error are

1. The uncertainty of the Z–R relationship related to the variability of the DSD.
2. The possible existence of along path attenuation when the radar beam goes through the most intense weather systems.
3. The difficulty of calibrating the radar by artificial targets.
4. The difficulty of identifing the type of hydrometeor.

Polarimetric radars have been developed in an attempt to overcome these difficulties. The foundations of the polarimetric technique are discussed in depth in Chap. 5. For the purpose of this section, we may just recall that due to the aerodynamical stress during their fall, raindrops adopt the shape of oblate spheroids with an axis ratio increasing with their equivalent diameter (Beard and Chuang, 1987; Andsager et al., 1999). When probing the rainy medium with a radar, this has two consequences: first, a backscattering effect which translates in particular as a differential returned power between horizontal and vertical polarisation (characterised by the differential reflectivity $Z_{\mathrm{DR}} = Z_{\mathrm{H}}/Z_{\mathrm{V}}$); second, a propagation effect which translates both in a differential attenuation between polarisations H and V, and in a differential phase shift between H and V returned signals that builds up through the precipitation medium.

The robustness of the inverse model parameterised by N_0^* was mentioned in Sect. 7.7.2. As illustrated in Fig. 7.11, after parameterisation by N_0^*, the rain relations are almost insensitive to the shape of the DSD. The oblateness law as a function of the drop diameter is an important element on which there are yet uncertainties. Because of the drop oscillations, the average shape (observed in a wind tunnel) differs from the Beard and Chuang (1987) equilibrium shape. The model presently used takes account of the most recent observations (Andsager et al., 1999). However, no recent data document the model beyond 4.5 mm diameter. Also, in the highly turbulent environment of very intense storms, it is possible that the disorganisation of the drop orientation affects the inverse model.

The application of spaceborne technology to polarimetric ground based weather radar was investigated theoretically by Testud et al. (2000) in which the three frequency bands, X, S, and C, were considered. Le Bouar and Testud (2001) then proceeded to a one month systematic test from the ground based C-band polarimetric radar of Darwin (Australia).

The general principle for application of the rain profiling algorithm to the ground based polarimetric radar lies in the quasi-linear relationship between A and K_{DP}. Assuming perfect linearity for simplification (in fact an eventual nonlinearity is taken into account in Testud et al., 2000), this relation may be written:

$$A = \gamma \cdot K_{\text{DP}} \quad \text{(independent of } N_0^*\text{)}. \tag{7.39}$$

Denoting PIA_{i-1}^i the path integrated attenuation between the two ranges r_{i-1} and r_i along the beam, we may estimate PIA from Φ_{DP} as:

$$\text{PIA}_{i-1}^i = \int_{r_{i-1}}^{r_i} A(s)\text{d}s = (\gamma/2). \left[\Phi_{\text{DP}}(r_i) - \Phi_{\text{DP}}(r_{i-1})\right]. \tag{7.40}$$

As a consequence, the attenuation profile $A(r)$ between the same boundaries may be derived from (7.18) using r_i as the reference range. $A(r_i)$ is itself given by

$$A(r_i) = \frac{Z_{\text{a}}^\beta(r_i)\left(\exp\left\{0.23\beta[\Phi_{\text{DP}}(r_i) - \Phi_{\text{DP}}(r_{i-1})]/\gamma\right\} - 1\right)}{I(r_{i-1}, r_i)}. \tag{7.41}$$

Thus, with the polarimetric radar, the analysis may be segmented along the beam, and this segmentation may be optimised in order to take into account that different types of rain (stratiform or convective) may be met along the beam with distinct values of N_0^*. The N_0^* estimator along each segment is similar to (24), except that it should account for the along path attenuation due to the previous segments (see more details in Le Bouar et al., 2001).

From the primary products of the analysis, $A(r)$ and $N_0^*(r)$ (this last parameter being defined segment by segment), $R(r)$ is determined using an A–R relationship like (7.23). The fact that two observables, Z_a and Φ_{DP}, are used to build the estimate $R(r)$, justifies the name of the algorithm: ZPHI. But the utilisation of the fully consistent inverse model depicted by Fig. 7.11 allows deriving *also* the differential attenuation profile $A_{\text{DP}}(r)$, from which a correction of the apparent differential reflectivity Z_{DRa} may be achieved in order to estimate the "true" Z_{DR}, as

$$Z_{\text{DR}}(r) = Z_{\text{DRa}}(r) \exp\left[0.46 \int_0^r A_{\text{DP}}(s)\text{d}s\right]. \tag{7.42}$$

This correction leads to an alternate rain rate estimator that combines A and Z_{DR}:

$$R/A = e\left(Z_{\text{DR}}\right)^f. \tag{7.43}$$

Figures 7.12 to 7.14 illustrate the potential of algorithm ZPHI in an operational exploitation of a ground based polarimetric radar. Figure 7.12ab illustrates the main products delivered by ZPHI: the rain rate field and the N_0^* retrieval. It is interesting to note that

1. ZPHI is able to retrieve the rain rate at the original radar resolution without amplifying the radar speckle noise (Fig. 7.12a);

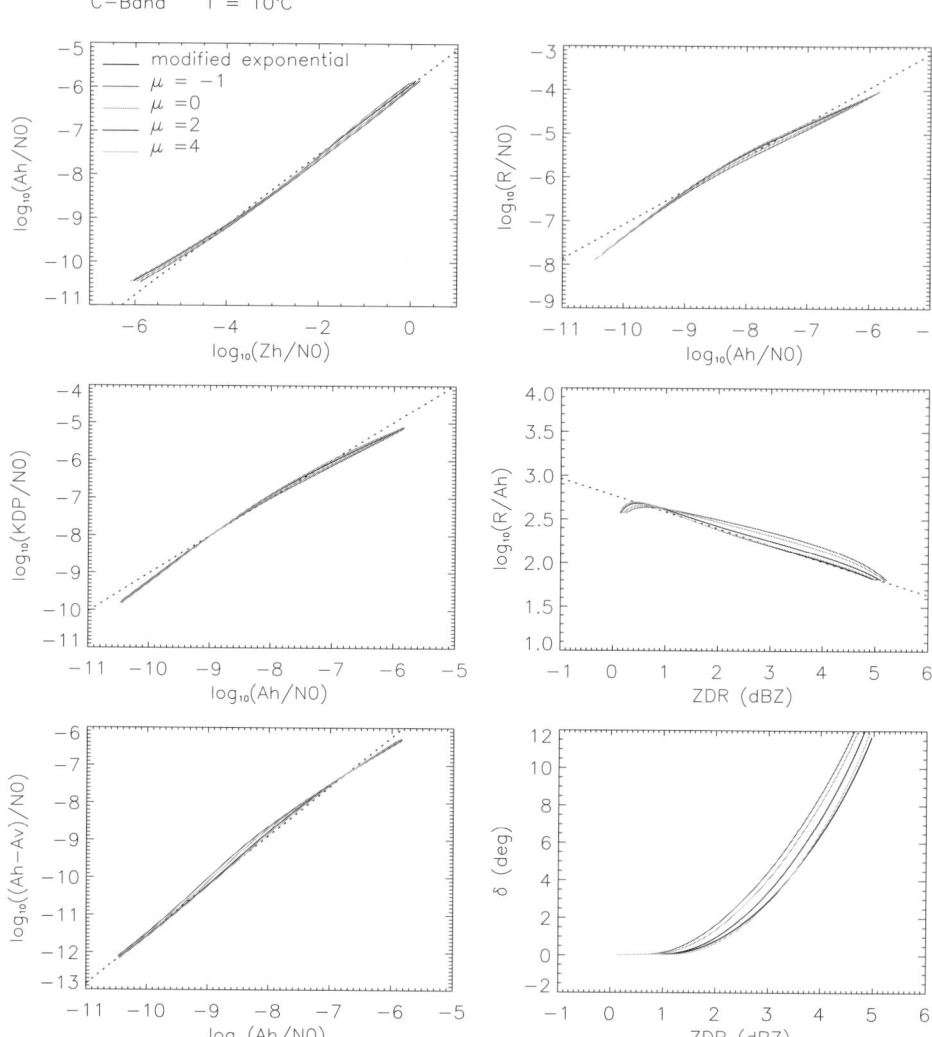

Fig. 7.11. Inverse model for a C-band ground based polarimetric radar (10°C, horizontal incidence). Left diagrams from top to bottom: normalised specific attenuation A_H/N_0^* against normalised radar reflectivity Z_H/N_0^*; normalised specific differential phase shift K_{DP}/N_0^* against A_H/N_0^*; normalised specific differential attenuation $A_{DP}/N_0^* = (A_H - A_V)/N_0^*$ against A_H/N_0^*. Right diagrams from top to bottom: normalised rain rate R/N_0^* against A_H/N_0^*; ratio R/A_H against Z_{DR}; back-scattering differential phase shift δ against Z_{DR}. In each diagram, various shapes of the DSD are considered: gamma DSD with μ between -1 and $+4$, or "modified exponential." The *dotted line* represents the best power law fit to the functional relationship corresponding to the "modified exponential"

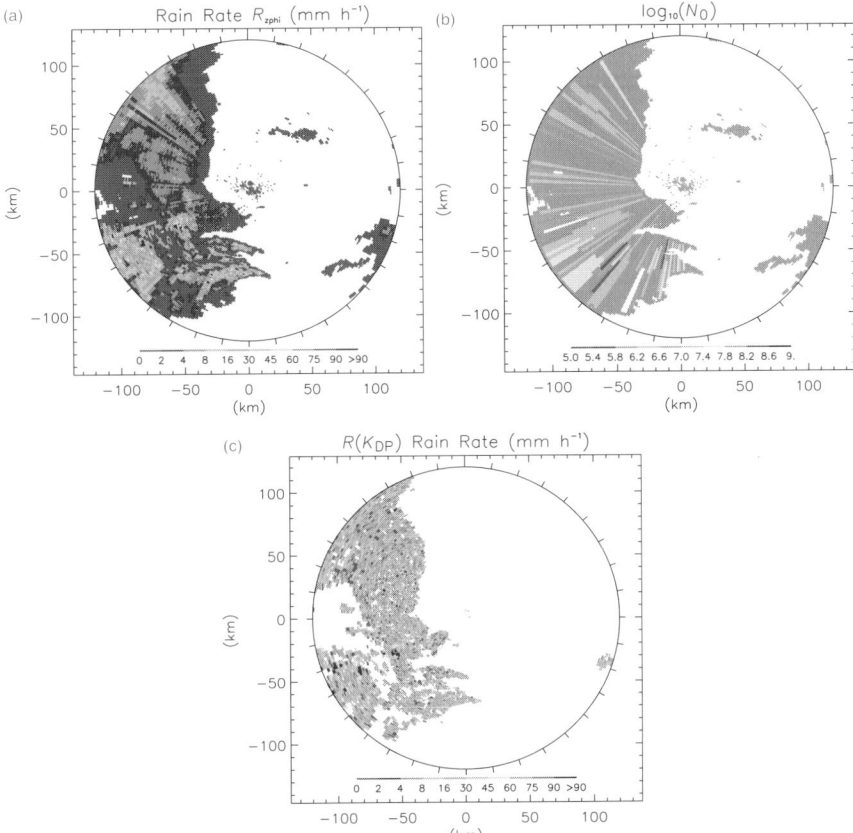

Fig. 7.12. PPI of the Darwin radar (at 1.6° elevation, 19 January 1998 around 2240 UTC). Top left is the rainfall rate, as estimated by ZPHI. Top right is the corresponding field of N_0^* (each ray corresponds to a segmentation; in the dark gray areas, ZPHI does not work, and the classical estimate is used instead). In the bottom a standard estimate of the rain rate based on K_{DP} is shown. The color Fig. 21 on page 330 shows structures more clearly

2. The N_0^* retrieval provides the right tendency between stratiform rain (NW quadrant) and convective rain (SW quadrant); for example, N_0^* is about 10 times higher in convective than in stratiform, in agreement with ground based disdrometer observations (Le Bouar et al., 2001).
3. In comparison, the standard estimate of R based on K_{DP} (Fig. 7.12c) is much more noisy and produces insignificant results when the rain rate is weak or moderate (in particular in stratiform areas).

Figure 7.13 illustrates one of the two calibration techniques that are available from ZPHI: the technique based upon the stability of the N_0^* histogram. It consists of adjusting the radar calibration by bringing to coincidence the histogram observed with the ground based disdrometer and that derived by the radar. The

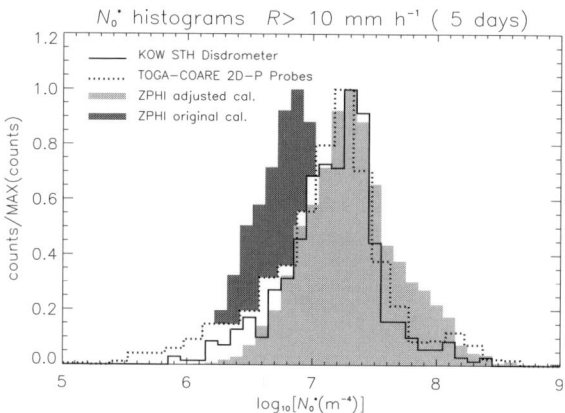

Fig. 7.13. Histograms of $\log10(N_0^*)$ for rainfall rate greater than 10 mm h-1, derived (i) from ZPHI (*dark and light grey-filled*), (ii) from the Kowandi-South BMRC disdrometer for 5 days of January 1998 (*full line*), and (iii) from 2-D-P probes during TOGA COARE (*dotted line*). The light-grey filled histogram is from ZPHI after a calibration adjustment in Z_a of -1 dBZ, while the dark filled one is for ZPHI with the original Z_a, see text

second technique is purely internal to the radar: the radar calibration is tuned to obtain the best correlation between the current R estimate by ZPHI [based on A and N_0^* through (7.23)] and the alternate estimate given by (7.43) (based on A and $Z_{\rm DR}$). In the one month exercise (January 1998) with the Darwin radar data, the two calibration techniques led to a consistent correction of -1 dBZ of the original Z data.

Figure 7.14 illustrates a validation of the ZPHI algorithm using the Darwin radar and a rain-gauge network. Radar- and rain-gauge have quite different time and space resolutions and sampling strategies. The time resolution of the rain gauge network is one minute, and the spacing between rain gauges is typically 10–20 km. The radar operates conical scanning at various elevations with revisit time of 12 minutes. Only the scan at 1.6° elevation is selected for comparison with rain gauges. To compensate for these different sampling characteristics, the radar estimate is averaged (i) in space, within a circle of 2 km radius centred at rain gauge site, and (ii) in time over three scans within ± 15 minutes. The corresponding rain gauge data is averaged over ± 15 minutes.

Figure 7.14a is a scatter plot representing a "point-by-point" comparison, covering seven rainy days of January 1998, between the rain rate estimate by ZPHI (noted $R_{\rm ZPHI}$) and that by the rain gauges (G). For reference, Fig. 7.14c displays the same scatter plot, but with the "classical" estimate $R(Z_a)$ (standard Z–R relationship without consideration of along path attenuation). Figure 7.14b shows the same scatter plot, but with an estimate denoted $R(A)$ based on the retrieved attenuation only, at fixed N_0^*.

Figure 7.14c confirms that the classical estimate is severely biased with respect to the rain gauge. With the classical estimate, the slope of the linear corre-

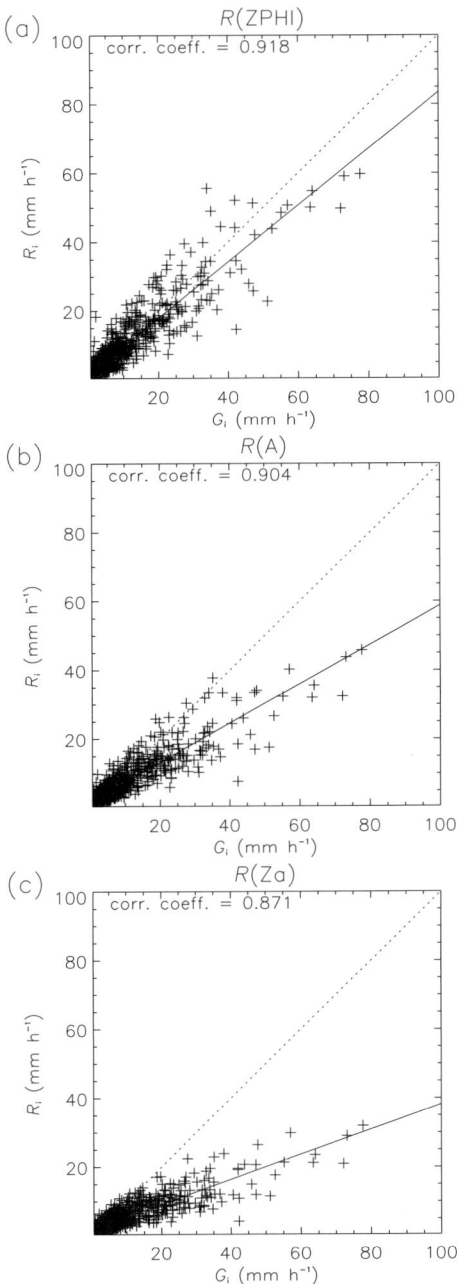

Fig. 7.14. "Point by point" comparison, during seven days of January 1998, of gauge rainfall rate G with collocated radar-derived estimates: (a) current ZPHI estimate, (b) R–A relation with fixed N_0, (c) classical" estimate R–Z_a. The rain-gauge data are averaged over ± 15 min, and the radar data are averaged over ± 15 min and within a circle of 2 km radius about the rain gauge

lation is 0.37, which means that on average, the classical estimate $R(Z_a)$ is only 37% of the gauge estimate G. The use of the $R(A)$ estimate with an N_0^* fixed to the Marshall–Palmer value ($0.8 \times 10^7 \mathrm{m}^{-4}$) sensibly reduces the bias. The slope increases to 0.58, which is a serious improvement with respect to the "classical estimate." With the R_{ZPHI} estimate, where the $R(A)$ relationship is tuned for the retrieved N_0^*, the slope further increases to 0.84. This demonstrates that not only the correction for attenuation, but also the dynamical adjustment, ray by ray, of N_0^*, is critical to retrieve a good rainfall rate estimate at C-band. Another important statistical feature is the regular improvement of the "linear correlation coefficient" from Fig. 7.14c to 7.14a: $\rho = 0.871$ with the classical estimate; $\rho = 0.904$ with the $R(A)$ estimate with fixed N_0^*; $\rho = 0.918$ with R_{ZPHI}.

7.7 The Rain Profiling Algorithm in Future Operational Applications

The rain profiling algorithm associated with the inverse model parameterised by the "normalised intercept parameter" N_0^*, reveals itself as a remarkable tool to correct the observed reflectivity for attenuation and to achieve a rain rate estimate immune with respect to DSD variability. Its robustness and mathematical simplicity make it a good candidate in many operational applications in radar meteorology.

Its capability in data analysis for a space borne precipitation radar is already demonstrated from the almost four year experience of the Tropical Rainfall Measurement Mission. The success of TRMM opens the field to the future GPM (Global Precipitation Mission), a preoperational satellite constellation aiming at monitoring the precipitation field on a global scale with a three-hour revisit time.

This rain profiling algorithm is of great value for ground based operational weather radars, too. Although not as prone to attenuation as spaceborne and airborne systems, this concept will enhance the accuracy of rain measurements. Associated with the polarimetric technique, it should constitute a breakthrough for accurate rainfall monitoring, and thus it should renew the interest of hydrologists in using weather radar data for quantitative precipitation measurements. Moreover this technique and algorithm allows reconsidering the realisation of "compact" polarimetric X-band radars for operational applications.

References

1. Andsager, K., K.V. Beard, and N.F. Laird, 1999: Laboratory measurements of axis ratios for large raindrops. J. Atmos. Sci., **56**, 2673–2683.
2. Bauer, P., 2001: Over-ocean rainfall retrieval from multi-sensor data of the Tropical Rainfall Measuring Mission (TRMM)-Part I: Development of inversion databases. J. Atmos. Oceanic Technol., **18**, 1315–1330.
3. Bauer, P., P. Amayenc, C.D. Kummerow, and E.A. Smith, 2001: Over-ocean rainfall retrieval from multi-sensor data of the Tropical Rainfall Measuring Mission

(TRMM)-Part II: Algorithm implementation. J. Atmos. Oceanic Technol., **18**, 1838–1855.
4. Beard K.V. and C. Chuang, 1987: A new model for equilibrium shape of raindrops, J. Atmos. Oceanic Technol. **44**, 1509–1524.
5. Chong, M., and J. Testud, 1983: Three-dimensional wind field analysis from dual Doppler radar data. Part III – The boundary condition: an optimum determination based on a variational concept. J. Climate Appl. Meteorol., **22**(7), 1227–1241.
6. Chong, M., J. Testud, and F. Roux, 1983: Three-dimensional wind field analysis from dual Doppler radar data. Part II – Minimizing the error due to temporal variation, J. Climate Appl. Meteorol., **22**(7), 1216–1226.
7. Chong, M., and J. Testud, 1996: Three-dimensional air circulation in a squall line from airborne dual-beam Doppler radar, 1996: a test of coplan methodology software. J. Atmos. Oceanic Technol. **13**(1), 36–53.
8. Ferreira F., P. Amayenc, S. Oury, and J. Testud, 2001: Study and tests of improved rain estimates from the TRMM precipitation radar, J. Appl. Meteorol., **40**, 1878–1899.
9. Hildebrand, P.H., W.-C. Lee, C.A. Walther, C. Frush, M. Randall, E. Loew, R. Neitzel, R. Parsons, J. Testud, F. Baudin, and A. Le Cornec, 1996: The ELDORA/ASTRAIA airborne Doppler weather radar: Design and observation from TOGA COARE. Bull. Am. Meteorol. Soc., February 1996, 213–232.
10. Hitschfeld, W., and J. Bordan, 1954: Errors inherent in the radar measurement of rainfall at attenuating wavelengths. J. Meteorol., **11**, 58–67.
11. Iguchi, T. and R. Meneghini, 1994: Intercomparisons of single frequency methods of retrieving a vertical rain profile from airborne or space-borne radar data. J. Atmos. Oceanic Technol., **11**, 1507–1516.
12. Iguchi, T. and R. Meneghini, 1995: Differential equations for dual-frequency radar returns. *Proc. 27th Conf. Radar Meteorol.*, AMS, pp. 190–193.
13. Iguchi, T., T. Kozu, R. Meneghini, J. Awaka, and K. Okamoto, 2000: Rain profiling algorithm for the TRMM Precipitation Radar. J. Appl. Meteorol., **39**, 2038–2052.
14. Jorgensen D.P., T. Matejka and J.D. Dugranrut, 1996: Multibeam technique for deriving wind fields from airborne Doppler radar. Meteorol. Atmos. Phys., **59**, 85–104.
15. Kabèche, A., and J. Testud, Stereoradar meteorology, 1995: A new unified approach to process data from airborne or ground-based meteorological radars. J. Atmos. Oceanic Technol. **12**(4), 783–799.
16. Kozu, T. and K. Nakamura, 1991: Rainfall parameter estimation from dual radar measurements combining reflectivity profile and path-integrated attenuation. J. Atmos. Oceanic Technol., **8**, 259–270.
17. Kozu, T., K. Nakamura, R. Meneghini, and W.C. Boncyk, 1991: Dual-parameter radar rainfall measurement from space: a test result from an aircraft experiment. IEEE Trans. Geosci. Remote Sensing, **29**, 690–703.
18. Kummerow, C., W. Barnes, T. Kozu, J. Shiue, and J. Simpson, 1998: The Tropical Rainfall Measuring Mission (TRMM) sensor package. J. Atmos. Oceanic Technol., **15**, 809–817.
19. Le Bouar E., J. Testud and T. Keenan, 2001: Validation of the rain profiling algorithm " ZPHI " from the C-band polarimetric weather radar in Darwin. J. Atmos. Oceanic Technol, **18**, N11, 1819–1837.
20. Marzoug, M., and P. Amayenc, 1994: A class of single- and dual-frequency algorithms for rain-rate profiling from a space-borne radar. Part I: Principle and tests from numerical simulations. J. Atmos. Oceanic Technol., **11**, 1480–1506.

21. Meneghini, R., and K. Nakamura, 1990: Range profiling of the rain rate by an airborne weather radar. Remote Sensing Environ., **31**, 193–209.
22. Oury S., J. Testud et V. Marcal, 1999: Estimate of precipitation from the dual beam airborne radar in TOGA-COARE. Part 1: The K–Z relationships derived from the stereo- and quad beam analysis. J. Appl. Meteorol. **38**(2), 156–174.
23. Oury S., J. Testud and X.-K. Dou, 2000: Estimate of precipitation from the dual beam airborne radars in Toga-Coare. Part 2: Precipitation efficiency in convective cells. Case study of 9th February 1993, J. Appl. Meteorol. **39**, 2371–2384.
24. Protat, A., Y. Lemaitre, and G. Scialom, 1998: Thermodynamic analytical fields from Doppler radar data by means of the MANDOP analysis. Q. J. R. Meteorol. Soc., **124**, 1633–1669.
25. Scialom G. and Y. Lemaitre, 1990: A new analysis for the retrieval of the three-dimensional wind field from multiple Doppler radars. J. Atmos. Oceanic Technol., **7**, 640–665.
26. Sekhon R.S. and R.C. Srisvastava, 1971: Doppler radar observations of dropsize distributions, J. Atmos. Sci., **28**, 983–994.
27. Sempere Torres, S.D., J.M. Porr and J.-D. Creutin, 1994: A general formulation for raindrop size distribution, J. Geophys. Res. (D), **103**, 1785–1797.
28. Testud, J. and M. Chong, 1983: Three-dimensional wind field analysis from dual Doppler radar data. Part I - Filtering, interpolating, and differentiating the raw data, J. Climate Appl. Meteorol., **22**(7), 1204–1215.
29. Testud J., E. Le Bouar, E. Obligis, M. Ali Mehenni, 2000: The rain profiling algorithm applied to polarimetric weather radar. J. Atmos. Oceanic Technol. **17**(3), 332–356.
30. Testud J. et S. Oury: 1997: Algorithme de correction d'atténuation pour radar météorologique, C.R. Acad. Sci. Paris, **324**, série2a, 705–710.
31. Testud, J., P. Amayenc, and M. Marzoug, 1992: Rainfall rate retrieval from spaceborne radar: comparison between single frequency, dual-frequency and dual-beam techniques. J. Atmos. Oceanic Technol., **9**(5), 599–623.
32. Testud J., S. Oury, P. Amayenc and R. Black, 2001: The concept of "normalized" distribution to describe raindrop spectra: a tool for cloud physics and cloud remote sensing. J. Appl. Meteorol., **40**(6), 1118–1140.
33. Ulbrich C.W.,1983: Natural variations in the analytical form of the drop size distribution. J. Climate Appl. Meteorol., **22**, 1764–1775.
34. Willis P.T., 1984: Functional fits to some observed dropsize distributions and parameterization of rain, J. Atmos. Sci., **41**, 1648–1661.

8 Radar Sensor Synergy for Cloud Studies; Case Study of Water Clouds

Herman Russchenberg[1] and Reinout Boers[2]

[1] Delft University of Technology, The Netherlands
[2] Royal Netherlands Meteorological Institute, The Netherlands

8.1 Introduction

Since the middle of the 1990s the use of radar for cloud profiling gained a lot of momentum because of the large need for reliable data to study the role of non-precipitating clouds in the climate system. High-frequency systems especially, for example, at 35 and 94 GHz, corresponding to about 0.9 and 0.3 cm wavelengths, were exploited for their feasibility to measure the structure of non-precipitating clouds. As a result of several experiments and field campaigns, it was soon realised that radar, although very useful, was not sufficient in revealing all the necessary information. It had to be combined with other sensors, like lidars or radiometers. A particular example is the observation of a low-level water cloud. The water droplets at the cloud base are in many cases too small to be detected by radar, and the radar will only measure part of the vertical profile of the cloud. Combination with an optical instrument like a lidar will then fill the gap. The lidar signal, however, is often absorbed before it reaches the cloud top, whereas the radar has no difficulty in observing the cloud particles there. Radar and lidar are in this case truly complementary instruments.

What is sensor synergy? Sensor synergy is the combination of different sensors in such a way that the observation process gives more information than what would have been acquired with the instruments separately. The fundamental principle underlying the benefit of sensor synergy lies in the fact that the observation of natural phenomena depends very much on the instruments used. For instance, depending on the frequency chosen, different scattering mechanisms of electromagnetic radiation come into play that also depend differently on the physical constitution of the scattering object. Combining different frequencies in one observation technique opens the option to retrieving more parameters of physical processes, as, for instance, cloud geometry and composition. Sensor synergy is crucial when it comes to creating a thorough overview of the physical state of the atmosphere and the processes therein.

In this chapter, the synergy of cloud radars and other remote sensing instruments, like lidar and a microwave radiometer, will be discussed and applied to the observation of water clouds.

8.2 Particle Scattering

The rationale for radar related sensor synergy follows from the different interaction mechanisms of electromagnetic waves with media when different sensor types are used. It is therefore important to understand this interaction process. In this section, the scattering process will be described and more specifically, the difference between scattering at radar wavelengths and optical wavelengths. This section is not intended to give a complete overview of the scattering theory. Only the relevant issues will be highlighted.

8.2.1 Scattering Mechanisms

In contrast to wavelengths of some centimetres that we have been dealing with up to now in this book, for shorter wavelengths, the backscattered radar signals generally are due to spatial and temporal fluctuations in the refractive index of the scattering medium. These fluctuations can be due to spatial variations in the concentration of water vapour, temperature, air pressure and atmospheric particles like hydrometeors. Three mechanisms can be distinguished:

Incoherent particle scatter. Assuming Rayleigh scattering, the backscattered power is described by the reflectivity factor Z:

$$Z = \int n(D) D^6 \mathrm{d}D, \tag{8.1}$$

with $n(D)$ the drop size distribution and D the diameter of the particles. Incoherent backscatter occurs when the number concentration of the particles is large and the (moving) particles are randomly distributed in space. In most cases, incoherent scattering dominates over other scattering mechanisms.

Coherent particle scatter. In turbulent environments, scattering may be due to coherent scattering by particles separated by turbulent eddies on a scale of half the radar wavelength. This mechanism may be dominant at centimetre-wavelengths in developing cumuli, although further research is still needed to confirm this. The reflectivity factor due to coherent particle scatter is approximately given by Erkelens et al. (2001) and de Wolf and Russchenberg (2000) as

$$Z = 2.44 \times 10^{-2} L_o^{-2/3} \beta^2 N^2 D^6 \lambda^{11/3} \tag{8.2}$$

with λ the radar wavelength and β the ratio of the standard deviation and mean of the spatial distribution of the particle mass; L_o is the macroscale[1] of the turbulence spectrum. A monodisperse drop-size distribution is assumed; N and D are, respectively, the number concentration and the size of the particles in the radar volume. The assumption of monodisperse distribution, although not representing real clouds, is sufficient for predicting the right order of magnitude of scattering.

---

[1] This is the characteristic scale of the mean wind flow before it breaks into smaller turbulent eddies.

Coherent air scatter. Turbulent environments with inherent time and spatial fluctuations of humidity and temperature may also lead to coherent backscatter of radar waves. The reflectivity factor is given by Erkelens et al. (2001) as

$$Z = 1.36 \cdot 10^{-3} C_n^2 \lambda^{11/3}, \qquad (8.3)$$

with C_n^2 the structure constant describing the fluctuations of the refractive index. This scattering mechanism can be dominant at centimetre-wavelengths in the convective boundary layer under clear air conditions and in the presence of scattered cumuli.

Attenuation. At high frequencies, attenuation of the radar signal along the propagation path has to be taken into account. For nonprecipitating water clouds, the attenuation can be of the order of 3 dB/km at 94 GHz for a liquid water content of 0.3 gm^{-3}. At 35 GHz, the attenuation becomes approximately 0.3 dB/km. At 3 GHz, the attenuation is negligible (Ulaby et al., 1982).

Optical extinction. In the optical limit, where the wavelength is much shorter than the particles, the extinction cross section equals twice its geometrical cross section (van den Hulst, 1981). For most water cloud types, the optical limit can be used for wavelengths shorter than 1 µm. The total extinction cross section per unit of volume is given by

$$\sigma = 0.5\,\pi \int n(D) D^2 \mathrm{d}D. \qquad (8.4)$$

For water clouds, the *optical* extinction is much larger than for mm and cm wavelengths. This implies that lidar signals will not penetrate as deeply into the clouds as the radars do. The optical depth is defined as the path-integrated value of σ.

The large droplet issue. Apart from cumuliform clouds in the boundary layer, where coherent backscatter may dominate the radar signal, the radar reflectivity factor of clouds is given by the sixth moment of the drop size distribution. Microphysical cloud properties are proportional to lower moments: in particular,

- the effective diameter to the 1st moment
- the optical extinction cross section to the 2nd moment
- the liquid water content to the 3rd moment.

This strong dependence of the radar reflectivity factor Z on drop size complicates the retrieval of microphysical quantities. Just a few relatively large drops present in the radar resolution volume can dominate the radar signal, while these droplets do not contribute significantly to the lower moments on the distributions and therefore to the microphysical quantities of interest.

It is illuminating to compare the sensitivity of radar and lidar to particle size. Table 8.1 gives the drop concentration that is needed to cause the reflectivity level given in the first column, assuming a uniform drop size distribution. It shows that the radar reflectivity is very sensitive to the drop size. A reflectivity of -30 dBZ can be caused by a billion drops of 10 μm, which is of the order of magnitude occurring in stratus clouds, or by one droplet of 316 μm. This means that in case of precipitating or nearly precipitating clouds, the reflectivity may be dominated by a few large drops. Table 8.1 also gives the optical depth that corresponds to the radar reflectivity, for the given diameter and concentration, and assuming a cloud thickness of 500 m. The optical depth is calculated in the so-called optical limit.

Table 8.1. Overview of relationships between droplet size, number concentration, radar reflectivity and optical depth. The bold numbers in the table denotes optically very thin clouds. Note that they are still very well distinguishable with the radar.

	Diameter (μm)					
	5	10	20	30	40	$N_t=1$ (m^{-3})
Reflectivity (dBZ)						
-40	$\frac{6.4 \cdot 10^9}{125}$	$\frac{10^8}{\mathbf{7.9}}$	$\frac{1.5 \cdot 10^6}{\mathbf{0.5}}$	$\frac{1.4 \cdot 10^5}{\mathbf{0.1}}$	$\frac{2.4 \cdot 10^4}{\mathbf{0.03}}$	$\frac{215}{\mathbf{3.6 \cdot 10^{-5}}}$
-30	$\frac{6.4 \cdot 10^{10}}{1250}$	$\frac{10^9}{79}$	$\frac{1.5 \cdot 10^7}{5}$	$\frac{1.4 \cdot 10^6}{\mathbf{1}}$	$\frac{2.4 \cdot 10^5}{\mathbf{0.3}}$	$\frac{316}{\mathbf{7.8 \cdot 10^{-5}}}$
-20	$\frac{6.4 \cdot 10^{11}}{12500}$	$\frac{10^{10}}{790}$	$\frac{1.5 \cdot 10^8}{50}$	$\frac{1.4 \cdot 10^7}{10}$	$\frac{2.4 \cdot 10^6}{\mathbf{3}}$	$\frac{464}{\mathbf{1.7 \cdot 10^{-4}}}$
-10	$\frac{6.4 \cdot 10^{12}}{125000}$	$\frac{10^{11}}{7900}$	$\frac{1.5 \cdot 10^9}{500}$	$\frac{1.4 \cdot 10^8}{100}$	$\frac{2.4 \cdot 10^7}{30}$	$\frac{681}{\mathbf{3.6 \cdot 10^{-4}}}$
	$\frac{\text{Drop concentration (m}^{-3}\text{)}}{\text{Optical depth}}$					$\frac{\text{Diameter (μm)}}{\text{Optical depth}}$

This *large droplet issue* has led to the notion that radar alone is not sufficient to derive microphysical properties with the accuracy that is necessary for the study of clouds in the climate system. Different sensors have to be combined. Especially the combination of radar with lidar and microwave radiometry is useful in this respect. A measured example of the significance of the drop size dependence is given in Fig. 8.1. To obtain this figure, cumulative drop size distributions, measured in stratocumulus clouds, were calculated and integrated over 2 hours. It shows that the majority of particles are smaller than 20 μ*m*. Larger particles do not contribute much to the number concentration any longer: the cumulated concentration hardly changes. This is, however, not true for the radar reflectivity: it is mainly determined by the relatively few large particles.

Figure 8.2 shows how large the scatter can be when large droplets are present in the clouds: for a given radar reflectivity, the retrieved liquid water content can vary by several orders of magnitude. It is impossible to retrieve the liquid water content from a radar measurement alone. It has to be combined with other instruments; Fig. 8.3 shows the ratio of the radar reflectivity and extinction of the

8 Radar Sensor Synergy for Cloud Studies   239

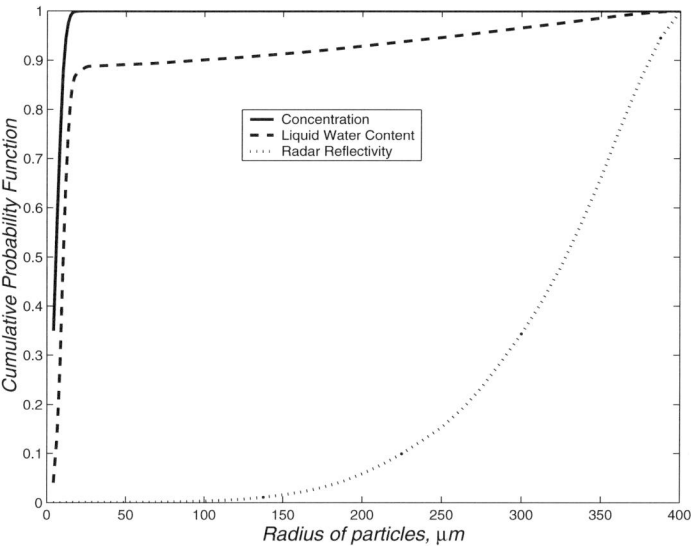

Fig. 8.1. An example of cumulative distributions of the number density, liquid water content and the radar reflectivity factor obtained in drizzling stratocumulus. It appears that particles larger than 20 micron diameter do not contribute much to the liquid water content, whereas their contribution to the radar reflectivity factor only starts at particles larger than 75 micron. Data was obtained during the Clare'98 campaign in the UK (October 7) (Clare'98 workshop, 1999)

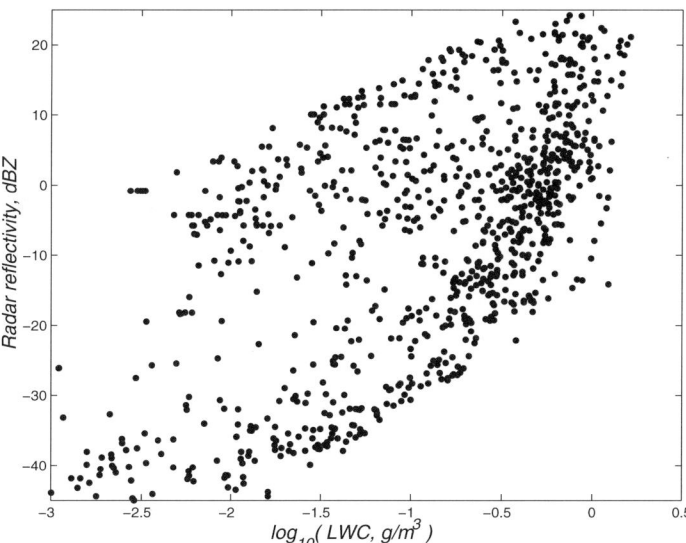

Fig. 8.2. Simulated radar reflectivity factor versus liquid water content. Data obtained during the CLARE'98 campaign in the United Kingdom (Clare'98 workshop, 1999)

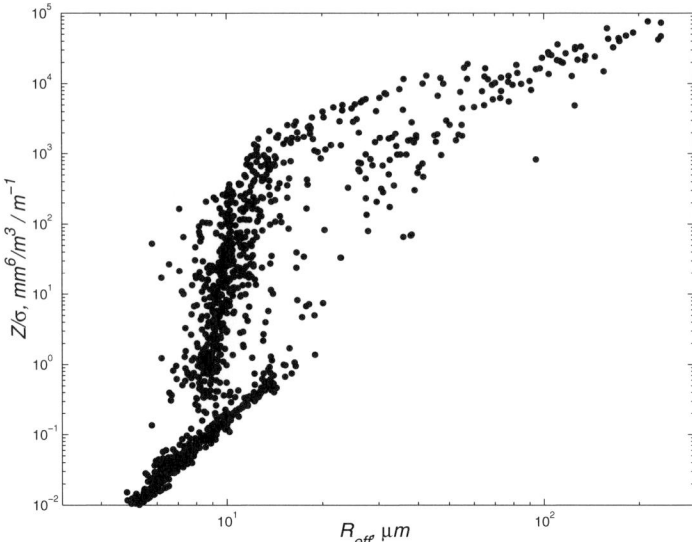

Fig. 8.3. The ratio of radar reflectivity factor and optical depth versus the effective radius in water clouds; data obtained during the Clare'98 campaign (Clare'98 workshop, 1999)

lidar signal as a function of the effective radius. The figure is based on simulations with measured drop size distributions in stratocumulus. Three regions can be distinguished: $\frac{Z}{\alpha} \leq 0.1$, due to small droplets alone; $0.1 \leq \frac{Z}{\alpha} \leq 10^4$, due to a mixture of small and large cloud droplets; and finally, $\frac{Z}{\alpha} \geq 10^4$, due to the predominance of large droplets (Baedi et al., 2000).

Figure 8.3 reveals that the combination of radar and lidar can be used to classify clouds in those with and without large droplets; as a result, the liquid water content can be retrieved more accurately (Krasnov and Russchenberg, 2002; Baedi et al., 2000).

8.3 Sensor Synergy

8.3.1 Observations of the Vertical Structure of Stratocumulus

Figure 8.4 shows an example of the radar reflectivity and the lidar backscatter as a function of time. The radar used for this measurement was the Delft Atmospheric Research Radar (Ligthart and Nieuwkerk, 1980), operating at 3.315 GHz.[2] At this frequency, backscattering can be due to the three earlier discussed scattering mechanisms. Throughout the measurement, a cloud layer is observed between 1500 and 1700 m. In the second half of the measurement, the convective activity

---

[2] This is not a typical frequency for a cloud radar, but it will be shown later that the data are sufficient when used in combination with other instruments.

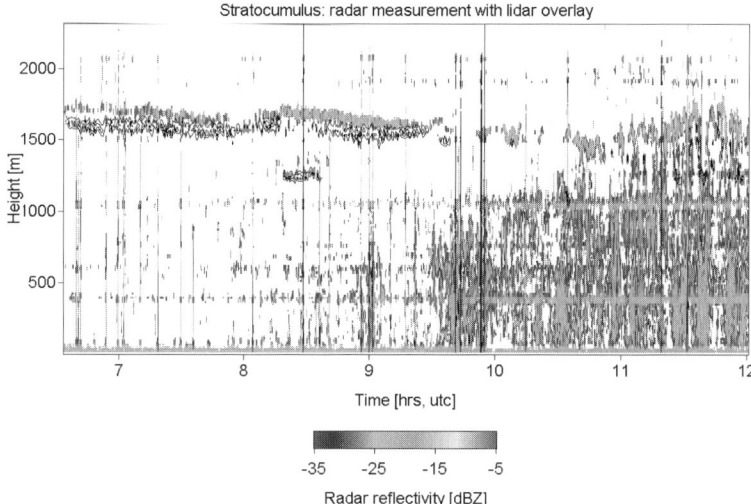

Fig. 8.4. Height–time diagram of radar reflectivity with contours of the lidar backscatter on April 19, 1996; stratocumulus, later growing cumuli after 10 UTC. The lidar backscatter has a maximum backscatter of 3000 (1000 s rad km) $^{-1}$. Radar resolution: 15 m, 5.12 s

in the boundary layer is increasing due to heating of the surface of the earth. This results in the enhanced speckle patterns in the radar plot due to backscattering by the containing water vapour, rising air. Isolated cumuli form out of the stratiform cloud layer The lidar backscatter has a maximum backscatter of 3000 (1000 s rad km) $^{-1}$; the radar reflectivity varies between -15 and -30 dBZ. To some extent, the radar signal in the cloud may be due to scattering by spatial inhomogeneities of the refractive index (caused, e.g., by entrainment); the quantification of this phenomenon is still subject to research.

The differences between the radar and lidar are intriguing. They give different cloud heights. The radar sees a cloud that is approximately 75 m higher than that the lidar sees. A small cloud layer at approximately 1200 m is observed by the lidar between 8 and 9 UTC; this cloud is optically too thick for the lidar to penetrate: the layer at 1600 m is not seen any longer. The radar did not measure the lower cloud.

8.3.2 Comparison with in Situ Measurements

During the observations shown by Fig 8.4, the cloud droplet size distribution was measured with a Forward Scattering Spectrometer Probe FSSP (Dye and Baumgardner, 1984). From the in situ data, the averaged radar reflectivity and lidar return were calculated. The results are given in Fig. 8.5; the simulations are in agreement with the actual observations of Fig. 8.4: the lidar gets a peak reflection from the cloud base, whereas the radar receives the maximum power from the cloud top. The sensitivity of the radar in this particular case is probably

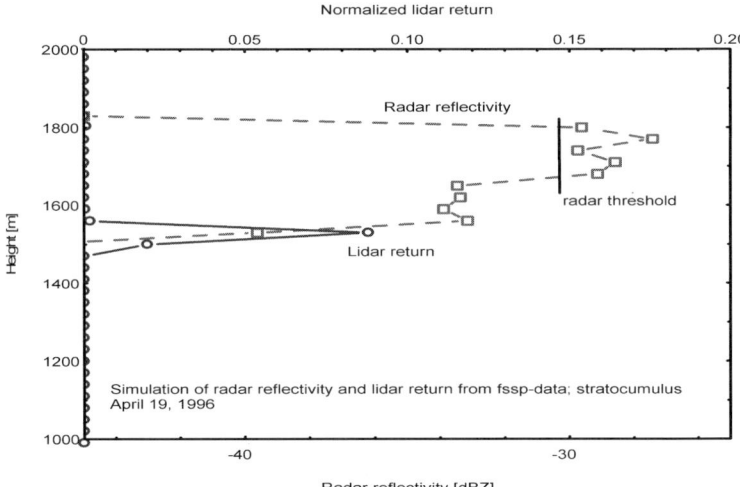

Fig. 8.5. Simulated height profiles of radar reflectivity and lidar return, based on in situ observations of the cloud droplet spectra. The lidar backscatter and extinction values were calculated using the optical approximation of scattering by spheres; applying exact Mie calculations will not give significant changes. The radar reflectivity is calculated with the Rayleigh approximation; at the frequency used here (3.315 GHz), the attenuation is negligible. Multiple scattering is ignored

not good enough to observe the cloud base. The aircraft was flying along horizontal tracks in the cloud field at different heights, not always close to the radar. The sampling volumes of the radar, lidar and FSSP were not coinciding, which means that the comparison of the in situ measurements with the radar/lidar observation can only be made qualitatively.

The observations of Fig. 8.5 can to some extent be generalized. In stratocumulus, the liquid water content as well as the mean drop size is often seen to increase with height (Martin et al., 1994; Gerber, 1996; Boers et al., 1994). This will have an impact on the vertical profiles of the lidar and radar signals due to differences in the scattering mechanisms in the microwave and optical range. In the absence of attenuation, which is the case for centimetre-wavelengths, the radar will receive its maximum power from the top of the cloud. The lidar signal, however, will have a peak in the lower part of the cloud: extinction prevents the lidar from penetrating into the cloud toward the top.[3]

The actual situation is, of course, more complex than these idealized clouds. The droplet concentration is never constant, and the height dependence of the liquid water content is never truly linear, but the idealization clearly shows a trend.

The radar technique at centimetre and milimetre wavelengths is well suited to detect the cloud boundaries, although the applicability is constrained by the

---

[3] From space, however, the situation will be different: both the radar and the lidar will see a maximum near the cloud top.

Table 8.2. Overview of usefulness of radar for cloud base detection

Cloud type	Radar for cloud base?
Stratus	Yes
Stratocumulus (no drizzle)	Yes
Stratocumulus (with drizzle)	No
Fair weather cumulus	Yes
Precipitation	No
Mid level clouds	Yes
High level clouds	Yes

cloud type. These constraints are a consequence of the D^6-dependence of the radar reflectivity factor. Large particles, whether they are precipitating or not, can obscure the radar reflection by the cloud base; in most cases the cloud top is unambiguously detectable. In Table 8.2, an overview of cloud types and the feasibility of radar for cloud base detection is given. In troublesome cases, the radar should be combined with lidar for proper identification of the cloud boundaries.

For nondrizzling stratus and stratocumulus, high radar sensitivity (-40 dBZ) is necessary. In case of drizzle and other types of precipitation, large particles may obscure the cloud base. Furthermore, at centimetre-wavelengths, scattering by fair weather cumulus and stratocumulus may be dominated by coherent scatter.

8.4 Case Study: Retrieval of the Microstructure of Water Clouds

8.4.1 Introduction

The three most important parameters linking clouds to climate are cloud optical depth, cloud droplet effective radius and cloud droplet concentration. Optical depth is a measure of the reflective capacity of a cloud and is linked to cloud albedo. It is a function of the cloud liquid water path and the cloud droplet effective radius (Stephens, 1978, 1990). Cloud droplet effective radius is inversely proportional to the one-third power of the droplet concentration (Twomey et al., 1984), which in turn is determined by local availability of anthropogenic and naturally occurring cloud condensation nuclei. So, local perturbations in observed effective radius may be influenced by industrial activities, population centers and continental – maritime transitions.

In this section, a new technique is outlined to obtain droplet concentration from groundbased remote sensing. It requires the synergism of three colocated remote sensing devices, namely, a cloud radar, a lidar and a microwave radiometer. Using a limited data set, it is demonstrated that this technique yields results that are in broad agreement with in situ cloud observations.

The remainder of this section is organised as follows. First, the theory will be presented whereby droplet concentration can be obtained from cloud dimension, liquid water path and (partial) extinction profiles. Next, the different remote sensing instruments will be introduced as well as the methods to derive the relevant cloud parameters. Then, the method will be tested using selected observations made during the CLARA campaign in The Netherlands (Clara report, 2001). We close with an error analysis to assess the validity of the method.

8.4.2 Theory

The present technique is based on the inherent relation between cloud optical extinction, cloud liquid water and cloud droplet concentration. In Boers et al. (1994), this relation is discussed in detail. They followed well-established thermodynamic principles to describe the shape of extinction and liquid water content profiles with height above cloud base. They used the fact that cloud droplet distributions can be adequately described by gamma distributions. We do not repeat their analysis but merely state the results, namely, that the extinction σ can be formulated as

$$\sigma = \pi Q A(\alpha) \left(\frac{4}{3} \frac{\rho_w}{\rho_0} \right)^{-\frac{2}{3}} A_d^{\frac{2}{3}} (1-D)^{\frac{2}{3}} N^{\frac{1}{3}} (z-z_B)^{\frac{2}{3}}, \qquad (8.5)$$

where Q is the extinction efficiency at visible wavelengths (~ 2 μm), A_d is the vertical gradient of liquid water content under adiabatic conditions, D is a parameter describing the deviation from adiabatic conditions (Betts, 1983), ρ_w is the density of liquid water, ρ_0 is the density of dry air (assumed constant within the cloud), z is height, z_B is cloud base height, and $A(\alpha)$ is defined as

$$A(\alpha) = \left[\frac{(\alpha+2)(\alpha+1)}{(\alpha+3)^2} \right]^{\frac{1}{3}}, \qquad (8.6)$$

where α is the exponent of size parameter r in the gamma distribution that describes the droplet size distribution. As indicated in Boers et al. (1994), normally an error of about 6–8% is introduced by assuming a constant value for N. However, the data to be shown later suggest that, on average, N was constant with height. In this model, it is further assumed that α is constant as well. This is a good assumption, as can be confirmed by calculations of this parameter from the in situ droplet size distributions, although it may vary slightly from day to day.

The value of A_d is a function of temperature, and care has to be taken to make a proper choice for it in individual applications. The best way is to remotely estimate the cloud temperature from an infrared radiometer, which was available during CLARA. However, in case this is not available, it can be estimated from radiosondes or by extrapolation of the surface temperature to cloud base using the adiabatic assumption.

The liquid water content (LWC) is assumed to vary linearly with height as

$$\text{LWC} = A_d (1-D)(z-z_B), \qquad (8.7)$$

so that the liquid water path is defined as

$$\text{LWP} = \frac{1}{2} A_\text{d}(1-D) h^2, \qquad (8.8)$$

where h is cloud depth.

The procedure we propose to follow is to determine the cloud depth h from a combination of lidar and radar data and to obtain LWP from the microwave radiometer data. Then (8.8) is used to obtain values of the parameter D. Essentially, this procedure yields values of the deviation of the cloud from its adiabatic state. If $D= 0$, this means that the layer is adiabatic. For D larger than 0, it means a progressive deviation of the cloud liquid water from the adiabatic value. Lidar extinction profiles can be obtained by treating the signal returns using a profile inversion technique developed by Klett (1983). Once these profiles have been obtained, the computed value of D is used to obtain best estimates of N by means of a least-squares regression of the functional form of (8.5) to the cloud base extinction profiles.

In practice, this technique is very simple to implement. However, to our knowledge, it has never been applied. We suspect that the reason is that there have been only very few groundbased remote sensing observations that could detect enough cloud parameters for this technique to work.

The technique is structurally similar to other techniques used to obtain effective radius (or droplet concentration) from a combination of optical depth and liquid water path (Boers, 1997), or liquid water path from a combination of optical depth and effective radius, as is common for satellite radiometry. This last technique relies on the classical relation between the optical depth τ (which is the height integrated extinction) and the liquid water path, such as described in many studies (Stephens, 1978):

$$\tau = \frac{3}{2} \frac{\rho_\text{w}}{\rho_0} \frac{\text{LWP}}{r_\text{eff}}. \qquad (8.9)$$

In our case, we use information on the height profiles of liquid water and optical extinction, rather than their height integrated values, to produce the same information.

8.4.3 Measurements

For the wavelength of the Delft Atmospheric Research Radar, the scattering from clouds is in the Rayleigh regime, so that larger precipitating particles (100–200 µm) contribute several orders of magnitude more to the scattering return than suspended cloud droplets (10–30 µm). As a result, the detection of cloud base, where suspended cloud droplets are small and precipitation droplets are large, is often not possible. Near cloud top, the cloud droplets are the largest, and precipitation particles mostly absent. Therefore, detection of cloud top is more straightforward. As the wavelength of DARR is considerably longer than that of the more frequently used 94-GHz cloud radars, temperature and humidity fluctuations can contribute to the backscatter returns when stratocumulus of cumulus clouds are probed. Fortunately, most temperature and humidity fluctuations occur near cloud top (except in the case of cumuli where mixing occurs during growth). The reason is that cloud top infrared cooling is the primary process responsible for the onset of turbulence underneath the overlaying temperature and humidity discontinuity associated with the top of the boundary layer.

The procedure for extracting cloud height from radar returns is to locate the maximum power of the range corrected radar returns probing from the top downward. Once the maximum has been located, the height of the 10% level of the maximum reflectivity (above the maximum) is designated as cloud top. This technique has been tested for several case studies during CLARA with good results (Russchenberg et al., 1998).

The short wavelengths associated with the lidars, in this experiment a 905-nm Vaisala CT75, imply that they operate in the Mie-scattering regime. Stable inversion techniques are available to convert the raw attenuated backscatter profiles into extinction profiles (Fernald, 1984; Klett, 1983). They rely on the assumption that the backscatter and extinction are proportional to each other. This assumption is excellent for water clouds, as any Mie calculation will attest. It then becomes possible to transform the lidar signal return equation into a homogeneous Ricatti equation, which has an analytic solution for the extinction factor.

The most serious problem associated with this technique is the inaccuracy of the calculated extinction profiles due to the unknown contribution of multiple scattering. There are now many papers in the literature that propose treatments for this problem (Young, 1995), none of which come up with a completely satisfactory solution. In general terms, multiple scattering is a function of the field of view of the lidar (FOV), the distance of the lidar to the cloud, the phase function of the particles, the concentration of droplets and the optical depth of the cloud. For some applications, multiple scattering can seriously affect the inversion of lidar profiles. In Spinhirne et al. (1989), a downward pointing lidar used on an aircraft flying in the stratosphere to probe boundary layer clouds, estimated that the contribution of multiple scattering was of the order of 50%. In our application, we expect the contribution of multiple scattering to be small on the basis of three arguments: (1) For the lidar used, the field of view was 0.5 mrad which is small. (2) The clouds were quite close to the receiver (generally within 1000–1800 m). And (3), we simulated single-scattered lidar profiles using the in situ aircraft observations of April 19, when the clouds were farthest away (around 2 kilometer) from the receiver. The resulting average simulated lidar profile was almost identical in shape to the average observed lidar profile, which suggests that the multiple scattering contribution must have been small. Therefore, we neglect its effect on the lidar profiles and extinction curves.

The procedure to detect cloud base starts by performing a profile inversion, as described above, yielding partial extinction profiles. Cloud base is determined as the height level on the digitalised extinction curve (30 m resolution) immediately above which the signal extinction exceeds the value of 2. Therefore, this process yields both a first estimate of cloud base as well as part of the extinction profiles directly above cloud base. Typical height at which total extinction of lidar signals ensues is 90 to 150 m above cloud base. The range resolution of 30 m implies that 3 to 5 reading points of measured extinction per profile can be used to regress against (8.5).

Further refinements to the procedure to detect cloud base are implemented when droplet retrieval is performed. At that point, the functional form of the extinction, as represented by (8.5), has to be fitted through the measured extinction profile. Given the value of D and a set of points on the extinction profile, the two free parameters are z_B and N. Their two-dimensional parameter space is searched to find the optimum solution that reduces the rms error between the theoretical and measured extinction curves. The restriction used here is that at least three points are necessary on the extinction curve.

Figure 8.6 demonstrates that excellent fits are obtained between the theoretical expression and the observations. The error bars indicate the shape change in the theoretical extinction curve for a 30% change in droplet concentration. One further iteration is performed. The new cloud base value (which is now located between two points on the original extinction profiles separated by 30 m) is used to update the value of D using (8.8), leaving LWP constant. Using this new value of D, which is typically between 0.01 and 0.07 different from the original value, a final regression is performed of the theoretical extinction curve to the observations.

Returning to Fig. 8.5, it becomes clear that the radar is able to detect cloud top but not cloud base, while the lidar is able to detect cloud base but not cloud top. So, both instruments are essential for the detection of cloud depth.

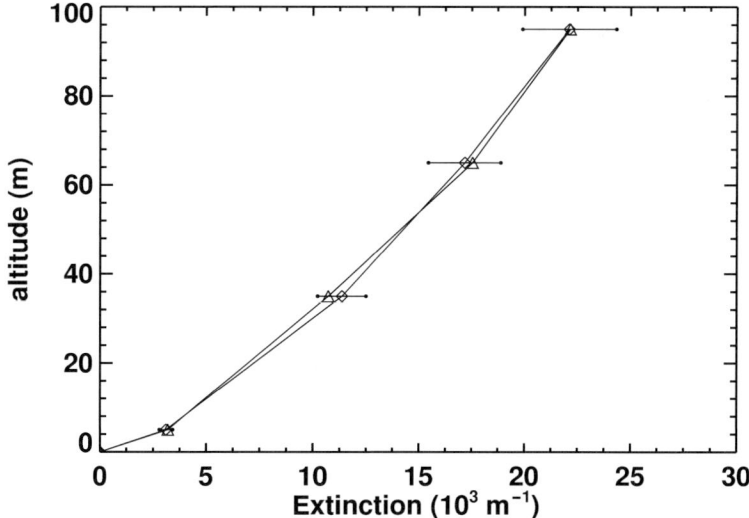

Fig. 8.6. Measured (triangles) and estimated (diamonds) extinction profiles

On selected days, it was possible to fly an instrumented aircraft near the central site of the experiment. The aircraft was equipped with a FSSP sensitive to particles with diameter between 1 and 47 μm. The probe also detected the number of particles exceeding the upper range limit but could not size those particles. In the absence of probes to detect and size the larger droplets, the measurements reported here only represent those droplets that are suspended in the clouds. The fifteen size bins of the FSSP were used to derive liquid water content, effective radius, mean radius and droplet concentration at a 1-s time resolution. They essentially form the verification material for any of the remote sensing measurements.

Values for $A(\alpha)$ can be evaluated from α, which in turn can be calculated in a straightforward manner from the fraction of the mean and effective radius $r_{\text{mean}}/r_{\text{eff}} = (\alpha+1)/(\alpha+3)$ using the properties of the gamma distribution which is used to represent the droplet spectra. The value of α varied somewhat with height, but, on average, it was around 5 for the days we analyzed. So under the range of observed clouds, A varies tightly between 0.869 and 0.896. In the subsequent analysis, we use the above value for α.

A narrow field of view (1 mrad) infrared radiometer was located near the other remote sensing instruments and used to detect cloud base temperature. The value of the cloud base temperature is important, as it fixes the value of the vertical gradient of the adiabatic liquid water content A_d.

8.4.4 Results

The observations to be shown were taken on 4 September 1996. On this day, a weak cold front was located over the Netherlands north of Delft. The line of the front was roughly from west–south–west to east–north–east. Satellite imagery indicated that clouds broke up entirely 50 to 100 km south of the front. However, just south of the front, a thick band of clouds persisted in the morning hours (4:00 to 6:00 UTC) while Delft was engulfed in fog. Winds were from the northwest for the entire time for which observations were recorded (4:00 – 24:00 UTC), with little or no perceptible shift in wind direction at the time of frontal passage over Delft (just before 12:00 UTC).

Figure 8.7a shows cloud base and cloud top, as derived from the lidar/radar combination. As the front approached, light drizzle fell between 7:00 and 8:00 UTC after which the cloud base, *as observed by the lidar*,[4] lifted rapidly from the surface to about 600 m. From the time of frontal passage onward there was a period of 4 hours during which cloud cover was not completely 100%. From about 16:00 UTC onwards cloud cover is almost 100%, and cloud base slowly lifted to 950 m after 20:00 UTC. The radar cloud tops coincide quite well with a temperature inversion, as determined from radiosondes. These points are indicated by pluses in Fig. 8.7a. There are several times when it was impossible to determine cloud top, in particular, around the time of frontal passage

---

[4] The lidar basically observes the lower boundary of drizzle and fog, which slowly disappears with time.

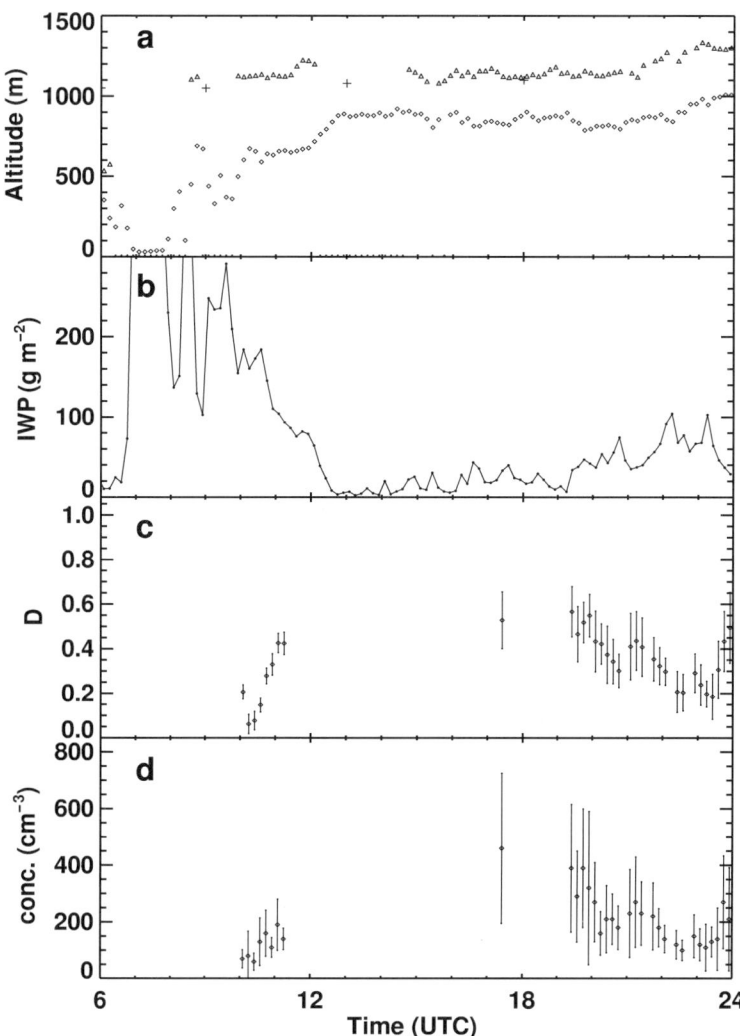

Fig. 8.7. Time series of observed and retrieved parameters on September 4 1996. (a) Time series of cloud base determined by lidar, and cloud top determined by radar. Individual points are 10 min averages. (b) Time series of microwave radiometer liquid water path. (c) Retrieval of parameter D. (d) Retrieval of droplet concentration

when clouds were thin and broken. After 16:00 UTC, cloud top lifted as well, so that the average cloud thickness remained constant. Figure 8.7b shows the trace of the liquid water path, derived from the microwave radiometer data; the procedure is described in Clara report (2001). The cloud liquid water path is very small between 12 and 15 UTC after which it appears to increase slightly as cloud cover thickens slightly. Figure 8.7cd will be described below but are included here with the other two time series to facilitate comparison.

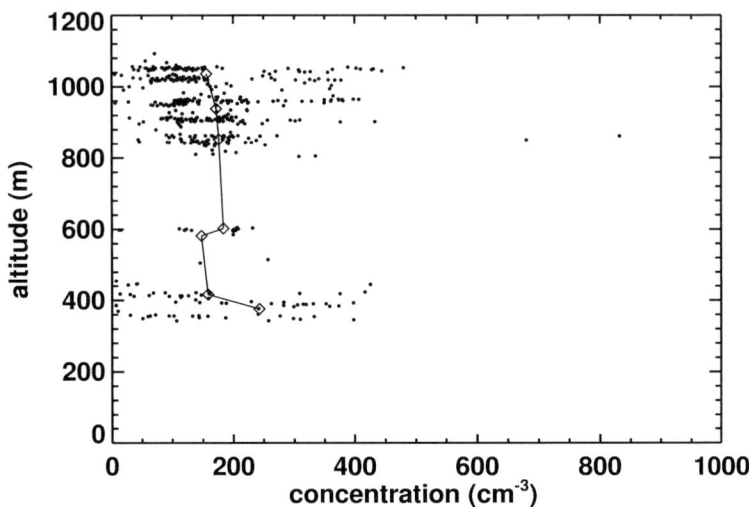

Fig. 8.8. 100 m averaged droplet concentrations taken south of 52.1 ° N on 4 September 1996. *Small dots* are 10-s averages

Between 8:00 and 11:00 UTC, the *lidar-observed* cloud base height showed great variations. Some of these variations were caused by the presence of precipitation which lowered the apparent cloud base. However, some of these are also associated with the presence of multiple cloud layers. This is especially evident just before 10:00 UTC.

Aircraft observations were carried out between 6:00 and 9:00 UTC, originating from the airport of Zestienhoven located 7 km to the southeast of Delft. Several flight passes were carried out in the close vicinity of Delft. Figure 8.8 shows the droplet concentration as a function of height from aircraft measurements. The measurements were scattered over two cloud layers, one between 400 and 600 m, the other between 800 and 1000 m. There were large variations in droplet concentration, but the 10 hPa averages fall within a tight range of 180–250 cm^{-3}. Also shown on the plot are the 10 s average values of the concentration to indicate the scatter around the mean. Clearly, the distribution of concentration values is somewhat asymmetric with the arithmetic mean slightly elevated above the median values. Figure 8.9 shows the liquid water content for the same time series. Also shown is the adiabatic liquid water content as the thin straight line. The slope of this line is more important than its exact location on this graph as it is difficult to ascertain the exact value of cloud base. Clearly, the actually measured liquid water content profile remains below the adiabatic value, implying that the layer is mixed with dry air.

Returning now to Figure 8.7, Figure 8.7c shows the time series of the 10 min averaged retrieved parameter D. After the initial small values near 10:00 UTC, it increased rapidly to about 0.5 after 11:00 UTC. As discussed above, just before 10:00 UTC, there were multiple cloud layers, and (evaporating) precipitation existed below cloud base. So, under these conditions, the idealized assumption

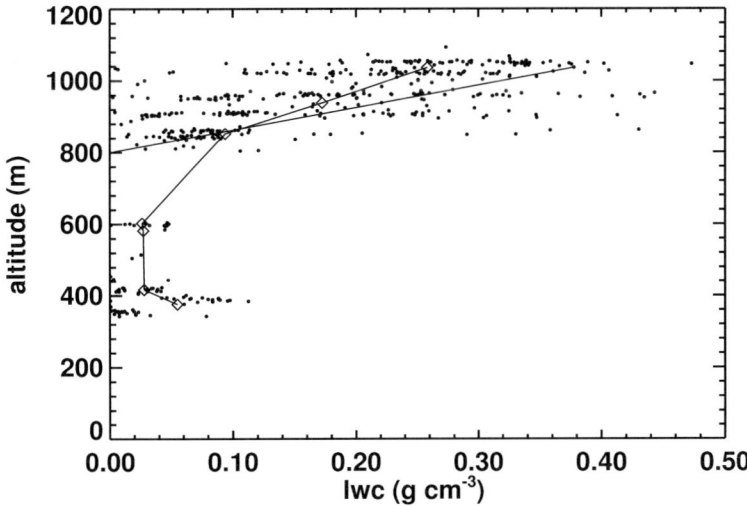

Fig. 8.9. 100 m averaged liquid water contents taken south of 52.1 ° N on 4 September 1996. *Small dots* are 10-s averages. The *thin straight line* is the adiabatic liquid water content

used to perform the retrieval, namely, a single nonprecipitating cloud deck, is not valid. Therefore, before frontal passage, the only valid results were obtained from 10:00 UTC onward. Near noontime, the liquid water path decreases rapidly clouds break up temporarily. Therefore, the errors become unacceptably large and no D or N retrieval is possible, until well after 19:00 UTC, except for one occasional retrieval near 17:30 UTC. These computations imply that during most of the day the cloud contained only roughly 50% of its possible adiabatic amount of liquid water. This is hardly surprising. Entrainment of unsaturated parcels into the cloud dilutes the air so that adiabatic conditions are the exception and not the rule. There are few, if any, similar measurements reported in the literature (Albrecht at el., 1990). We note that these results are consistent with the average in situ liquid water contents. These appear to depart from their adiabatic values by a similar amount.

Figure 8.7d shows the 10-min averaged retrieval of the droplet concentration. This graph shows some similarity to Fig. 8.7c due to the dependence of N on the parameter D. On average, concentrations are restricted to values under 400 cm^{-3}. The first values on the plot after 10:00 UTC were measured very close to the time of overflights. The droplet retrievals range from 100 to 200 cm^{-3} and underestimate the mean of the aircraft data by about 50%. After 18:00 UTC, when the LWP values increase, the mean of the retrievals of N returned to the values ranging from 100 to 400 cm^{-3}.

8.4.5 Concluding Remarks

It was shown that in order to retrieve cloud parameters that are important for radiation transfer in stratocumulus, cloud radar has to be combined with other instruments, like a microwave radiometer and lidar. This chapter is not intended to give a complete overview of the work in this field. Although the case study only dealt with water clouds, and more specifically stratocumulus, the radar-lidar synergy has also been applied to other cloud types (Donovan and van Lammeren, 2001; Donovan et al., 2001).

Water clouds are best observed with high-frequency radars (35 and 94 GHz), as at these frequencies, coherent scattering mechanisms do not play a significant role. For the observation of other clouds types, there is no fundamental difference between the performance of centimetres and milimetres wavelengths radars.

The technique discussed in this chapter can be improved in several ways. First, the reliability of liquid water path estimates could be improved if the sensitivity of liquid water detection is increased at low liquid water path, for example by including an extra 94 GHz microwave channel. Second, the technique can be improved if better estimates of cloud dimensions are available and if the vertical radar profile through the cloud is incorporated. Third, the technique can be improved if a more powerful lidar would enable the retrieval of the extinction profile at a higher resolution. The more points are available on the lidar extinction curve, the easier it is to apply the curve fitting technique necessary to adjust the observed extinction curve to the theoretical extinction curve.

8.5 Acknowledgements

The authors thank the following persons for their contribution to the work described in this chapter: J.S. Erkelens, V.K.C. Venema, A.C.A.P van Lammeren, O. Krasnov, A. Apituley, R.P Baedi, J.J.M de Wit, H. ten Brink and S. Jongen. This work is the result of collaboration in The Netherlands between Delft University of Technology, the Royal Netherlands Meteorological Institute KNMI, Eindhoven University of Technology, the National Institute for Public Health and the Environment RIVM and the Netherlands Energy Research Foundation ECN.

References

1. Albrecht, B.A., C.W. Fairall, D.W. Thomson, A.B. White and J.B. Snider, 1990: Surface-based remote sensing of the observed and the adiabatic liquid water content of stratocumulus clouds. Geophys. Res. Lett., **17**, 89–92.
2. Baedi, R.P., J.J.M. de Wit, H.W.J. Russchenberg, J.S. Erkelens and J.P.V. Poiares Baptista, 2000: Estimating the effective radius from radar and lidar based on the Clare'98 data set. Phys. Chem. Earth (B), **25**, 1057–1062.
3. Betts, B.K., 1983: Thermodynamics of mixed stratocumulus: Saturation point budgets. J. Atmos. Sci, **40**, 2655–2670.

4. Boers, R. and R.M. Mitchell, 1994: Absorption feedback in stratocumulus clouds: Influence on cloud top albedo, Tellus. **46A**, 229–241.
5. Boers, R., 1997: Simultaneous retrievals of cloud optical depth and droplet concentration from solar irradiance and microwave liquid water path. J. Geophys. Res, **102**, 29881–29891.
6. CLARA report 2001: Clouds and radiation: intensive observational campaigns in The Netherlands, RIVM, NRP Program, Report 410 200 057.
7. Clare'98 report: *Cloud Lidar and Radar Experiment, Workshop Proceedings*, Sept. 1999, WPP-170, ISSN 1022-6656, ESTEC, Noordwijk, The Netherlands.
8. Donovan, D.P. and A.C.A.P. van Lammeren, 2001: Cloud effective particle size and water content profile retrievals using combined lidar and radar observations – 1. Theory and examples. J. Geophys. Res. Atmos., **106**, 27425–27448.
9. Donovan, D.P., A.C.A.P. van Lammeren, R.J. Hogan, H.W.J. Russchenberg, A.Apituley, P. Francis, J. Testud, J. Pelon, M. Quante and J. Goddard, 2001: Cloud effective particle size and water content profile retrievals using combined lidar and radar observations – 2. Comparison with IR radiometer and in situ measurements of ice clouds. Geophys. Res. Atmos., **106**, 27449–27464.
10. Dye, D. and D. Baumgardner, 1984: Evaluation of the Forward Scattering Spectrometer Probe. Part I: Electronic and optical studies. J. Atmos. Oceanic Technol., **1**, 329–344.
11. Erkelens, J.S., V.K.C. Venema, H.W.J. Russchenberg and L.P. Ligthart, 2001: Coherent scattering of microwaves by particles: Evidence from clouds and smoke. J. Atmos. Sci., **58**, 1091–1102.
12. Erkelens, J.S., S.C.H.M. Jongen, H.W.J. Russchenberg and M.H.A.J. Herben, 1999: Estimation of cloud droplet concentration from radar, lidar and microwave radiometer measurements. *Proc. Symp. Remote Sensing Cloud Parameters: Retrieval and Validation*, Delft, The Netherlands, pp. 107–111.
13. Fernald, F.G, 1984: Analysis of atmospheric lidar observations: Some comments. Appl. Opt., **23**, 652.
14. Gerber, H., 1996: Microphysics of marine stratocumulus clouds with two drizzle modes. J. Atmos. Sci., **53**(12), 1649–1662.
15. Hulst, H.C. van den, 1981: *Light Scattering by Small Particles*. Dover Publications, New York.
16. Klett, J.D., 1983: Stable analytical inversion solution for processing lidar returns. Appl. Opt., **20**, 211–220.
17. Krasnov, O. and H.W.J. Russchenberg, 2002: The relation between the radar to lidar ratio and the effective radius of droplets in water clouds: An analysis of statistical models and observed drop size distributions. *11th Conf. Cloud Phys.*, Ogden, AMS, USA.
18. Lammeren, A.C.A.P. van, H.W.J. Russchenberg, A. Apituley and H. Ten Brink, 1997: CLARA: A data set to study sensor synergy. *Proc. Workshop on Synergy of Active Measurements in the Earth Radiation Mission*, Geesthacht, Germany, ESA EWP 1968 or GKSS 98/eE10.
19. Ligthart, L.P. and L.R. Nieuwkerk, 1980: FM-CW Delft Atmospheric Research Radar. IEE Proc., **127**, Pt F.
20. Martin, G.M., D.W. Johnson and A. Spice, 1994: The measurement and parametrization of effective radius droplets in warm stratocumulus clouds. J. Atmos. Sci., **51**, 1823–1842.
21. Peter, R. and N. Kampfer, 1992: Radiometric determination of water vapor and liquid water and its validation with other techniques. J. Geophys. Res., **97**, 18173–18183.

22. Russchenberg, H.W.J., 1992: *Ground-Based Remote Sensing of Precipitation Using a Multi-Polarized FM-CW Doppler Radar.* Delft University Press.
23. Russchenberg, H.W.J., V. K .C. Venema, A. C. A. P. van Lammeren and A. Apituley, 1998: Cloud measurements with lidar and a 3 GHz radar. Final report for ESA, contract PO151912.
24. Slingo, A., S. Nicholls and J. Schmeitz, 1982: Aircraft observations of marine stratocumulus during JASIN. Q. J. R. Meteorol. Soc., **108**, 833–856.
25. Spinhirne, J.D., R. Boers and W.D. Hart, 1989: Cloud top liquid water from lidar observations of marine stratocumulus. J. Appl. Meteorol., **28**, 81–90.
26. Stephens, G., 1978: Radiation profiles in extended water clouds. J. Atmos. Sci., **35**, 2111–2122.
27. Stephens, G. and Tsay, 1990: On the absorption anomaly. Q. J. R. Meteorol. Soc., **118**.
28. Twomey, S.A., M. Piepgras and T.L. Wolfe, 1984: An assessment of the impact of pollution on global cloud albedo, Tellus, **36B**, 356–366.
29. Ulaby, F.T., R.K. Moore and A.K. Fung, 1982: Microwave Remote Sensing. Addison-Wesley, Reading, Massachussets, p. 314.
30. Wolf, D.A. de and H.W.J. Russchenberg, 2000: Radar reflection from clouds: Gigahertz backscatter cross sections and Doppler spectra. IEEE Trans. Antenna Propagation, **48**, 254–259.
31. Young, S.A., 1995: Analysis of lidar backscatter profiles in optically thin clouds. Appl. Opt., **34**, 7019–7031.

9 Assimilation of Radar Data in Numerical Weather Prediction (NWP) Models

Bruce Macpherson,[1] Magnus Lindskog,[2] Véronique Ducrocq,[3] Mathieu Nuret,[3] Gregor Gregorič,[4] Andrea Rossa,[5] Günther Haase,[2] Iwan Holleman,[6] and Pier Paolo Alberoni[7]

[1] Met Office, London Road, Bracknell RG12 2SZ, UK
[2] Swedish Meteorological and Hydrological Institute,
   S-60176 Norrköping, Sweden
[3] Météo-France, CNRM/GMME, 42 avenue G. Coriolis,
   F-31057 Toulouse, France
[4] University of Ljubljana, Department of Physics, Jadranska 19,
   SI-1000 Ljubljana, Slovenia
[5] MétéoSwiss, Krähbühlstrasse 58, CH-8044, Zürich, Switzerland
[6] KNMI, Department of Satellite Data, Wilhelminalaan 10,
   NL-3730 AE De Bilt, Netherlands
[7] Arpa - Servizio Meteorologico Regionale, Viale Silvani 6,
   I-40122 Bologna, Italy

9.1 Introduction

Radar data have exciting potential for improving forecasts from operational numerical weather prediction (NWP) models. This potential, already partially realised, arises from a combination of developments. NWP models of the European National Meteorological Services (NMS) are now running routinely at the 10 km grid scale and in a few years will be moving to resolutions of the order of 2 km. Such high resolution models require correspondingly high resolution wind and moisture data for initialisation, which radar networks are well placed to deliver. Secondly, NWP data assimilation techniques have advanced considerably in the 1990s, with the arrival of techniques capable of extracting information from time sequences of observations only indirectly related to model prognostic variables. The first decade of the twenty-first century is likely to see further improvements in computing power, microphysical parametrisation and assimilation methods which will enable better exploitation of the information available from weather radars. Thirdly, developments in radar networking and processing around Europe are beginning to reach a maturity which makes feasible the routine operational delivery of quality controlled radar information of an accuracy sufficient for worthwhile NWP assimilation.

The COST-717 Action entitled "Use of Radar Observations in Hydrological and NWP models" (Rossa, 2000) has given a framework for European radar and NWP scientists, along with those from the hydrological community, to work together to realise this potential. COST-717 is divided into three working groups, the third of which is entitled "Using Radar Information for Assimilation into Atmospheric Models." The present chapter has arisen out of a review by members of Working Group 3, seeking to inform the radar and NWP communities about each other's disciplines and provide an accessible introduction to the literature for those working at the intersection of these fields.

Before reviewing the assimilation of radar data, it will be useful to outline some basic concepts and techniques in the field of NWP data assimilation, which is unfamiliar not only to radar scientists but also to a good number within the numerical modelling community.

9.2 Data Assimilation

9.2.1 Basic Concepts

A very helpful descriptive tutorial on atmospheric data assimilation for the non-specialist is found in Schlatter (2000), whose presentation has influenced this section. Data assimilation may be defined as the process of estimating meteorological conditions on a regular grid from two main sources of information: (1) observations from disparate sources, whenever and wherever taken, and (2) a numerical model, which incorporates through mathematical equations what is known about the atmosphere. The goal is to construct the best set of initial conditions, known as the analysis, from which to integrate the NWP model forward in time.

A fundamental concept is that of the *data assimilation cycle*. This means that if, for example, a 6-hour data assimilation cycle is used, then a 6-hour numerical model forecast is merged with observations during the assimilation procedure to produce an initial state for a new model integration. The 6-hour forecast launched from this new model state is itself merged 6 hours later with a new set of observations to produce a new initial state, and so on.

The available observations may be of various quantities related to the atmosphere, such as pressure, temperature, wind, humidity, cloudiness, satellite radiance and radar intensity. These need not be the same quantities carried as prognostic variables by the NWP model. All that is required for an observation to be suitable for assimilation is that it must be possible to estimate the observed quantity from variables carried by the model, via the so-called *forward model*, or *observation operator*. The success of this assimilation will, however, be influenced by the sophistication with which the model represents the physical processes giving rise to the measured quantity. For example, a model may have a simple diagnostic relationship to derive surface rainfall rate from prognostic variables, which could be used to assimilate observations of that quantity. But it may lack the detailed microphysics needed for a credible estimate of the reflectivity measured by a radar aloft, thus making direct assimilation of reflectivity impractical.

The observations are typically irregularly distributed in space and time, and the number of observations is in general small compared to the number of variables in the initial state of the forecast model. Observations alone are inadequate to initialise the model. This is why we need to introduce a priori information in the form of a short-range NWP model forecast. To combine the two kinds of information optimally, we need statistical knowledge about the errors of this short forecast and statistical knowledge about the observation errors. We may also exploit additional a priori information regarding so-called 'balance relationships' between (errors in) different forecast model variables. The best known example is probably the geostrophic relationship between wind and pressure gradient.

Given an observation at one location, assimilation schemes use statistical knowledge of the short-range forecast error structures and correlations between errors in different model variables to decide how to spread the information from that observation spatially (and temporally) and how to influence different model variables. These forecast error correlations and balances are known reasonably well and can be modelled empirically. The spatial patterns of forecast error correlations are usually referred to as *structure functions*. The assimilation is said to be univariate in the case where cross-correlations between different variables are zero, i.e. each variable is analyzed independently of the other variables. Multivariate assimilation is more general and in most cases, more physically justified. When the increments to different model variables are not in good multivariate balance, this can lead to an analysis which does not represent the atmosphere's slowly varying motions realistically. Instead, the model may respond by generating spurious high-frequency waves, which may also lead to some loss of observational information just added. Avoiding this by the construction of a well-balanced analysis is the *initialisation problem*. For a thorough grounding in these (and many other) aspects of data assimilation, along with its historical development, the reader is referred to Daley (1991).

Even with good knowledge of observation errors and forecast errors, an extra difficulty originates in the model's inability to model the physics behind the measurement perfectly and the limited spatial resolution of the model. These manifest themselves as another source of discrepancy between observations and model estimates termed the *error of representativeness*. For example, the temperature measurement inside a cumulonimbus cloud may be very accurate but quite unrepresentative of average conditions over a model grid-square of side, say 20 km, in a model which does not resolve thunderstorms explicitly. The error of representativeness is usually treated by increasing the assumed observation error for assimilation.

9.2.2 Techniques

Early data assimilation techniques were based on simple spatial interpolation. For example, polynomial fitting or distance weighting were used to adjust a 12-hour forecast locally to observed values in the vicinity of each model grid point. A first step toward more advanced methods was taken with the arrival of *statistical* interpolation or *optimum* interpolation. Essentially, this technique applies linear minimum variance estimation of analysis errors locally at each grid point. Simplified error covariance models, assuming, for example, horizontal homogeneity and isotropy, were applied for this assimilation technique. Three-dimensional multivariate statistical interpolation schemes dominated operational NWP during the period 1975–1995.

During this period, another popular and conceptually simple approach was developed, generally referred to as 'nudging.' It has been used extensively in research applications, but also in some operational systems. In this method, observations are 'nudged' into the model at each time step of the integration by adding an extra term to the prognostic equations, forcing the model toward the observations. The forcing term is proportional to the difference between the observed value and the model's estimate of the observed quantity, scaled by tunable 'nudging coefficients' which may depend on the relative error of the observation and the model estimate. One advantage of nudging is that the forecast model can adjust gradually to the observed information being added. This can help the model maintain a balanced state without generation of spurious high-frequency modes, thereby addressing the initialisation problem.

With the increasing variety of observations that are nonlinearly coupled to the forecast model variables, the linear spatial interpolation techniques were eventually regarded as unsatisfactory for optimal use of the data. A new generation of assimilation schemes has been based around the variational approach. At its roots, variational assimilation is an application of mathematical techniques going back to Gauss, but it is only in the last decade or so that practical operational assimilation systems have been developed in this framework, building on some foundational work in the mid-1980s. For more background, see Schlatter (2000) and references therein.

Variational assimilation provides a means of deriving the optimal analysis as a combination of observations and model forecast. A key concept is that of the *cost function* (or *penalty function*) which is a sum of terms, each measuring the 'distance' between the model state and the available information. One term measures the 'distance' or 'fit' to the observations, another the fit to the previous short-model forecast or *background field*. In each term, the distance is weighted by the inverse of the error covariance matrix for that information source. Minimisation of a cost function with this structure gives the model state with minimum expected analysis error. Additional terms can be added to the cost function, for example, to give constraints on gravity wave activity and alleviate the initialisation problem.

In three-dimensional variational assimilation (often known as *3D-Var*), the minimisation takes place at a single time, using observations valid at (or close to) a single time. In four-dimensional variational assimilation (*4D-Var*), observations distributed over a time window are analysed, and the minimisation consists of finding a model trajectory in phase space which best fits the available data. This can be very demanding computationally, relying as it does on an iterative technique and several integrations of the model before arriving at the minimum of the cost function. Since the analysis determined by 4D-Var is an actual space–time trajectory produced by the model, the initialisation problem is solved en route.

During the minimisation of the cost function in 3- or 4D-Var, the gradient of the cost function with respect to a change in its dependent variables must be estimated. Calculation of this gradient requires the *adjoint* of the forward model used to calculate model estimates of observed quantities. If we linearise the forward model and represent it as a matrix, then the adjoint is obtained from the transpose of that matrix. In 4D-Var, the forward model is the full prognostic model, and many papers refer to the associated *adjoint model*, whose derivation can be a complex exercise, particularly where physical parametrizations with on-off thresholds are concerned.

4D-Var is run operationally at the European Centre for Medium Range Weather Forecasts (ECMWF) at Météo-France and at the Japan Meteorological Agency. As we shall see later, it has also been used for a number of studies in radar data assimilation, where it promises to give a model state which best distributes the complex information present in, say, a reflectivity measurement amongst the various model variables: dynamical, thermodynamic and microphysical. 3D-Var assimilation of rainfall data is also possible, but has more limited potential than 4D-Var, since only through the time dimension can increments to thermodynamic and microphysical variables feed back into modifying the wind field which can sustain the desired precipitation structure within the model. Given its complexity and cost, however, 4D-Var is not the only current practical option for assimilation of radar data, and a variety of simpler techniques are still being applied.

9.3 Assimilation of Radar Precipitation Data in NWP Models

9.3.1 Introduction

The impetus for introducing precipitation data into NWP models came from tropical meteorology. In lower latitudes, there is a relative lack of conventional data from surface and radiosonde stations. Perhaps more influential was the fact that in the tropics there is no dynamical balance like that described by quasi-geostrophic theory in midlatitudes which links through to the moisture field via diagnosed vertical motions. Tropical dynamics are also known to be heavily impacted by latent heat release from convective precipitation. Tropical models tended to suffer from large errors in the analysed moisture field and consequently (since convection is a major forcing term) in the divergent wind field. There was therefore a rather slow equilibration over a few days of evaporation and precipitation on the global scale. Therefore, a procedure that could treat rainfall data held the prospect of improving tropical forecasts substantially. Rainfall estimates derived from satellite, rather than radar, were the focus for these studies. Techniques applied usually fell into the categories of 'physical initialisation' or 'latent heat nudging,' depending on the approach to dealing with the model's convective parametrisation. We review these methods separately.

In midlatitudes, there has, of course, been a need to improve precipitation forecasts, since mesoscale details are usually not represented well enough in the analysed fields. This is especially important in the first few hours of the forecast, when precipitation forecasts suffer from the spin-up problem, as the model dynamical and hydrological fields come into balance. Therefore, even a short-lived beneficial impact of precipitation assimilation on precipitation forecasts is very welcome for nowcasting and short-range forecasting. Researchers began to use radar data in a variety of assimilation systems and in models of increasing resolution. After looking at the achievements of physical initialisation and latent heat nudging, we survey work on variational assimilation, focussing finally on efforts by all means to assimilate radar data in meso-γ scale models. These are nonhydrostatic models with horizontal resolution ranging from 6 km to 500 m. Most previous studies have worked with surface rain rate estimates, but attempts to assimilate reflectivity directly are also covered.

9.3.2 Physical Initialisation

Physical initialisation (PI), as formulated in Krishnamurti et al. (1991), consists of two steps. First, there is a diagnostic calculation of surface fluxes and a humidity analysis consistent with observed precipitation rates. Secondly, a nudging of the diagnosed fluxes into the model takes place during a preintegration phase. The key concept behind the humidity analysis is an inversion of the planetary boundary layer and convective parameterization schemes so that diagnosed precipitation rates correspond to observed ones. First, fluxes of sensible and latent heat are calculated that correspond to measured precipitation rates. Then 'reverse similarity theory' is used to obtain temperature and humidity above the constant flux layer. Finally, a reverse convective parameterization scheme yields the vertical profile of humidity. PI has been found to reduce spin-up substantially and improve precipitation forecasts (Donner, 1988, Krishnamurti et al., 1993 and Treadon, 1996).

An assimilation method based partly on PI was developed for the Japan spectral model using precipitation observations from a radar–rain-gauge network over Japan (Aonashi, 1993; Matsumura et al., 1997). Within this scheme, a PI method calibrates the thermodynamic and dynamic variables of the objective analysis data in such a way that model precipitation rates agree with observations. Afterward, a nonlinear normal mode initialization (Daley, 1991) is conducted. The combination of PI and NMI makes diabatic heating in the analysis consistent with that produced in the model forecast. Experiments showed that the method reduces both spin-up and position errors in precipitation forecasts.

More recently developed is a kind of PI (Gregoric, 2001) where diagnostics from radar data are used to replace some closure assumptions in the convective parametrization. The locations and dimensions of convective cells are identified from the radar data and stored in lookup tables accessed by the convection scheme. Convection is only triggered in the model, where the radar confirms the existence of convective precipitation.

9.3.3 Latent Heat Nudging

Latent heat nudging (LHN) is a method of forcing an NWP model toward observed precipitation rates. It is based on the observation that since relatively little moisture is stored in clouds, the column integrated latent heating rate must be approximately proportional to the precipitation rate. The principle is to correct the model's latent heating at each time-step by an amount calculated from the difference between observed and model estimated precipitation. This extra heating then acts as a source term in the thermodynamic equation, which in turn brings about an adjustment in the model vertical velocity field that takes the model precipitation rate closer to that observed. LHN is simpler than PI in that it does not seek to intervene directly within the physical parametrisations to bring about the adjustment.

Since usually only the surface rainfall rate is known or available, one has to specify the vertical structure of applied heating. Vertical radar reflectivity profiles, however, increasingly are used for correcting rain rates at ground, as presented by Germann and Joss and Koistinen et al. in this book. So, if available in proper quality, they will give vertical structures. Two possibilities have been tried to date: idealised profiles may be constructed, or else a scaled model profile may be selected as in Manobianco et al. (1994) and Jones and Macpherson (1997), assuming that the model has the correct structure despite an incorrect intensity. This has some advantages: it ensures consistency with the model's parameterisations, and it allows evolution of the profile with time during the integration. It is assumed that the model's separation of explicit (stratiform) and implicit (convective) precipitation is correct. If one does not trust the model separation, it is possible to treat the convective part separately through closure assumptions in the parameterisation scheme.

A third option to construct heating profiles could be based on the work in Cartwright and Ray (1999), where an algorithm is proposed within a cloud resolving model to relate latent heating profiles to model reflectivities. The same algorithm could then be used to derive 'observed heating' from observed three-dimensional reflectivities from radar. There are plans to test this in an LHN context at Météo-Swiss.

The LHN algorithm runs into difficulty in the case when the model point is dry, yet precipitation is observed. In this case, the algorithm can include a search for nearby points where the model is raining. Heating profiles from there can then be applied at the point where it is desired to introduce rain, but this refinement has limited success. This situation of introducing rain absent in the model is a great challenge for precipitation assimilation.

LHN has been found beneficial in midlatitudes. Some authors report dramatic impact on quality of precipitation forecasts (Wang and Warner, 1988). Others working with radar assimilation (operational since 1996) in the mesoscale model at the Met Office in the UK have found smaller, yet noticeable positive impact in general, with occasional larger benefits (Macpherson, 2001), of which Fig. 9.1 is the best example. Recent results with the Met Office mesoscale model indicate that these larger benefits for short-period (6–12-hour) rainfall forecasts originate

262 B. Macpherson et al.

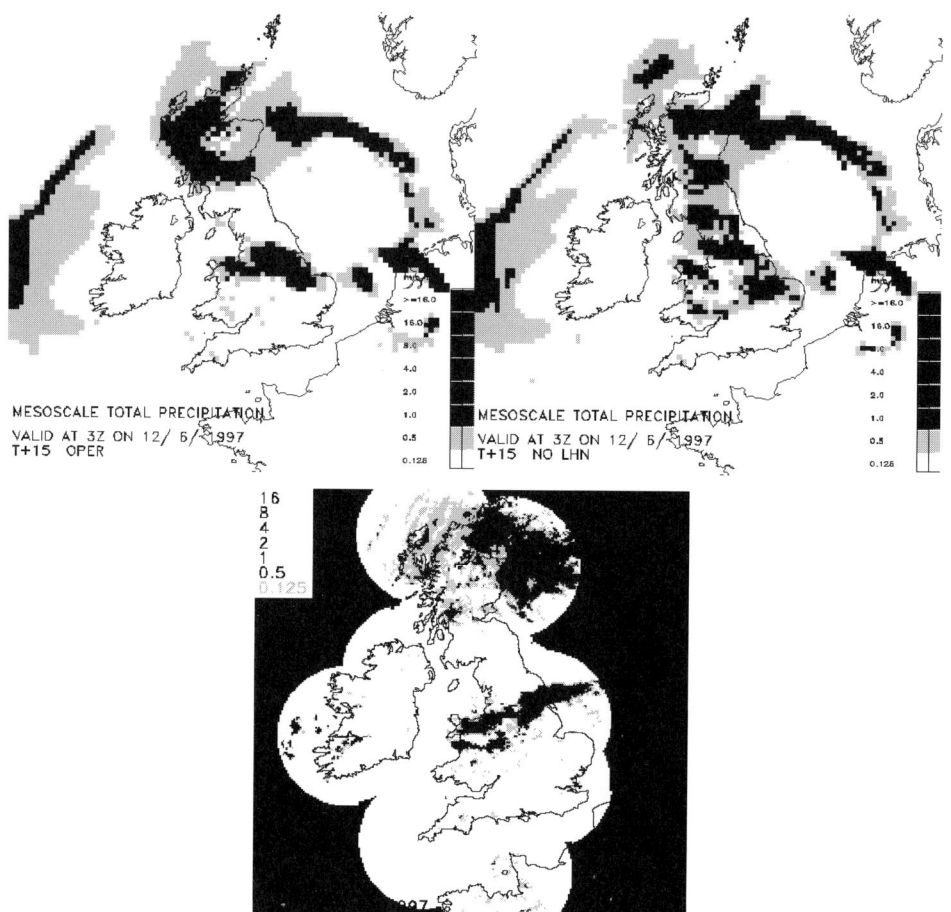

Fig. 9.1. An unusually long lasting forecast impact of radar rain rate assimilation in the Met Office (UK) mesoscale model, taken from Macpherson (2001). Bottom frame shows a radar picture for 03 UTC, 12 June 1997, with a northward moving rain band over central England. Top left frame shows an operational precipitation rate forecast at $t+15$ hours. Top right frame is from an experiment with NO assimilation of radar data. Darker tones imply heavier rain rates. This figure also appears as color Fig. 22 on page 331

just as frequently from radar data, as from other observation types such as aircraft, radiosonde and surface data.

A modified version of the LHN algorithm, including elements of physical initialisation, but still based on nudging, was implemented operationally in 2001 in the ETA model (grid length 22 km) at the National Centers for Environmental Prediction (NCEP) in the USA. The assimilation of observed precipitation is performed with a form of nudging that is applied during the 3-hour forecasts that occur between successive analyses in the data assimilation cycle. At

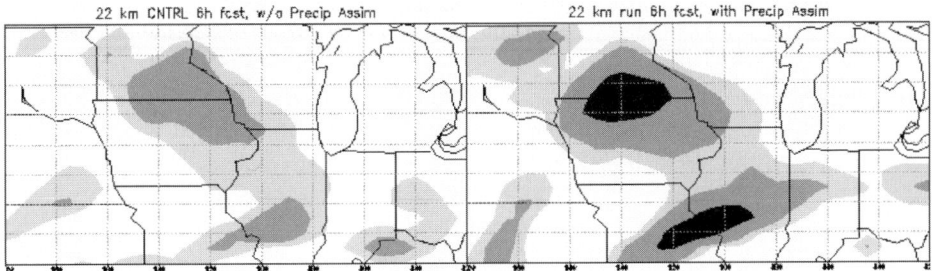

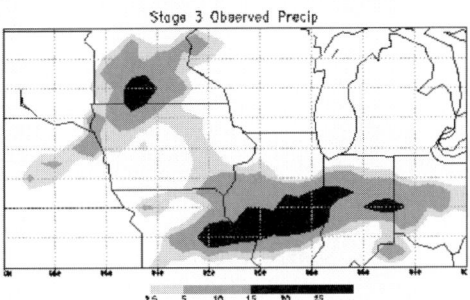

Fig. 9.2. Impact of precipitation assimilation on the 6-hour precipitation forecast from the NCEP ETA model for a case with data time 00 UTC, 11 April 2001, taken from Lin et al. (2001). Upper panels show accumulated precipitation in the first 6 hours of model forecast, on the left from a run without precipitation assimilation, and on the right with precipitation assimilation included. The lower panel shows observed accumulation from a precipitation analysis based on radar data and rain gauges. Darker tones imply larger rainfall amounts

each time step, the observed and forecast precipitation are compared; then the model's temperature, moisture, cloud and precipitation are mutually adjusted to produce a forecast value closer to the observed value. The observed values come from hourly precipitation analyses derived from 2500 automatic reporting gauges and hourly precipitation estimates from radars. This technique produces initial precipitation rates that match observations and also ensures that the soil moisture (coupled to the model precipitation through the land-surface physics package) evolves with the observed precipitation. The soil moisture, in turn, is very important in how the boundary layer evolves in the forecast. Some noticeable improvement in the first 6 hours of the precipitation forecast (Fig. 9.2) is reported, along with a modest overall improvement in longer term (24 hour) precipitation forecasts. For a fuller account, see Lin et al. (2001).

9.3.4 Variational Assimilation

Some pioneering studies (Zupanski and Mesinger, 1995; Zou and Kuo, 1996) gave early indication of the likely value of 4D-Var for assimilation of precipitation data. In a case study of strong convection with the NCAR/Penn State MM5 4D-Var system (Guo et al., 2000), the impact of various data types on the simulation was compared. Hourly surface precipitation data (not from radar) were important in preserving the precipitation structure of the squall line, although wind profile data had a more beneficial impact on precipitation accuracy.

On a finer scale, the adjoint of a cloud resolving model has been developed and applied to the assimilation of three-dimensional radar data (Sun and Crook, 1997, 1998) in a 4D-Var system that can be used to assimilate data from one or more Doppler radars. The horizontal resolution of the cloud scale model was 500 metres, and only the warm microphysical processes were parameterized. The thermodynamical and microphysical fields, as well as the three-dimensional wind field, were determined by minimizing a cost function defined by the difference between the observed radial velocities and the reflectivities (or rainwater mixing ratio) and their model counterparts. The derivation of the adjoint of the physical processes with on/off switches follows that in Zou et al. (1993), and the microphysical scheme was modified for the evaporation of rain and the rainwater fall velocity. It was found that assimilating the rainwater mixing ratio obtained from the reflectivity data resulted in a better performance of the retrieval procedure than directly assimilating the reflectivity. Differential reflectivity data were also used to produce a better estimate of the rainwater mixing ratio and hence, to improve the microphysical retrieval.

The same variational retrieval technique of Sun and Crook (1997, 1998) was applied in a flash flood case simulation with a cloud scale model, described briefly in Wilson et al. (1998). The numerical forecasts were found to significantly improve over persistence and extrapolation in the 60-minute time frame. Afterward, in Wu et al. (2000), the application of Sun and Crook (1997, 1998) was extended to convective storms where the ice phase plays an important role. As a complete ice microphysics parameterisation will have a complex adjoint model with poor convergence properties for the minimisation due to many nonlinearities, a simplified cold microphysical scheme was developed. This had no snow category and only one category for the non-precipitating species (cloud water and cloud ice). The differential reflectivity was used to discriminate between the rain and hail and allowed to employ phase-dependent $Z-M$ relationships. The results of this work were mixed: although the analysis system was able to retrieve all the main features of the storm, the simulations were unable to reproduce the evolution of the observed storm – the simple microphysical parameterisation was unable to follow the actual cloud physics.

On the global scale, an initial approach to the complexity of variational assimilation of precipitation data has been reported in Marecal and Mahfouf (2000). The authors assimilated data from the Tropical Rainfall Measurement Mission (TRMM) into the ECMWF global model. The data were rain rate estimates from the TMI microwave imaging radiometer which had been 'calibrated' by the TRMM precipitation radar. The distinctive feature of this study was a two-step assimilation: first, an initial 1D-Var retrieval of temperature and humidity profiles from the rainfall data, then a 4D-Var assimilation of the total column

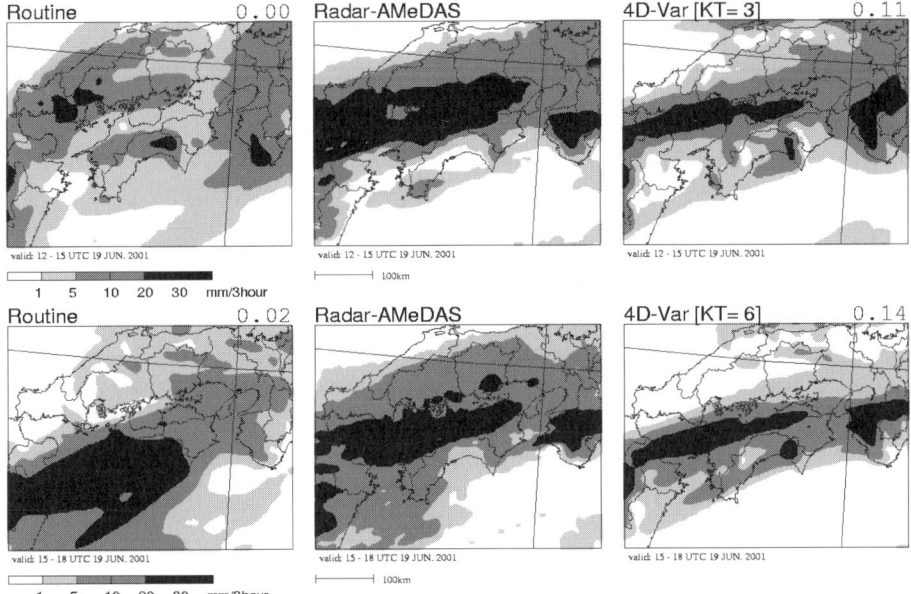

Fig. 9.3. Forecasts and observations of 3-hour precipitation amount. Centre column shows the Radar-AMeDAS observations of the Japan Meteorological Agency, right column shows forecasts starting from a 4D-Var analysis and left column shows the forecasts from a 'routine' analysis using PI for assimilation of rainfall data. Darker tones imply larger rainfall amounts. Initial time of forecasts is 12-UTC 19 June 2001. Top row is for forecast period 0–3 hours; bottom row is for period 3–6 hours. Taken from Ishikawa (2002). This figure also appears as color Fig. 23 on page 332

water vapour (TCWV) produced by the 1D-Var. TCWV is easier to assimilate within 4D-Var than rainfall rate itself. This study found a positive impact on the humidity and wind analysis, though little forecast impact. As with LHN, it relies for success on the presence of some rain in the model background field.

At the Japan Meteorological Agency, a 4D-Var mesoscale system (Ishikawa, 2002) has been implemented operationally, including assimilation of hourly precipitation amounts from a synthesis of radar and gauge data. The 4D-Var scheme has replaced a system with Optimum Interpolation for conventional data and a PI approach to rainfall assimilation. An example of the difference in impact on the 0–6 hour forecast period is given in Fig. 9.3.

In the USA, there have been tests of a 3D-Var rainfall assimilation scheme for TRMM TMI data in the NCEP global model. There was some improvement in the fit to pressure data with rainfall assimilation, and the impact on other fields was neutral; but the fit to other moisture data was degraded, perhaps because of some problem with the precipitation physics. Forecast impact was largely neutral, though an improvement in the track of a hurricane was noted.

A case study of 4D-Var precipitation assimilation in the NCEP ETA model was reported by Zupanski et al. (2002). The situation was a major snowstorm affecting the U.S. East Coast. Hourly rainfall observations (a combination of radar and gauge data) were found to have a beneficial impact as far as 36 hours into the forecast. This impact could be detected in the upper tropospheric wind field as well as the surface rainfall.

The basic problem for a variational approach to precipitation assimilation is the discontinuous ("on-off") nature of the precipitation physics schemes. In the case of precipitation observations, highly nonlinear parameterisation schemes must be linearised for development of the adjoint version of the model required by the minimisation procedure for the cost function. Alternatively, simpler physics schemes can be coded for the linear model, which are not a direct linearisation of the full model physics (i.e. the linear model is not *tangent linear* to the full model). It is not clear what may be lost in this simplification. Technical issues for the minimisation of the cost function arising from these characteristics of the physics are discussed by Bao and Kuo (1995), Xu (1996) and Zou (1997).

9.3.5 Meso–γ Scale Models

On the cloud-resolving (meso-γ) scale, the first assimilation procedure using radar data was developed in Lin et al. (1993), using the Colorado State University Regional Atmospheric Modeling System (RAMS). The numerical simulation was initialised with three-dimensional dynamical, thermodynamical and microphysical fields derived from multiple Doppler radar observations. The horizontal resolution of the simulation was 2 km, and only warm microphysical processes were parameterised in the numerical model. The procedure initialised the wind field with Doppler wind data and filled the data void region in order to provide a smooth transition from the observational domain to the background state. The pressure and potential temperature perturbations were obtained from a thermodynamic retrieval method following Hane and Ray (1985). The water vapour content was set to a saturated value where the radar had detected precipitation and above the lifting condensation level. The rainwater content was derived from the reflectivities by using a Z–M relationship, whereas the cloud water was assumed to be zero. After feasibility of the initialisation method was established with simulated storm data, it was tested with multiple Doppler radar observations from a tornadic storm. The very short range prediction (less than 15 minute) showed good agreement with the observations, although the modelled storm seemed to evolve faster than the observed storm.

The same approach of moisture and microphysical adjustments has been developed in Xue et al. (1998), Bielli and Roux (1999), Haase (2002), Ducrocq et al. (2000) and to some extent in Zhang (1999). All of these experiments except Haase (2002) used a cold microphysical scheme.

In Xue et al. (1998), reflectivity data were used to deduce the initial cloud water content and to moisten the initial state. A distinctive feature of this work is that the adjustments to the water vapour and cloud water fields were applied during an intermittent data assimilation period: in their experiments, the reflectivity and the radial velocity were assimilated at 15-minute intervals during the last hour of the assimilation period. They found that the assimilation of radar reflectivity had a large positive impact on the simulation of a squall line case.

In Bielli and Roux (1999), the production rate of precipitation was used to modulate the adjustment of water vapour content in the observed precipitation areas: the relative humidity was assumed to be 100%, except where the production rate of precipitation was negative. In some of the experiments, the cloud water content was set empirically inside the precipitation areas and also where the production rate of precipitation was positive. The Doppler-derived three-dimensional wind fields were also used to initialise the model. The results obtained from simulations of a tropical mesoscale convective system data set have shown that it was important to describe, even crudely, the saturated and unsaturated areas in connection with the updraft and the rear-to-front flow described by the Doppler winds. Initialising cloud water contents did not bring significant improvements. In the initialisation method of Haase (2002), radar reflectivities were used to modify the vertical wind as well as the specific humidity and temperature profiles.

The initialisation method in Ducrocq et al. (2000) is also based on cloud and precipitation analyses, but adapted to the French networks. The reflectivities are available only on a single PPI (Each PPI – Plan Position Indicator – is taken at a single, fixed elevation angle, and thus forms a cone of coverage in space). These were used to determine the rainy areas and to impose cloud where the reflectivities were greater than a given threshold. The vapour mixing ratio was set to saturation in cloudy regions, and a Z–M relationship was used with an empirical vertical distribution to initialise the rainwater mixing ratio. In some experiments, a constant cloud water value was imposed. The initialisation scheme has been applied to simulation of a real case of a convective system. It gave a large impact on the results : use of radar and satellite observations allowed the model to trigger the convection which is not the case when simulations start from a conventional large scale analysis. These results have been confirmed on another convective case also over flat areas. Inserting cloud water was found to have no significant impact.

A cloud analysis system was applied in Zhang (1999) to synthesize several data sources (radar and satellite observations) and to construct a three-dimensional cloud analysis. This system, called ADAS (Brewster, 1996; Zhang et al., 1998), is based on the LAPS cloud analysis (Albers et al., 1996) with several modifications. The three-dimensional radar reflectivities are used to impose clouds and to determine the hydrometeor type and mixing ratio. If the reflectivity exceeds a threshold, clouds are inserted in the radar echo region. The type of hydrometeor is determined from the wet bulb potential temperature, and hail is diagnosed when the three-dimensional radar reflectivity is above a given threshold. Then, the mixing ratio of the hydrometeors is derived from hydrometeor type-dependent Z–M relationships. The outputs of the cloud analysis system are then provided to a moisture and diabatic initialisation scheme. In this, the cloud water and ice mixing ratio are simply set to the analysed values. The rainwater, snow and hail mixing ratios are usually initialised to smaller values than the analysed values, as inserting the total amount of precipitate could prohibit, by its drag, the development of updrafts. Then, the thermal field is adjusted to account for the latent heating associated with the inserted cloud water. The relative humidity field is also modified, and cloudy regions are moistened. The impact of the initialisation on the meso-γ scale numerical prediction has been validated on simulated storm data.

So, to summarise the assimilation of radar reflectivity in cloud resolving models, reflectivity is always used in an indirect way to modify the moisture fields. In some studies, the reflectivity data are also used for initialisation, via Z–M relationships, of the contents of the nonprecipitating and/or precipitating species.

9.3.6 Summary

The three main approaches to assimilation of precipitation data have been reviewed – physical initialisation, latent heat nudging and variational assimilation. At this stage, there is no clearly preferred technique for assimilation of radar rainfall data on all scales for all weather systems. The most natural one seems to be 4D-Var, but this still faces various technical and scientific challenges before reliable and computationally viable schemes can be produced for widespread operational use.

9.4 Assimilation of Radar Wind Data

9.4.1 Introduction

The growth in high resolution limited area NWP models has brought a greater interest recently in the assimilation of Doppler radar wind data. The models require observations with high spatial and temporal resolution for determining the initial conditions, for which purpose radar data, such as presented by Gekat et al. (2003) and Koistinen et al. (2003), both this book, are particularly appealing. Wind is also a primary prognostic variable of the models.

Radar wind data, fully or partially preprocessed, have been assimilated into a number of atmospheric models, over a wide range of spatial resolutions. A lot of different assimilation techniques have been tried, which we review in approximate order of sophistication. Most of the work described here has so far been applied in research mode, but radar winds are also assimilated operationally. It should also be noted that some of the relevant studies on Doppler wind assimilation have already been discussed under the heading of precipitation assimilation on the meso–γ scale.

9.4.2 The Successive Correction Method

At NOAA's Forecast Systems Laboratory (FSL), the Local Analysis and Prediction System (LAPS) has been developed (McGinley, 1989). The analysis is performed on a resolution of approximately 10 km horizontally and 50 hPa vertically. The observation residuals are spread vertically using simple vertical structure functions. The analysis is then performed level by level and is based on the successive correction method, an early and more empirical forerunner of the 'optimum interpolation' technique. The analysis system uses information from various data sources, and radar wind data play a key role, being subject to some special treatment (Albers, 1995). Observational errors are assumed to be uncorrelated, and the background error structure functions are modelled.

The LAPS analysis system has also been used at Servizio Meteorologico Regionale in Bologna, Italy. Applied to Doppler radial wind data, it reportedly refines the wind analysis (Alberoni et al., 2000).

9.4.3 Optimum Interpolation

Radar wind information is assimilated operationally in the form of VAD (Velocity Azimuth Display) profiles within the multivariate Optimum Interpolation (OI) scheme used in the Rapid Update Cycle (RUC) atmospheric prediction system at NCEP in the USA. The RUC atmospheric prediction system is the operational version at NCEP of the Mesoscale Analysis and Prediction System (MAPS) developed at NOAA's Forecast Systems Laboratory (FSL). The RUC system is applied on a limited area with an intermittent data assimilation cycle.

The OI scheme of the RUC system (Benjamin et al., 1991) uses a hybrid vertical coordinate system that combines a terrain following coordinate system near the ground with a potential temperature coordinate system above. The analysis variables are the horizontal wind components, the Montgomery stream function and the condensation pressure. Due to the choice of vertical coordinate system and analysis variables, some preprocessing of observations is needed. The VAD wind profiles are treated as standard PILOT balloon observations. The observation errors are assumed to be uncorrelated. Multivariate structure functions are used for the background error correlations.

9.4.4 Methods Employing Thermodynamic Retrievals

The study of Liou (1990) employed an assimilation procedure which combined the thermodynamic retrieval technique (Gal-Chen, 1978) with so-called 'direct insertion' and wind adjustment methods. Simulated radial wind data were nudged into a dry version of the nonhydrostatic, fully elastic RAMS model. The fields were then adjusted variationally to fulfill the equation of continuity and some other constraints, and finally, thermodynamic fields were retrieved from the wind field.

The method was demonstrated through identical twin experiments. First, a control run was conducted to generate a time series of the east–west component of the wind, which served as simulated observational data. The control run was started from a perturbed initial state for the potential temperature. During the model integration, the perturbation developed to a thermal bubble which rose toward the upper boundary. A number of additional runs were performed from different initial states, without the initial perturbations in the potential temperature field. Pseudo-observations from the control run were then nudged into the integrations with different frequencies. It was shown that the potential temperature perturbation could not be completely recovered from wind data only. In addition, some detailed potential temperature data were needed. The success of the assimilation was also related to the insertion frequency of the observations. Assimilation of real data from Doppler radars was among the suggestions for future work.

9.4.5 4D-Var

VAD wind profiles can be assimilated by either OI or variational methods. Variational methods, however, are also ideally suited for assimilation of Doppler radar radial winds. The observation operator projects the model wind along the radar beam direction, and only a partial preprocessing of the data is needed.

An early attempt at using 4D-Var for assimilation of radar radial winds into an atmospheric model was described in Wolfsberg (1987). Unfortunately the assimilation encountered severe convergence problems. Some years later, further attempts at 4D-Var assimilation of radar wind data were reported in Kapitza (1991). Simulated radar radial wind observations were assimilated into a dry nonhydrostatic mesoscale model. The 4D-Var formulation used in this assimilation contained no background a priori information. A series of 'identical twin' experiments with the 4D-Var system were performed. In these, a reference model integration was started from an atmosphere at rest, but including a sub-area with a temperature excess of 1 K. The temperature excess caused a hot bubble of air rising through the initially neutrally stratified model atmosphere, as the integration proceded. During the first 200 seconds, all dependent variables were sampled at each time step and for all grid points. These data then served as observations in the 'twin' experimental runs. In one experiment, only the east-west wind component from the reference run was assimilated. This case represented a situation when radar radial winds were the only source of data. The assimilation only approximately managed to recover the thermal structure of the hot rising bubble. Significant improvements were achieved if temperature data from the reference run were also assimilated.

As mentioned in the discussion of 4D-Var precipitation assimilation, a four-dimensional variational Doppler radar analysis system (VDRAS), as in Sun and Crook (1997, 1998), has been developed to assimilate radial winds and reflectivities from single or multiple Doppler radars. Before assimilation, the data are interpolated from the original polar geometry to a Cartesian grid. The observation error correlations are neglected, and a relatively simple model is used for the background error covariances. The assimilation system employs univariate horizontal structure functions for the background error correlations. The VDRAS system has been applied to both simulated and real data. The application of the system to different stages of a convective storm demonstrated that the detailed structure of wind, thermodynamics and microphysics could be obtained with reasonable accuracy.

So far, all 4D-Var schemes that have been used to assimilate radial wind data suffer from the lack of a multivariate formulation of the structure functions. These ensure that the radial wind observations modify also the thermodynamic fields. Instead, in the 4D-Var schemes just described, the model equations and the time dimension are used to obtain the three-dimensional wind, as well as thermodynamic fields.

9.4.6 3D-Var

3D-Var has been used operationally from 1997–1999 in the NCEP ETA forecasting system in the USA to assimilate radar wind information in the form of VAD profiles. In this scheme, as in the OI scheme of the RUC system, the VAD profiles were treated as standard PILOT balloon observations. The data were later withdrawn pending improvements in quality control and readmitted to the operational system in March 2000. Also in the USA, VAD profiles and radial winds are assimilated into the 3D-Var version of the 'MAPS' system at FSL. VAD wind profiles are also assimilated into the operational models at the Met Office in the UK.

Radial winds in the form of radial wind 'superobservations' have been assimilated into research versions of the NCEP ETA forecasting system (Parrish and Purser, 1998), and the 3D-Var scheme of the HIgh Resolution Limited Area Model (HIRLAM) forecasting system in Sweden (Lindskog et al., 2000). In both these systems, the radial wind raw observations are spatially averaged to form 'superobservations' that are representative of the characteristic scale of the model. The calculation of the model counterpart of the radial wind superobservation involves a relatively simple projection of the horizontal wind along the radar beam line.

An interesting comparison was made with the HIRLAM system of the relative impact from assimilating VAD profiles and radial wind superobservations in the same model (Lindskog et al., 2002), from which Fig. 9.4 is taken. It shows that for almost all 24-hour forecasts in the ten day period, the scores of the runs using radar wind information in either form are superior to the ones not using radar wind data. Both forms of wind information were found to give similar benefits in the model forecast. This is encouraging for the potential benefits in exchanging VAD profiles within Europe for assimilation into the current generation of regional models, with resolutions of 10–20 km. The operational exchange of wind data is under discussion, see Gekat et al. (2003), this book and http://www.chmi.cz/OPERA/. It would seem likely, however, that radial wind data will be favoured in higher resolution models on the scale of a few kilometres or less.

The 3D-Var systems described above, as well as the OI scheme of the RUC system at NCEP, include multivariate structure functions, which ensure that

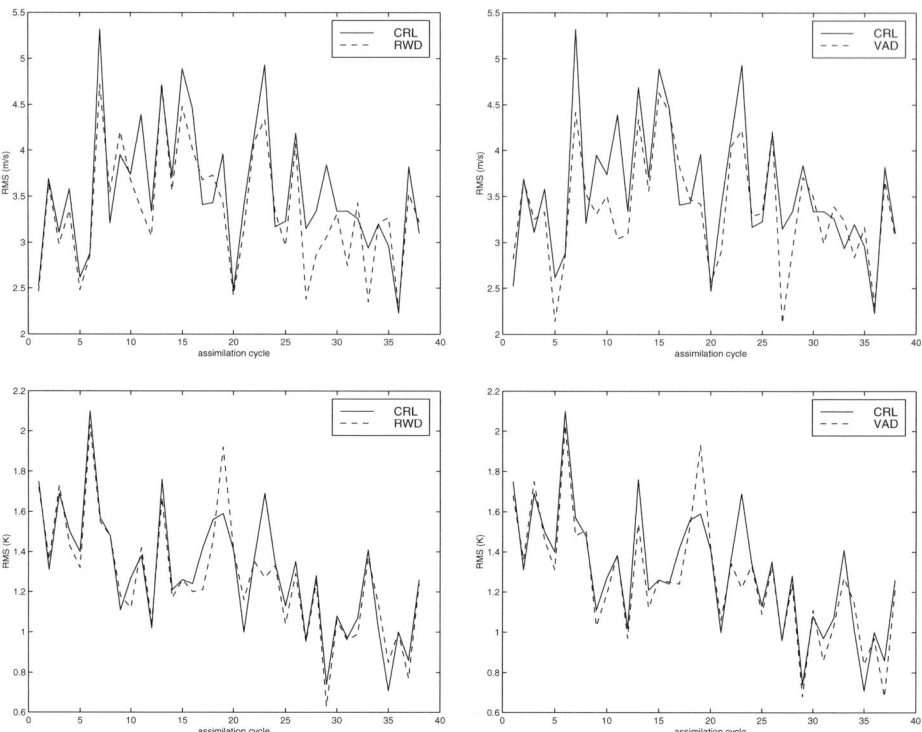

Fig. 9.4. Day-to-day variability of the 24-hour forecast 850 hPa wind speed (*upper row*) and temperature (*lower row*) rms scores for a run utilizing conventional observations only (CRL), for a run utilizing conventional observations and radar radial wind superobservations (RWD), and for a run utilizing conventional observations and radar VAD winds (VAD). The units are m/s and K, respectively. Taken from Lindskog et al. (2002)

the observed radar wind observations affect also the thermodynamic fields. The observation errors of the radar winds are assumed to be uncorrelated.

9.4.7 Summary

A variety of methods for assimilation of radar wind data into atmospheric models have been tested. One of the main differences between the various approaches is the way in which the thermodynamic fields are adjusted to agree with the analyzed wind field. The OI and 3D-Var schemes presented here use multivariate structure functions, which only permit linear relations between errors in the different model variables. In the methods based on nudging and 4D-Var, the governing equations of the model and the time dimension are used to ensure that the thermodynamics are consistent with the wind field. Nonlinear relations as well as nonhydrostatic processes may be implicitly included in 4D-Var and nudging approaches.

9.5 Quality Control of Radar Data for NWP

9.5.1 Precipitation Data

Assimilation in NWP models of surface rain rate estimates derived from radar is potentially hindered by all the well-known sources of error in these estimates, as presented and discussed in full length throughout this book. Accurate measurements of on-ground rainfall are troubled by measurement errors in the reflectivity values, assumptions about drop size distributions, clutter, anomalous propagation, attenuation and uncertainties introduced in the translation of reflectivity values to on-ground rainfall rates. These include range effects and various sources of gradients in the vertical reflectivity profile such as bright band and orographic enhancement.

In the Met Office Mesoscale Model in the UK, the system reported in Macpherson (2001) assimilates surface rain rate estimates produced by the Nimrod nowcasting system (Golding, 1998). Nimrod takes data from a network of 15 C-band radars around the UK and processes it with the help of a wide variety of other meteorological information, such as mesoscale NWP output, satellite imagery and rain gauge data. From analyses of rain rate, forecasts out to 6 hours ahead are generated by blending extrapolation techniques with NWP model forecasts. The Nimrod system has also been developed to analyse and predict variables such as precipitation type, cloud cover, snow probability, visibility and, most recently, soil moisture. Radar data have a rôle to play in helping to analyse several of these other variables.

Radar processing within Nimrod addresses a number of the above mentioned error sources. Further detail on the corrections applied for these effects is given in Harrison et al. (2000). Typical root-mean-square differences between radar estimates and rain gauges are of the order of a factor of two. The authors report that quality control and correction can reduce these differences by 30%, on average. Representativeness errors make up a good part of the residual.

It is worth noting that the impact in an NWP model of a given gross error in the input radar data will depend on the details of the assimilation scheme and the meteorological situation. For example, a spurious observation of rain in a high pressure system due to anomalous propagation may excite a temporary response in an NWP model through assimilation via latent heat nudging (LHN). However, the dynamical structures in the model will tend not to support rainfall in anticyclonic conditions and will be only weakly and briefly distorted by the incorrect data. A 4D-Var scheme, on the other hand, with stronger feedback from the observation to the model dynamics than with LHN, may develop a more intense and longer lasting evolutionary error in response to the bad data.

9.5.2 Wind Data

Radar wind data also give rise to quality control issues of concern to NWP assimilation. Radial winds can be aliased due to the limited maximum range and velocity measured by a Doppler radar. There are several algorithms available to dealias measured radial winds. The aliasing problems can, however, be overcome to a large extent by using different measurement techniques, like dual PRF (Pulse Repetition Frequency) or staggered PRT (Pulse Repetition Time). Radial wind data can be processed to obtain wind profiles at the radar site using a VAD-like or Volume Velocity Processing (VVP) algorithm, as described in Gekat et al. (2003), this book. Representativeness of the winds obtained from Doppler radars, for both the radial winds and the wind profiles, is the major problem when using these data. Problems can also originate in the signal from migrating birds, which leads to inconsistency in the wind data. Algorithms have been developed as part of the NWP processing at NCEP in the USA to control quality of radar wind profiles for a variety of errors, including bird contamination (Collins, 2001), see also Gekat et al. (2003), this book.

9.6 Treatment of Radar Data Errors in Assimilation

An extra challenge when merging observations with an NWP model arises when the measurements are associated with complicated error structures. A reflectivity or radial wind field measured by radar certainly comes in this category. To assimilate data with spatially correlated observation errors, the assimilation scheme should, in addition to structure functions for the background errors, include structure functions for observational errors. The first challenge is to accurately describe the observational error correlations. The second is that most assimilation schemes are not designed to handle these observational error correlations easily, although it is theoretically possible to do so. These difficulties have generally been circumvented in data assimilation by applying data selection and preprocessing algorithms that are assumed to remove the observational error correlations or else by giving reduced weight to information from data sources with correlated errors.

Some results in this area have been presented by Lin et al. (2000), coming from a number of 'identical twin' experiments, using a version of the VDRAS-system of Sun and Crook (1997, 1998). It was observed that the velocity field obtained with the applied smoothness constraint was insensitive to spatially correlated errors. More research is needed.

Assimilation schemes working with surface rain rate have usually included some simple treatment of the estimated errors in rainfall rate derived from the radar processing and extrapolation to ground level. In the Met Office Mesoscale Model in the UK, the initial version of Jones and Macpherson (1997) assumed that rain rate estimates were of uniformly high quality within 100 km of the radar site and of decreasing quality out to maximum range. The assumed radar errors were used to weight the combination of model forecast and observed rates to produce the analysed rain rate used as the target for nudging. Later, a more sophisticated quality measure was introduced (Macpherson, 2000) which takes account of lower accuracy in derived surface rain rate when the beam is above the freezing level. The assimilation was found, however, to be rather insensitive to this upgraded error formulation.

In the assimilation of TRMM data into the ECMWF model (Marecal and Mahfouf, 2000), tests were conducted on the sensitivity to the assumed rain rate error of the TRMM TMI data. When it was increased from 25% to 50% of the observed rain rate, the assimilation showed only a weak response.

In working to improve quality control of radar data for assimilation, one should bear in mind that the assimilation system itself can be an effective weapon in detecting errors. This is particularly true for the wind field on larger scales, where a variety of other good data give a good quality analysis independent of the radar data. Regular monitoring of differences between the model and radar VAD profiles on a monthly timescale can detect systematic errors in radar processing which may be missed in a single analysis. These can then be reported back to the data producer for investigation. An example is given in Rinne and Fortelius (2001) of occasional large random errors in radar processing which led to an average difference between model and radar profiles of 2 m/s at the radar site. Of course, this kind of monitoring is harder to do on smaller scales where the model analysis is more suspect, and is unlikely to be possible for precipitation data until models improve substantially.

The errors in the forward model required for variational assimilation of radar data will also need to come under closer scrutiny in future high resolution systems. An example of a tool which could help with this is the Radar Simulation Model (RSM) developed at the University of Bonn (Meetschen et al., 2000). This allows quite a detailed simulation of the reflectivity field that would be 'seen' by placing a radar at a given location in the model atmosphere. By varying parameters of the RSM, one could build up a picture of the sensitivity of the forward model errors to, for example, details of the microphysics and aspects of the synoptic situation.

All the above studies show that the treatment of observation errors in assimilation of radar data is at a relatively immature stage. One should not conclude too soon that specification of the error characteristics (beyond removal of gross errors) is unimportant. It is acknowledged that, as assimilation schemes improve, so they become more vulnerable to bad data. This suggests that the next generation of 4D-Var schemes for reflectivity and radial wind assimilation will become more demanding of effective quality control to remove gross errors in radar data, together with a realistic appraisal of the magnitude and structure of the residual errors in 'good' data.

9.7 Future Prospects

Building on the work reviewed above, the next few years promise to be stimulating for the development of radar data assimilation and the international exchange of radar data for NWP. Progress will depend on good cooperation between radar and NWP scientists. Those nations and institutes which strengthen such links will be better placed to capitalise on the promise of the next generation NWP systems to deliver improved short-period weather forecasts.

References

1. Alberoni, P.P., P. Mezzasalma, S. Costa, T. Paccagnella, P. Patruno, and D. Cesari, 2000: Doppler radar wind data assimilation in mesoscale analysis. Phys. Chem. Earth (B), **25**, 1263–1266.
2. Albers, S.C., 1995: The LAPS wind analysis. Weather and Forecasting, **10**, 342–352.
3. Albers, S.C., J.A. McGinley, D.L. Birkenheuer and J.R. Smart, 1996: The local analysis and prediction system (LAPS): analysis of clouds, precipitation and temperature. Weather and Forecasting, **11**, 273–287.
4. Aonashi, K., 1993: An initialization method to incorporate precipitation data into a mesoscale numerical weather prediction model. J. Meteorol. Soc. Japan, **71**, 393–406.
5. Bao, J.W. and Y.-H. Kuo: 1995. On-off switches in the adjoint method: step functions. Mon. Weather Rev., **123**, 1589–1594.
6. Benjamin, S.G., K.A. Brewster, R. Brümmer, B.F. Jewett, T.W. Schlatter, T.L. Smith, and P.A. Stamus,1991: An isentropic meso-scale analysis system and its sensitivity to aircraft and surface observations. Mon. Weather Rev., **117**, 1586–1603.
7. Bielli, S. and F. Roux, 1999: Initialization of a cloud-resolving model with airborne Doppler radar observations of an oceanic tropical convective system. Mon. Weather Rev., **127**, 1038–1055.
8. Brewster, K., 1996: Implementation of a Bratseth analysis scheme including Doppler radar. *Proc. 15th Conf. Weather Anal. Forecasting*, AMS, 1996, pp. 92–95.
9. Cartwright, T.J. and P.J. Ray, 1999: Radar-derived estimates of latent heating in the subtropics. Mon. Weather Rev., **127**, 726–742.
10. Collins, W.G., 2001: The quality control of Velocity Azimuth Display (VAD) Winds at the National Centers for Environmental Prediction. *Proc. 11th Symp. Meteorol. Observations Instrum.*, AMS, pp. 317–320.
11. Daley, R.: *Atmospheric Data Analysis*. Cambridge University Press, Cambridge, 1991.
12. Donner, L., 1988: An initialization for cumulus convection in numerical weather prediction models. Mon. Weather Rev., **116**, 377–385.
13. Ducrocq, V., J. Lafore, J. Redelsperger, and F. Orain, 2000: Initialisation of a fine scale model for convective system prediction: A case study. Q. J. R. Meteorol. Soc., **126**, 3041–3066.
14. Gal-Chen, T., 1978: A method for the initialization of the anelastic equations: Implications for matching models with observations. Mon. Weather Rev., **106**, 587–602.
15. Golding, B.W., 1998: Nimrod: a system for generating automated very short range forecasts. Meteorol. Appl. **5**, 1–16.
16. Gregoric, G., 2001: Diagnostics of deep convection using radar data and mesoscale numerical model. *Proc. 30th Conf. Radar Meteorol.*, AMS, pp. 177–178.
17. Guo,Y.-R., Y.-H. Kuo, J. Dudhia, D. Parsons and C. Rocken: 2000: Four-dimensional variational data assimilation of heterogeneous mesoscale observations for a strong convective case. Mon. Weather Rev., **128**, 619–643.
18. Haase, G., 2002: A physical initialization algorithm for non-hydrostatic weather prediction models using radar derived rain rates. Ph.D. Thesis, Meteorologisches Institut der Universität Bonn.

19. Hane, C.E and P.S. Ray, 1985: Pressure and buoyancy fields derived from Doppler radar data in a tornadic thunderstorm. J. Atmos. Sci., **42**, 18–35.
20. Harrison, D.L., S.J. Driscoll, and M. Kitchen, 2000: Improving precipitation estimates from weather radar using quality control and correction techniques. Meteorol. Appl., **6**, 135–144.
21. Ishikawa, Y., 2002: 'Meso-scale analysis.' In: *'Outline of the Operational Numerical Weather Prediction at the Japan Meteorological Agency'*, JMA pp. 26–31.
22. Jones, C.D. and B. Macpherson, 1997: A latent heat nudging scheme for the assimilation of precipitation data into an operational mesoscale model. Meteorol. Appl., **4**, 269–277.
23. Kapitza, H., 1991: Numerical experiments with the adjoint of a non-hydrostatic mesoscale model. Mon. Weather Rev., **119**, 2993–3011.
24. Krishnamurti, T.N., J. Xue, H.S. Bedi, K. Ingles and D. Oosterhof., 1991: Physical initialization for numerical weather prediction over the tropics. Tellus, **43AB**, 53–81.
25. Krishnamurti, T.N., H.S. Bedi and K. Ingles, 1993: Physical initialization using SSM/I rain rates. Tellus, **45A**, 247–269.
26. Lin, C.-L., T. Chai, and J. Sun, 2000: Adjoint retrieval of wind and temperature fields from a simulated convective boundary layer. *Proc. 14th Symp. Boundary Layer Turbulence*, AMS, pp. 106–107.
27. Lin, Y., M.E. Baldwin, K.E. Mitchell, E. Rogers and G.J. DiMego, 2001: Spring 2001 changes to the NCEP Eta Analysis and Forecast System: assimilation of observed precipitation. *Proc. 18th Conf. Weather Anal. Forecasting and 14th Conf. Numerical Weather Prediction*, Fort Lauderdale, FL, AMS, pp. J92–J95.
28. Lin, Y., P.S. Ray and K.W. Johnson, 1993: Initialization of a modeled convective storm using Doppler radar-derived fields. Mon. Weather Rev., **121**, 2757–2775.
29. Lindskog, M., H. Järvinen, and D.B. Michelson, 2000: Assimilation of radar radial wind in the HIRLAM 3D-Var. Phys. Chem. Earth (B), **25**, 1243–1250.
30. Lindskog, M., H. Järvinen, and D.B. Michelson, 2002: Development of Doppler radar wind data assimilation for the HIRLAM 3D-Var. HIRLAM Technical Report 52, HIRLAM-5 Project.
31. Liou, Y.-C., 1990: Retrieval of three-dimensional wind and temperature fields from one component wind data by using the four-dimensional data assimilation technique. Master's thesis, University of Oklahoma.
32. Macpherson, B., 2000: Assimilation of precipitation in the Met Office Mesoscale Model. *Proc. ECMWF/EuroTRMM Workshop on Assimilation of Clouds and Precipitation*, pp. 405–414.
33. Macpherson, B., 2001: Operational experience with assimilation of rainfall data in the Met Office Mesoscale Model. Meteorol. Atmos. Phys., **76**, 3–8.
34. Manobianco, J., S. Koch, V.M. Karyampudi and A.J. Negri: 1994: The impact of assimilating satellite derived precipitation rates on numerical simulations of the ERICA IOP 4 cyclone. Mon. Weather Rev., **122**, 341–365.
35. Marecal, V. and J.-F. Mahfouf, 2000: Four dimensional variational assimilation of total column water vapour in rainy areas. Technical Memorandum 314, ECMWF.
36. Matsumura, T., I. Takano, K. Aonashi and T. Nitta, 1997: Improvement of spin-up of precipitation calculation with use of observed rainfall in the initialization scheme. In: *Numerical Methods in Atmospheric and Ocean Modelling*, Canadian Meteorological and Oceanographic Society / NCR Research Press, pp. 353–368.

37. McGinley, J., 1989: The local analysis and prediction system. *Proc. 12th Conf. Weather Anal. Forecasting*, AMS, pp. 15–20.
38. Meetschen, D., S. Crewell, P. Gross, G. Haase, C. Simmer and A. v. Lammeren, 2000: Simulation of weather radar products from a mesoscale model. Phys. Chem. Earth (B), **25**, 1257–1261.
39. Parrish, D.F. and J. Purser, 1998: Anisotropic covariances in 3D-Var: Application to hurricane doppler radar observations. *Proc. HIRLAM 4 Workshop on Variational Analysis in Limited Area Models*, Météo-France, Toulouse.
40. Rinne, J. and C. Fortelius, 2001: Data assimilation and problems in the quality control of Doppler winds. *Proc. 30th Conf. Radar Meteorol.*, AMS, pp. 182–183.
41. Rossa, A.M., 2000: The COST 717 action: Use of radar observations in hydrological and NWP models. Phys. Chem. Earth (B), **25**, 1221–1224.
42. Schlatter, T., 2000: Variational assimilation of meteorological observations in the lower atmosphere: a tutorial on how it works. J. Atmos. Solar Terrestrial Phys., **62**, 1057–1070.
43. Sun, J. and N.A. Crook, 1997: Dynamical and microphysical retrieval from Doppler radar observations using a cloud model and its adjoint: Part I. Model development and simulated data experiments. J. Atmos. Sci., **54**, 1642–1661.
44. Sun, J. and N.A. Crook, 1998: Dynamical and microphysical retrieval from Doppler radar observations using a cloud model and its adjoint: Part II. Retrieval experiments of an observed Florida convective storm. J. Atmos. Sci., **55**, 835–852.
45. Treadon, R.E., 1996: Physical initialization in the NMC global data assimilation system. Meteorol. Atmos. Phys., **60**, 57–86.
46. Wang, W. and T.T. Warner, 1988: Use of four-dimensional data assimilation by Newtonian relaxation and latent-heat forcing to improve a mesoscale model precipitation forecast: A case study. Mon. Weather Rev., **116**, 2593–2613.
47. Wilson, J., N.A. Crook and C. Mueller, 1998: Nowcasting thunderstorms: A status report. Bull. Amer. Meteorol. Soc., **79**, 2079–2099.
48. Wolfsberg, D., 1987: Retrieval of three-dimensional wind and temperature fields from single-Doppler radar data. CIMMS Report No. 84, Cooperative Institute for Mesoscale Meteorological Studies.
49. Wu, B., J. Verlinde and J. Sun, 2000: Dynamical and microphysical retrievals from Doppler radar observations of a deep convective cloud. J. Atmos. Sci., **57**, 262–283.
50. Xu, Q., 1996: Generalized adjoint for physical processes with parameterized discontinuites. Part I: Basic issues and heuristic examples. J. Atmos. Sci., **53**, 1123–1142.
51. Xue, M., D. Wang, D. Hou, K. Brewster, and K.K. Droegemeier, 1998: Prediction of the 7 May 1995 squall line over the central U.S. with intermittent data assimilation. *Proc. 12th Conf. Numerical Weather Prediction*, AMS, pp. 191–194.
52. Zhang, J., 1999: Moisture and diabatic initialisation based on radar and satellite observations. Ph.D. Thesis, University of Oklahoma.
53. Zhang, J., F. Carr and K. Brewster, 1998: ADAS cloud analysis. *Proc. 12th Conf. Numerical Weather Prediction*, AMS, pp. 185–188.
54. Zou, X., I. Navon and J. Sela,1993: Variational data assimilation with moist threshold processes using the NMC spectral model. Tellus, **45A**, 370–387.
55. Zou, X. and Y.-H. Kuo, 1996: Rainfall assimilation through an optimal control of initial and boundary conditions in a limited-area mesoscale model. Mon. Weather Rev., **124**, 2859–2882.
56. Zou, X., 1997: Tangent linear and adjoint of 'on-off' processes and their feasibility for use in 4-dimensional variational data assimilation. Tellus, **49A**, 3–31.

57. Zupanski, D. and F. Mesinger, 1995: Four-dimensional variational assimilation of precipitation data. Mon. Weather Rev., **123**, 1112–1127.
58. Zupanski, M., D. Zupanski, D.F. Parrish, E. Rogers and G. DiMego, 2002: Four-dimensional variational data assimilation for the Blizzard of 2000. Mon. Weather Rev., **130**, 1967–1988.

Glossary

Compiled with use of

Collier, C. (Ed.): Applications of Weather Radar Systems, 2nd Ed., John Wiley, Chichester 1996

Rinehard, R.E.: Radar of Meteorologists, 3rd Ed. Rinehart Publishing, Grand Forks, ND 1997

DoC/NOAA: Fed. Met. Handbook No. 11, Doppler Radar Meteorological Observations, Part A-D, DoC, Washington D.C. 1990–1992

ACU
Antenna Control Unit.

A/D converter
ADC. Analog-to-digital converter. The electronic device which converts the radar receiver analog (voltage) signal into a number (or count or quanta).

ADAS
ARPS Data Analysis System, where ARPS is Advanced Regional Prediction System.

Aliasing
The process by which frequencies too high to be analyzed with the given sampling interval appear at a frequency less than the Nyquist frequency.

Analog
Class of devices in which the output varies continuously as a function of the input.

Analysis field
Best estimate of the state of the atmosphere at a given time, used as the initial conditions for integrating an NWP model forward in time.

Anomalous propagation
AP. Anaprop, nonstandard atmospheric temperature or moisture gradients will cause all or part of the radar beam to propagate along a nonnormal path. If the beam is refracted downward (superrefraction) sufficiently, it will illuminate the ground and return signals to the radar from distances further than is normally associated with ground targets.

Antenna
A transducer between electromagnetic waves radiated through space and electromagnetic waves contained by a transmission line.

Antenna gain
The measure of effectiveness of a directional antenna as compared to an isotropic radiator, maximum value is called antenna gain by convention.

Antenna reflector
The portion of an antenna system which reflects the energy from the radiating element into a focused beam generally circular parabolas for weather radars.

ARMAR
Airborne Rain MApping Radar.

A-scope
A deflection-modulated display in which the vertical deflection is proportional to target echo strength and the horizontal coordinate is proportional to range.

ASL
Above sea level.

Assimilation
The process of estimating meteorological or hydrological conditions on a regular grid from observations and a numerical model.

Assymmetric Digital Subscriber Line
ADSL. A technology allowing transmission rates of up to 2 Mbit/s in WANs by applying different rates for uplinks and downlinks.

ATC
Air Traffic Control.

Attenuation
Any process in which the flux density (power) of a beam of energy is dissipated.

Attenuator
A device or network that absorbs part of a signal and transmits the remainder with a minimum of distortion.

Autocorrelation
A measure of similarity between displaced and undisplaced (in time, space, etc.) versions of the same function.

Automatic gain control
AGC. Any method of automatically controlling the gain of a receiver, particularly one that holds the output level constant regardless of the input level.

Average power
Pulsed radars transmit over a very low duty cycle, i.e. many intense but short and widely separated pulses. The average power is a radar's peak power, its PRF, its pulse length.

Azimuth
A direction in terms of the 360° compass, north at 0°, east at 90°, south at 180°, west at 270°, etc.

B-scope
An intensity-modulated rectangular display with azimuth angle as the horizontal coordinate and range as the vertical coordinate or vice versa.

Backing wind
A change in wind direction with height in a counterclockwise sense representing cold air advection.

Background field
Latest short forecast from an NWP model, used as the starting point for assimilation of observations to produce the next analysis.

Backscatter
That portion of power scattered back in the incident direction.

Backscattering cross section
Equivalent area required for an isotropic scatterer to return to a receiver the power actually received.

Backscattering phase shift
Phase shift between H and V polarised waves induced by the backscattering process.

BALTEX
Baltic Sea Experiment.

Bandwidth
The number of cycles per second between the limits of a frequency band.

Band-pass filter
A filter whose frequencies are between given upper and lower cutoff values, while substantially attenuating all frequencies outside these values (this band).

Bandwidth, 3 dB
The frequency span between the points on the selectivity curve at which the insertion loss is 3 dB greater than the minimum insertion loss. Also called 3 dB passband.

Beam filling
The measure of variation of hydrometeor density throughout the radar sampling volume. If there is no variation in density, the beam is considered to be filled.

Beam width
Angular width of antenna pattern. Usually that width where the transmitted (i.e. one-way) power density is one-half that on the axis of the beam. (Half-power or 3 dB point).

Bias
A systematic difference between an estimate of and the true value of the parameter.

Bin
Radar sample volume.

Bipolar video
The phase detection of the Doppler radar backscattered signal using the transmitted signal as a reference. Also known as the I (in-phase) and Q (quadrature phase) component of the Doppler radar signal.

Bispectral analysis
See multispectral analysis.

Bistatic radar
A radar which uses separate antennas for transmission and reception. Usually, the transmitter and receiver are at different locations. Bistatic radars depend upon forward and sideward scattering of the signal from the target to the receiver.

Bistatic radar network
A network comprising one transmitting pencil-beam radar and one or more passive, low-gain, nontransmitting receivers at remote sites.

BMRC
Bureau of Meteorology Research Centre (Australia).

Boundary layer
The layer of a fluid adjacent to a physical boundary in which the fluid motion is affected by the boundary and has a mean velocity less than the free-stream value.

Bounded weak echo region

BWER. A core of weak equivalent reflectivity in a thunderstorm that identifies the location of a strong updraft. The updraft is so strong that large precipitation particles do not have time to form in the lower and mid-levels of the storm and are prevented from falling back into the updraft core from above. The weak echo region is bounded when, in a horizontal section, the weak echo region is completely surrounded or bounded by higher reflectivity values. See also weak echo region.

Bow echo

A radar echo pattern in which a line of echoes, typically around 100 km in length, bulges forward under the influence of a corridor of strong winds taking the shape of a bow.

Bragg scattering

Scatter from small-scale fluctuations (i.e. turbulence) in the refractive index of the atmosphere. Bragg scatter comes from fluctuations which are wavelength/2.

Bright band

A region of enhanced reflectivity in the vicinity of the freezing/melting level caused by the complex interaction of small ice crystals (poor reflectors) coalescing into larger clumps and melting from the outside (water coated ice is a strong reflector) then collapsing into raindrops (smaller cross section) and falling more rapidly (more dispersed droplets).

Brightness temperature

(also called the equivalent blackbody temperature). The temperature of a blackbody emitting the same amount of radiation as the object being viewed.

BUFR

Binary Universal Form for the Representation of meteorological data.

CAPE

Convectively available potential energy, a measure of the instability of the atmosphere.

CAPE

Convection And Precipitation Electrification experiment.

CAMPR

CRL Airborne Multiparameter Precipitation Radar.

CAPPI

Constant altitude plan position indicator, a display of radar data at a constant altitude, constructed from data obtained at several elevation angles. See also PPI.

Cartesian coordinates
The familiar "x–y" coordinate system, in which the axes are at right angles to each other. Raw radar data, often in polar coordinates, can always be converted to Cartesian coordinates.

C-Band
Radar wavelength of about 5 cm.

CETP
Centre d'Etudes des Environnements Terrestres et Planétaires (Vélizy, France).

CLARA
CLouds And RAdiation Experiment.

Clear-air echoes
Radar returns from cloud- and precipitation-free (optical clear) air. Clear-air echoes are caused either by Bragg scatter or by returns from discrete targets such as insects or birds. Bragg scatter comes from small-scale fluctuations of the refractive index (i.e. turbulent fluctuations of humidity).

Cloud base
For a given cloud or cloud layer, it is the lowest level in the atmosphere where cloud particles are visible.

Clutter
Echoes that interfere with observation of desired signals on a radar display. Usually applied to ground targets.

Coherency matrix
A matrix giving the generally polarized electromagnetic radiation backscatter from precipitation particles.

Coherent radar
A radar that utilizes both signal phase and amplitude to determine target characteristics such as velocity, spectrum width. See Doppler radar.

COHO
Coherent oscillator.

Cold front
A narrow transition zone separating advancing colder air from retreating warmer air. The air behind a cold front is cooler and typically drier than the air it is replacing.

Combined sewer system
Dirty water (sewage) and rainwater drained or processed in the same system.

Complex index of refraction
$m = n + i \cdot k$, where n is the normal index of refraction, i is sqrt(-1) and k is the absorption coefficient.

Complex signal
A signal containing both amplitude and phase information.

Condensation
The process by which water vapor becomes a liquid; the opposite of evaporation, which is the conversion of liquid to vapor. Sublimation is the process by which a solid forms directly from vapor.

Confluence zone
A region where streamlines of the atmospheric flow at any level come closer together, indicating acceleration and corresponding ageostrophic flow.

Convection
In general, mass motions within a fluid resulting in transport and mixing of properties of that fluid. In meteorology, convection is referred to the atmospheric motions that are predominantly vertical, such as rising air currents produced by surface heating. Forced convection is the ascent of air induced by some external force. An example of this convection is the lifting of lighter, warmer air by an advancing cold front. Other examples include convection resulting from orographic lifting and convergence. Free convection is the rising of heated air and the sinking of cooler air without the need of external forces.

Convergence
A principal air flow relative to a moving synoptic scale midlatitude weather system. Often used flows are, for example, warm conveyor belt (WCB), cold conveyor belt (CCB) and dry intrusion. Quite often cloud and precipitation bands or dry zones, generated by the vertical air motions of the conveyor belts, are also called conveyor belts.

Conveyor belt
A principal air flow relative to a moving synoptic scale midlatitude weather system. Often used flows are, for example, warm conveyor belt, cold conveyor belt and dry intrusion. Quite often cloud and precipitation bands or dry zones, generated by the vertical air motions of the conveyor belts, are also called conveyor belts.

COST
COoperation in Science and Technology.

Cost function
Sum of terms measuring the 'distance' between a model state and the available information – also known as the penalty function.

Covariance
A measure of the degree of association between two variables. In Doppler radars, the argument (or angle) of the covariance of the complex signal is a measure of the Doppler frequency.

CPI
Coherent processing interval.

CRL
Communication Research Laboratory (Japan).

Crosssection (of radar targets)
The area intercepting that amount of power, which, if scattered isotropically, would return to the radar receiver an amount of power equal to that actually received.

CSR
Clutter-to-signal ratio.

dB
Decibel. A unit of gain equal to ten times the common logarithm of the ratio of two power levels or 20 times the common logarithm of the ratio of two voltage levels.

dBm
Decibels related to 1 mW. The standard unit of power level used in microwave work.

dBZ
A logarithmic expression for reflectivity factor, referenced to $(1 \text{ mm}^6/1 \text{ m}^3)$. $\text{dBZ} = 10 \log (Z/1 \text{ mm}^6/1 \text{ m}^3)$. See dB.

Dealiasing
Process of correcting for aliases in the velocity measurement. Also known as unfolding.

Default value
A setting or value that will be used in a given software program unless changed.

Dielectric constant
Term in the radar equation $K = (m^2 - 1)/(m^2 + 1)$, where $\mid K \mid$ is the magnitude of the expression for the complex index of refraction m. For water, the dielectric factor is $\mid K \mid^2 = 0.93$; for ice, $\mid K \mid^2 = 0.197$.

Differential reflectivity
ZDR. A radar reflectivity parameter formed using the reflectivities measured at both horizontal (Z_H) and vertical (Z_V) linear polarizations; ZDR = 10 log Z_H/Z_V. See ZDR.

Diffluence zone
A region where the streamlines of the atmospheric flow at any level spread out horizontally or sideways.

Diffraction
The process by which the direction of radiation is changed so that it spreads into the geometric shadow region of an opaque or refractive object that lies in a radiation field.

Disaggregation / downscaling
The process of transferring precipitation estimates or other parameters over large areas (grid squares) to smaller areas (grid squares). Often used in the context of NWP precipitation output being reduced using a model for input to a hydrological model. The opposite process is known as aggregation.

Disdrometer
Equipment that measures and records the size distribution of raindrops.

Display resolution
The area or two-dimensional product of the X and Y coordinates represented by one picture element (pixel) of a raster scan display.

Distributed models
River or urban drainage catchment models using input data on a grid at each point of which are formulated process equations.

Divergence
A measure of the expansion of a vector field. In meteorology, usually wind or mass divergence. When observed near the ground, it is associated with downdrafts.

DLR
German Aerospace Center, Deutsches Zentrum für Luft- und Raumfahrt.

Doppler dilemma
$V_{\max} \cdot R_{\max} = c \cdot l/8$, where $V_{\max}$ is maximum unambiguous velocity, $R_{\max}$ is maximum unambiguous range, c is speed of light, and l is wavelength.

Doppler effect
If an electromagnetic source moves relative to an observer, there is a shift in the observed frequency. If the source is receding from the observer, the observed frequency will appear to decrease.

Doppler frequency shift
$f = 2 \cdot V/l$, where V is radial velocity of the target, l is the wavelength.

Doppler radar
A type of radar that measures not only the intensity of a returned signal, but also its Doppler shift, and hence the radial velocity of the target.

Doppler shift
The change in frequency between a transmitted radar signal and that returned from a moving target.

Doppler velocity
Reflectivity-weighted average velocity of targets in the pulse volume. Determined by phase measurements from a large number of successive pulses. Also called radial velocity. Gives only the radial component of the velocity vector. It is generally assumed that raindrops and other particles are advected with the wind and have no own motion except their falling velocity. Motions toward the radar are negative, motions away from the radar are positive.

Downburst
A severe thunderstorm downdraught that spreads out radially from the storm at the ground causing sudden large changes in the boundary layer wind flow.

Downconverter
Integrated assembly of components required to convert microwave signals to an intermediate frequency range for further processing. Generally consists of an input filter, local oscillator filter, if filter, mixer and frequently an lo frequency multiplier and one or more stages of if amplification. May also incorporate the local oscillator, AGC/gain compensation components and RF preamplifier.

Drop Size Distribution
DSD. Number concentration of raindrops by class of equivalent diameter (in $mm^{-1} \cdot m^{-3}$ or m^{-4}).

Dual Doppler
The use of two Doppler radars to measure two different radial velocities. With some mathematics, these two radial wind components can be synthesized to a spatial distribution of fully 2-D (horizontal) winds.

Ducting
The phenomenon by which the radar signal propagates along the boundary of two dissimilar air masses. The radar ranges with ducted propagation are greatly extended holes in the coverage. Ducting occurs when the upper air is exceptionally warm and dry in comparison with the air at the surface. Ducting occurs when $dN/dh <= -157N$-units/km. N is refractivity.

Duplexer
A device in the waveguide which protects the sensitive receiver from the full power of the transmitter. It usually contains one or more TR (transmit-receive) tubes.

Dwell time
Time over which a signal estimate is made. For operational radars, usually the time required for the antenna to transverse one degree.

Dynamic range
The range from the minimum, which is at a level at or below the amplifiers' internally generated noise, to a maximum input signal level that a component can accept and amplify without distortion.

Echo
Energy backscattered from a target, as seen on the radar display.

Echo tops
The maximal height of a defined nonzero reflectivity contour.

ECMWF
European Centre for Medium Range Weather Forecasts.

EDOP
ER2 Doppler radar.

Effective (equivalent) radar reflectivity
The summation of returned power per unit volume proportional to the sixth power of the diameters of spherical water drops in the Rayleigh scattering region.

Effective (excess) rainfall
The amount of rain that finds its way into a river after the rain has made up catchment moisture deficits and surfaces and soil are saturated.

ELDORA-ASTRAIA
ELectra DOppler RAdar - Analyse STéréoscopique par RAdar á Impulsions Aéroporté.

Elevation
The vertical pointing angle of the antenna; 0° is horizontal, 90° is vertical.

Elevation scan
The process of the radar completing a full 360° rotation in azimuth for a specific elevation angle.

Ensemble prediction
Forecasts based upon the use of input data having slightly different characteristics or values.

ETH
Eidgenössische Technische Hochschule (Swiss Federal Institute of Technology, Zürich).

EUMETSAT
European Organisation for the Exploitation of Meteorological Satellites.

Evaporation
General: The physical process by which a liquid is transformed into a gaseous state. Meteorology: The transfer of water from a wet surface to the atmosphere. The surface may be the sea, a lake, bare soil or vegetation. In the case of vegetation, the term evapotranspiration is often used to refer to the total water loss due to soil evaporation and plant transpiration into the atmosphere.

FAST
Fore and Aft Scanning Strategy.

FASTEX
Fronts and Atlantic Storm-Track Experiment.

FEH
Flood Estimation Handbook.

FFT
Fast Fourier transform.

Flare echo
On radar images, a patch of anomalous echoes associated with hail not too far from the transmitter. It is caused by multiple scattering from the precipitation and the ground. Also referred to as a hail spike.

Flood forecasting
The total system for warning of inundation of land and property involving the use of observational data, such as precipitation, flow and catchment

morphology, NWP forecasts, models of fluvial, urban and coast systems and dissemination and alarm procedures.

FM-CW
Frequency-modulated continuous wave: radar principles in which the distance of the target is obtained from a frequency change in the signal.

Folding
Doppler radar velocities are said to be folded when they lie outside the Nyquist interval of twice the Nyquist velocity, V_a, where $V_a = \pm \frac{\text{PRF} \cdot \lambda}{4}$.

Forward model
Model that estimates observed parameters from NWP model variables – also known as the observation operator.

Frequency
The number of recurrences of a periodic phenomenon per unit time. Electromagnetic energy is usually specified in hertz (Hz), which is a unit of frequency equal to one cycle per second. Weather radars typically operate in the gigahertz range (GHz). See also wavelength.

Frequency band
A range of frequencies between some upper and lower limit.

Frequency carrier
The fundamental transmitted microwave frequency between 2700 and 3000 MHz. It is modulated so that it exists for a few microseconds each pulse.

Front
The transition zone between two distinct air masses. These air masses could be different in temperature or moisture content. The most commonly known examples of fronts are the cold front and the warm front.

FSL
Forecast Systems Laboratory.

FSR
Flood Studies Report.

Gain
A change in signal power, voltage or current. Usually applied to a change greater than one and expressed in decibels. See, for example, antenna gain.

Gate
See range gate.

Gating (range gating)
The use of electric circuits in radar to eliminate or discard the target signals from all targets falling outside certain desired range limits.

GHz
Gigahertz (billions of hertz).

GPM
Global Precipitation Measurement.

GPS
Global Positioning System. A network of satellites which provide extremely accurate position and time information. Useful in remote locations or for moving platforms.

Graupel
A lightly rimed ice aggregate often found in vigorous storms. Formed when an ice aggregate collects supercooled liquid water droplets.

Ground clutter
The pattern of radar echoes from fixed ground targets.

Gust front
The boundary between the horizontally propagating cold air outflow from a thunderstorm and the surrounding environmental air.

Gyro
A device used for measuring changes in direction. Often used in antenna stabilisation.

Hail spike
See flare echo.

Heterodyne
In heterodyne systems, a known local oscillator (LO) is combined (in a mixer) with the signal you want to measure to produce a beat frequency (called the IF) which is at a lower frequency than the original signal (and thus easier to work with).

HIRLAM
High Resolution Limited Area Model.

Hook echo
A pendant or hook on the right rear of an echo that often identifies mesocyclones on the radar display. The hook is caused by precipitation drawn into a cyclonic spiral by the winds, and the associated notch in the echo is caused by precipitation-free, warm, moist air flowing into the storm.

Hydrograph
   Plot of river flow or level as a function of time.

Hydrological (or Water) cycle
   The movement of water from the oceans to the atmosphere and back to the oceans, sometimes via the land. This cycle operates on all scales from local through regional to global.

Hydrology
   The study of the distribution of water on or under the earth's surface.

Hydrometeor
   A particle of condensed water (liquid, snow, ice, graupel, hail) in the atmosphere.

Hydrometeorology
   The application of meteorology to hydrological problems.

Hz
   Hertz. Unit of frequency equal to one cycle per second.

IF
   Intermediate frequency. In superheterodyne receiving systems, the frequency to which all selected signals are converted for additional amplification, filtering and eventual direction.

Index of refraction
   See refractive index.

In-phase or 'I' component
   The component of a complex signal along the real axis in the complex plane.

Insertion loss
   The transmission loss measured in dB at that point in the passband which exhibits the minimum value.

ISDN
   Integrated Services Digital Network. A switched network providing end-to-end digital connectivity for simultaneous transmission of voice and data over multiplexed communication channels.

Isolated storm
   An individual cell or group of cells that are identifiable and separate from other cells in a given geographic area.

Isolation
The ratio of the power level applied at one port of a component to the resulting power level at the same frequency appearing at another port.

Isotropic
Having the same characteristics in all directions, as with isotropic antennas. Directional or focused antennas are not isotropic.

JPL
Jet Propulsion Laboratory.

KDP
Specific differential phase shift (degree/km) induced by the different propagation characteristics of H and V waves.

Klystron
Radar amplifier tube operating at high power (up to 2 MW) from which the same phase is maintained over many emitted pulses.

Kriging (named after D. G. Krige)
A suite of interpolation techniques that uses regionalized variable theory to incorporate information about the stochastic aspects of spatial variation when estimating interpolation weights.

LAN
Local Area Network.

LAPS
Local Analysis and Prediction System.

Latent Heat Nudging
LHN. Assimilation method to force an NWP model toward observed precipitation rates by correcting the model's latent heating rate based on observed precipitation rates.

LDR
Linear Depolarization Ratio. Ratio between the reflectivity received at vertical polarization, but transmitted with horizontal polarization, and the reflectivity at horizontal polarization. Depolarization of the horizontal polarized pulse is caused by asymmetric particles which are not aligned horizontally. Depolarization of rain is normally very small (approx -40 dB). It is high (-20 to -10 dB) for melting snow and water coated hail or graupel.

Lidar
Optical equivalent of a radar (Light Detection and Ranging).

LNA
Low Noise (Microwave) Amplifier.

Local oscillator
An oscillator used in superheterodyne receiver which when mixed with an incoming signal results in a sum or difference frequency equal to the intermediate frequency of the receiver.

LPATS
Lightning Position and Tracking System.

Lumped models
Hydrological model which relates the characteristics of the river hydrograph to the physiographic factors usually employing the unit hydrograph or transfer function approach.

LWC
Liquid Water Content, mostly in g per cubic meter.

MAP
Mesoscale Alpine Programme.

Magnetron
A self-exciting oscillator tube generating pulses, the phase of which is random. Used to produce the radio frequency signal transmitted by some radars. It utilizes a strong magnetic field to help induce the RF signal generated.

Main lobe
The envelope of electromagnetic energy along the main axis of the beam as formed by the antenna reflector.

MAPS
Mesoscale Analysis and Prediction System.

Maximum unambiguous range
The maximum range to which a transmitted pulse wave can travel and return to the radar before the next pulse is transmitted. $R_{\max} = c/(2 \cdot \text{PRF})$, where c is the speed of light, PRF is pulse repetition frequency.

Maximum unambiguous velocity
The maximum Doppler radial velocity, V_a, which can be measured by a radar without the occurrence of folding where $V_a = \pm \text{PRF} \cdot \lambda/4$. The velocity may be increased by using a radar which can transmit two pulse repetition frequencies.

MCC
Mesoscale Convective System.

MCS
Mesoscale Convective Complex.

MDS
Minimum Detectable Signal.

Mean Doppler velocity
Reflectivity-weighted average velocity of targets in a given volume sample. Usually determined from a large number of successive pulses. Also called mean radial velocity. Doppler velocity usually refers to spectral density first moment, radial velocity to base data.

Mean radial velocity
The component of motion of the target toward or away from the radar.

Melting level
That height in the atmosphere at which ice melts to water i.e., at which the temperature is 0°C. This may range from 0–5 km above the earth's surface.

Melting Layer
ML. That height interval of the atmosphere, where snow particles, during fall, melt to water drops.

MESAN
SMHI's Mesoscale Analysis.

Meso-gamma scale
Scale at which NWP models can resolve cloud features explicitly (without subgrid scale parametrisation), of the order a few kilometres or less.

Microwaves
High-frequency radio waves lying roughly between infrared waves and radio waves. Microwaves are generated by electron tubes, such as the klystron and the magnetron, or solid-state devices with built-in resonators to control the frequency or by oscillators. Microwaves have many applications for radio, television, radar, test and measurement communications, distance and location measuring, and more.

Mie scattering
Scattering of electromagnetic radiation which occurs when the diameter of the target scatterers is comparable to the wavelength of the incident radiation (e.g., for hail).

MM5
Mesoscale Model, fifth generation (NCAR).

Monostatic radar
A radar that uses a common antenna for both transmitting and receiving.

MSG
Meteosat Second Generation.

MTI
Moving target indicator, often used in ATC radars to aid the removal of unwanted radar echoes, in this case those from precipitation.

Multicell storm
A storm that consists of a cluster of single cells that are often short-lived.

Multiparameter radar
Radar having more than one wavelength or polarization characteristic.

Multiple Doppler analysis
The use of more than one radar (and hence more than one look angle) to reconstruct spatial distributions of the 2-D or 3-D wind field, which cannot be measured from a single radar alone. Includes dual Doppler, triple Doppler and overdetermined multiple Doppler analysis.

NASA
National Aeronautics and Space Administration.

NASA-GSFC
National Aeronautics and Space Administration - Goddard Space Flight Center, Greenbelt, Maryland.

NASDA
National Space Development Agency of Japan.

NCAR
National Center for Atmospheric Research, Boulder, CO., USA.

NCEP
National Centers for Environmental Prediction.

NMS
National Meteorological Service.

NOAA
National Oceanic and Atmospheric Administration.

NOAA-ERL
National Oceanic and Atmospheric Administration - Environmental Research Laboratory.

Noise figure
The ratio (in dB) between the signal-to-noise ratio applied to the input of the microwave component and the signal-to-noise ratio measured at its output. It is an indication of the amount of noise added to a signal by the component during normal operation. Lower noise figures mean less degradation and better performance.

NORDRAD
NORDic country RADar network.

Normal Mode Initialization
NMI. Technique to remove spurious high-frequency waves from Numerical Weather Prediction models.

Nudging
NWP. Assimilation technique in which model equations are extended with a forcing term which depends on the difference between observations and model estimates of the observed quantities.

NWP
Numerical Weather Prediction.

Nyquist frequency
The highest frequency that can be determined in data that have been discretely sampled. For data sampled at frequency f, this frequency is $(f/2)$. Doppler radar sampling frequency (rate) is equal to the pulse repetition frequency (PRF).

Nyquist velocity, or interval
The maximum unambiguous velocity that can be measured by a Doppler radar.

Optimum interpolation (OI)
Technique based on linear minimum variance estimation of analysis errors at each grid point of a numerical model – also known as statistical interpolation.

Oscillator
The general term for an electric device that generates alternating currents or voltages. The oscillator is classified according to the frequency of the generated signal.

Parabolic antenna
An antenna with a radiating element and a parabolic reflector that concentrates the radiated power into a beam and also concentrates the returned signal.

PIA
Path integrated attenuation.

Peak power
The amount of power transmitted by a radar during a given pulse. Note that because these pulses are widely spaced, the average power will be much smaller.

Pedestal
A generic term for the structure supporting the antenna dish. Usually includes the drive motors and one end of the servo loop.

Phase
A particular angular stage or point of advancement in a cycle; the fractional part of the angular period through which the wave has advanced, measured from the phase reference.

Phase noise
The short-term frequency variations in the output frequency which appear as energy at frequencies other than the carrier. It is usually expressed in terms of dBc or as an rms frequency deviation in a specified frequency removed from the carrier.

Phase shift
The angular difference of two periodic functions.

PhiDP
ϕ_{DP}. Differential phase, the difference of phase shifts for horizontal and vertical polarizations.

Physics-based models
These hydrological models use quantities which are, in principle, measurable and use parameters which have physical significance. The model equations represent the small-scale physics of homogeneous systems.

Physical initialisation
PI. Technique to impose consistency between precipitation rates diagnosed from model physical parametrisation schemes and observed rain rates.

Pixel
A term derived from 'picture elements.' It is the smallest resolvable part of a satellite or radar image.

PMF
Probable Maximum Flood.

PMS
Particle Measuring System.

Polarization radar
A radar which takes advantage of ways in which the transmitted waves' polarization affect the backscattering. Such radars may alternately transmit horizontal and vertically polarized beams and measure differential reflectivity and depolarisation ratios.

POLDIRAD
Polarisation-Diversity Radar, C-band research radar of DLR Oberpfaffenhofen.

Power
$P = I \cdot V = V \cdot V/R = I \cdot I \cdot R$, where I is current (amps), V is voltage (volts), R is resistance (ohms), P is power (watts). I is not to be confused with the 'I' of 'I and Q', the in-phase and quadrature components.

Power divider
A passive resistive network which equally divides power applied to the input port between any particular number of output ports without substantially affecting the phase relationship or causing distortion.

PPI
Plan-Position Indicator. An intensity-modulated display on which echo signals are shown in plan view with range and azimuth angle displayed in polar coordinates, forming a map-like display. Each PPI is taken at a single, fixed elevation angle, and thus forms a cone of coverage in space. PPIs may be run in sequence, for a number of elevation angle creating a "volume scan.".

Precision
The accuracy with which a number can be represented, i.e. the number of digits used to represent a number.

Probable maximum flood
PMF. The theoretical greatest depth of precipitation for a given duration that is physically possible over a given size of storm area at a particular geographic location at a certain time of year.

Propagation
Transmission of electromagnetic energy as waves through or along a medium.

Pulse
A single short duration transmission of electromagnetic energy.

Pulse duration
Time occupied by a burst of transmitted radio energy. This may also be expressed in units of range (pulse length) or time (pulse duration). Also called pulse width.

Pulse Forming Network
Series of high voltage condensers whose discharges via a transformer trigger the magnetron or Clystron and define pulse length.

Pulse length
$h = c \cdot t$, where t is the duration of the transmitted pulse, c is the speed of light, h is the length of the pulse in space. Note, in the radar equation, the length $h/2$ is actually used for calculating pulse volume because we are only interested in signals that arrive back at the radar simultaneously.

Pulse radar (or pulsed radar)
A type of radar, designed to facilitate range measurement, in which the transmitted energy is emitted in periodic brief transmissions, or pulses.

Pulse repetition frequency
PRF. The number of pulses transmitted per second. Also called pulse repetition rate. Typical PRFs may range from 300–1200 Hz. See also Nyquist frequency.

Pulse repetition rate
See Pulse repetition frequency.

Pulse repetition time
PRT. The time interval from the beginning of one pulse to the beginning of the next succeeding pulse.

Pulse resolution volume
The volume in which the radar data for one range bin are measured. Defined by the width of the radar beam (approx 1°) and half the length of the transmitted pulse. 1 μs pulses are 150 m deep, 2 μs pulses are 300 m deep.

Pulse width
The time occupied, or the linear distance in range occupied by an individual broadcast from a radar.

Pulse-pair processing
Nickname for the technique of mean velocity estimation by calculation of the signal complex covariance argument. The calculation requires two consecutive pulses, hence "pulse-pair."

Quadrature
The component of the complex video signal which, in Doppler radar, is 90° out of phase with the reference oscillator or the in-phase component.

Quantitative precipitation forecasting
QPF. The preparation of forecasts of precipitation amount, timing and spatial distribution. This may involve the use of real-time data and numerical models providing deterministic or stochastic forecasts.

Radar
Radio Detection and Ranging.

Radar cross section
The area of a fictitious perfect reflector of electromagnetic waves that would reflect the same amount of energy back to the radar as the actual target.

Radar equation
Relationship between the power received by a radar system and the sum of the cross sections of individual scatterers. The equation involves the transmitted power, antenna gain, radar wavelength and the range of the target.

Radar reflectivity
The sum of all backscattering cross sections (e.g., precipitation particles) in a pulse resolution volume divided by that volume. The radar reflectivity can be related to the radar reflectivity factor through the dielectric constant term $\mid K \mid^2$ and the radar wavelength.

Radar reflectivity factor
Z = the sum (over i) of $(N_i \cdot D_i^6)$, where N_i is the number of drops of diameter D_i in a pulse resolution volume. Note that Z may be expressed in linear or logarithmic units. The radar reflectivity factor is simply a more meteorologically meaningful way of expressing the radar reflectivity and therefore mostly just called "reflectivity".

Radial velocity
The component of motion of the target toward or away from the radar. Doppler radars only detect radial components of velocity.

Radiometer
Passive device which measures the radiation emanating from an object (e.g. a cloud).

RAMS
Regional Atmospheric Modeling System.

Range
Distance from the radar antenna.

Range bin
See range gate.

Range folding
Apparent range placement of a multiple return. A multiple return appears at the difference of the true range and a multiple of the unambiguous range.

Range gate
The discrete point in range along a single "radial" (beam) of radar data at which the received signal is sampled. Range gates are typically spaced at 100–1000 meter intervals. A "radial" of radar data is composed of successive range gates, out to the maximum unambiguous range.

Range height indicator
RHI. An intensity-modulated display with height as the vertical axis and range as the horizontal axis for a given azimuth-angle. A "vertical cross section" in a plane passing through a radar.

Rayleigh scattering
Scattering of electromagnetic radiation which occurs when the diameter of the sperical target scatterers is small (one-tenth) compared to the wavelength of the radiation (e.g., for most raindrops).

Real time
A description of how close a process such as data collection or analysis takes place to the instant that a geophysical process occurs. Usually, procedures designated as taking place in real time occur within a few seconds of the occurrence of a geophysical process.

Receiver
The electronic device which detects the backscattered radiation, amplifies it and converts it to a low-frequency signal which is related to the properties of the target.

Reflectivity
A measure of the fraction of radiation reflected by a given surface defined as the ratio of the radiant energy reflected to the total that is incident upon that surface. Lazy radar meteorologists and others working with radar data from storms frequently say "reflectivity" when they should say radar reflectivity factor or equivalent radar reflectivity factor.

Reflectivity factor
Integral over the backscatter cross section of the particles in a pulse volume. For particles small compared to the wavelength the scatter cross section is D^6, where D is the diameter of the particle. Radars are calibrated in a way to give directly (assuming the dielectric constant of water) the reflectivity

factor from the received backscattered energy. Units for the reflectivity factor are mm^6 m^{-3} or the logarithmic value of this in dBZ.

Refraction
The process in which the direction of energy propagation is changed as a result of a change in the speed of propagation caused by changes in density within the medium or as the energy passes through the interface representing a density discontinuity between two media.

Refractive index
A measure of the amount of refraction. Numerically equal to the ratio of wave velocity in a vacuum to the wave speed in the medium, i.e. $n = c/v$ where v is actual speed, and c is speed of light in a vacuum.

Refractivity
$N = (n-1) \cdot 10^6$, where n is refractive index and N is a function of temperature, pressure and vapor pressure (in the atmosphere).

Relative humidity
The amount of water vapor in the air compared to the amount of water vapor the air can actually hold at a given temperature and pressure. When air has a relative humidity of 100 %, it is saturated.

RF
Radio Frequency. In radar work, generally used for the transmitted or received signal as opposed to a signal translated to a different frequency, for example, the intermediate frequency (if) or the detected video signal.

RHI
Range Height Indicator. An intensity-modulated display on which echo signals are shown in plan view with range as the horizontal axis and height as the vertical axis. Each RHI is taken at a single fixed azimuth-angle.

RUC
Rapid Update Cycle, an atmospheric prediction system at NCEP.

RSM
Radar Simulation Model.

Sample volume
The volume of the atmosphere which is being instantaneously sampled by a radar; the power returned at any one instant which is the total backscatter from a volume of atmosphere equal to 1/2 the pulse length multiplied by the beam diameter.

S-Band
Radar wavelength of about 10 cm.

Scatterer
Any object capable of reflecting a radar signal.

Scattering cross section
See cross section.

Second-time-around echoes
See second-trip echoes.

Second-trip echoes
Radar echoes from previous pulses which occur in the interpulse period between successive radar pulses.

Sensitivity time control
STC. Time adjusted attenuation of the linear receiver to compensate for the range dependent signal reduction.

Sensor synergy
Combined use of instruments to acquire more information than with individual use of the instruments.

Shear
The rate of change of the vector wind in a specified direction normal to the wind direction. Vertical shear is the variation of the horizontal wind in the vertical direction.

Side lobe
Secondary radiated energy "away from" the radar main beam. Typically contains a small percent of energy compared to the main lobe but may produce erroneous echoes.

Signal-to-noise ratio
SNR. A ratio that measures the comprehensibility of data, usually expressed as the signal power divided by the noise power, expressed in decibels (dB).

Spectral density
The distribution of power by frequency.

Spectrum width
A measure of dispersion of velocities within the radar pulse resolution volume. Standard deviation of the velocity spectrum. It contains components from wind shear, the dispersion of drop terminal velocities, turbulence and the antenna scan rate.

Sphere calibration
Reflectivity calibration of a radar by pointing the dish at a metal sphere of (theoretically) known reflectivity. The sphere is often tethered to a balloon.

Spurious signal
Undesired signals produced by an active microwave component, usually, at a frequency unrelated to the desired signal or its harmonics. Spurious outputs are both harmonically and nonharmonically related signals.

Signal quality index
SQI. A threshold to eliminate signals either too weak or of a too large spectrum width. It varies between 0 for white noise and 1 for a noise-free zero-width signal. Needs experimental adjustment.

Squall line
Any line or narrow band of active thunderstorms.

STALO
Stabilized local oscillator of radar.

Standard deviation
The positive square root of the signal variance. The velocity standard deviation is called spectrum width.

Stochastic models or physics
The use of statistical representations to formulate a model of environmental processes or the physics of such processes.

Structure functions
Spatial patterns of forecast error correlations in an NWP model, i.e. correlations between errors in one variable at one location and errors in the same (or different) variable at another location.

Supercell
A large, long-lived (up to several hours) cell consisting of one quasi-steady downdraft couplet that is generally capable of producing the most severe weather (tornadoes, high winds, and giant hail).

Supercooled liquid water
In the atmosphere, liquid water can survive at temperatures colder than $0°C$. Many vigorous storms contain large amounts of supercooled liquid water at cold temperatures. Important in the formation of graupel and hail.

Superobservations
Pseudo-observations produced by spatially averaging finer scale radar observations to the larger characteristic scale of the model.

Superrefraction
A condition of atmospheric refraction when radar waves are bent more than normal. Superrefraction occurs when $dN/dh < -39.2N$-units/km. N is refractivity.

TCWV
Total Column Water Vapour.

Time–height display
An intensity-modulated display which has height as the vertical coordinate and time as the horizontal coordinate, usually used for vertically pointing antennas only.

TMI
TRMM Microwave Imager.

TOGA-COARE
Tropical Ocean and Global Atmosphere – Coupled Oceanic and Atmospheric Response Experiment.

Transmitter
The equipment used for generating and amplifying a radio-frequency (rf) carrier signal, modulating the carrier signal with intelligence, and feeding the modulated carrier to an antenna for radiation into space as electromagnetic waves. Weather radar transmitters are usually magnetrons or klystrons.

Tretyakov wind shield
One type of the shield structure surrounding a rain gauge, aimed to eliminate measurement errors due to wind deflection. Shields improve gauge catch of snowfall, on average by 20–70 %, depending on the wind velocity, vegetation and human structures around each gauge.

Triple Doppler
Since any wind has three components (say, in the x, y and z directions) and a single radar measures in only one direction (radial), a single radar cannot give the 3-D winds everywhere it samples. However, if three different radars view a storm from three different locations, the three measured radial velocities can be transformed into the actual 3-D wind field.

TRMM
Tropical Rainfall Measurement Mission.

TRMM PR
TRMM precipitation radar.

Unambiguous range
  The range to which a transmitted pulse wave can travel and return to the radar before the next pulse is transmitted. See Maximum Unambiguous Range.

Uniform wind technique
  UWT. A simplified VVP under the assumption that the wind field is constant across sector elements of about $15° \cdot 10$ km.

VAD
  Velocity Azimuth Display. The VAD technique is used to retrieve the vertical wind profile at the radar site with a Doppler radar. PPI scans at some elevation cover various heights with one scan.

Variational assimilation
  Technique involving minimisation of a 'cost' or 'penalty' function in order to retrieve the best estimate of a model state.

3D-Var
  Variational assimilation applied in three spatial dimensions for observations valid at, or close to, a single time. The minimisation finds a state with minimum expected analysis error variance.

4D-Var
  Variational assimilation applied in four dimensions (three spatial dimensions plus time) for observations distributed throughout a time window. The minimisation finds the model trajectory which best fits the available data.

VDRAS
  Variational Doppler Radar Analysis System.

Veering wind
  A change in wind direction with height in a clockwise sense representing warm air advection.

Velocity aliasing (folding)
  Ambiguous detection of radial velocities outside the Nyquist interval. On a radial velocity display, this shows up as regions where the radial velocity suddenly switches from one extreme to its opposite (e.g., max toward to max away).

Vertical shear
  The rate of change of wind speed or direction, with a given change in height.

VHF
Very high frequency, 30–300 MHz.

Volume scan
The process of completing a series of specified scans in a specific sequence in order to sample a whole atmospheric volume.

VORTEX
Verification of the Origins of Rotation in Tornadoes EXperiment.

VPR
Vertical Profile of Reflectivity.

VVP
Volume Velocity Processing. A way to guess the large-scale two-dimensional winds, divergence and fall speeds from one-dimensional radial velocity data. Essentially a multivariate regression which fits a simple wind model to the observed radial velocities. Similar to VAD, except it uses different functions for the fit.

WAN
Wide Area Network.

Warm front
A narrow transition zone separating advancing warmer air from retreating cooler air. The air behind a warm front is warmer and typically more humid than the air it is replacing.

Wavelength
The distance a wave will travel in the time required to generate one cycle. The distance between two consecutive wave peaks (or other reference points) in space. Weather radar wavelengths typically range from 1 mm to 50 cm.

Weak echo region
WER. Within a convective echo, a localized minimum of equivalent reflectivity associated with a strong updraft region. The weak echo region is bounded when in a horizontal section, the weak echo region is completely surrounded or bounded by larger reflectivity values.

Wind Shear
The rate of change of wind speed or wind direction over a given distance.

WMR
Weighted Multiple Regression Analysis.

X-band
Radar wavelength of about 3 cm.

ZDR
Differential Reflectivity. Ratio between the reflectivity of a horizontal polarized pulse and the reflectivity of a vertical polarized pulse. ZDR depends on the asymmetry of the shape, the orientation and the falling behavior of the particles. ZDR is positive for oblate raindrops, zero or slightly negative for hail and graupel. Note that ZDR is strongly biased by differential attenuation during the passage of the radar pulse through heavy rainfall.

ZPHI
Algorithm for Rain Retrieval combining the radar reflectivity Z and the differential phase shift ϕ_{DP}.

Z-R relationship
An empirical relationship between radar reflectivity factor Z (in mm^6/m^3) and rain rate R (in mm/hr), usually expressed as $Z = AR^b$ where A and b are empirical constants.

Color Plates

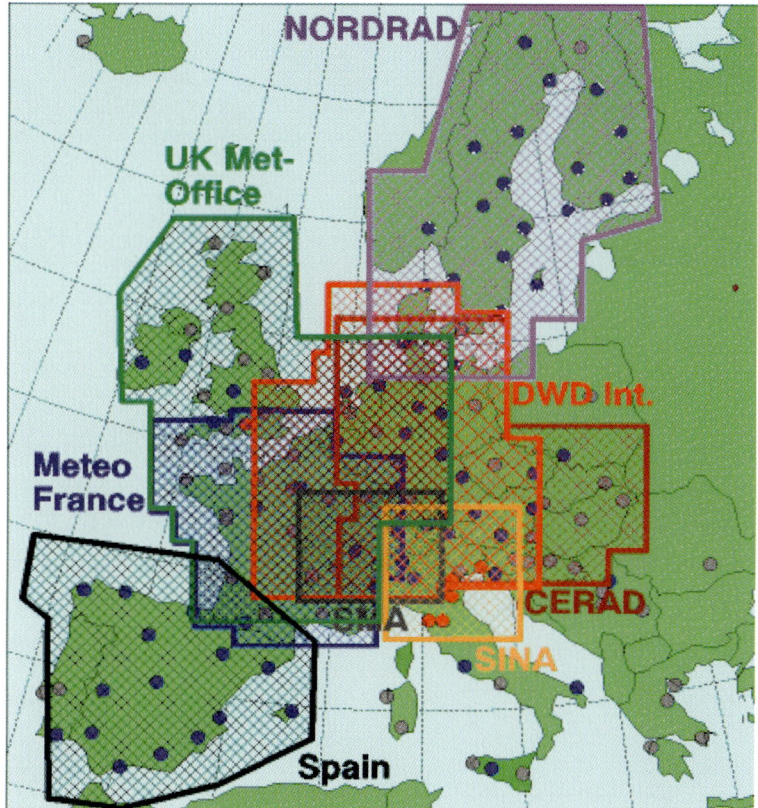

Fig. 1. European weather radar networks for which composite products are operationally available. DWD Int. is the international composite of the German Weather Service DWD. Embedded are the Swiss radar network and part of the north Italian network. CERAD, the Central European Radar composite, extends the coverage to eastern European countries. Networking for southern Europe is under discussion

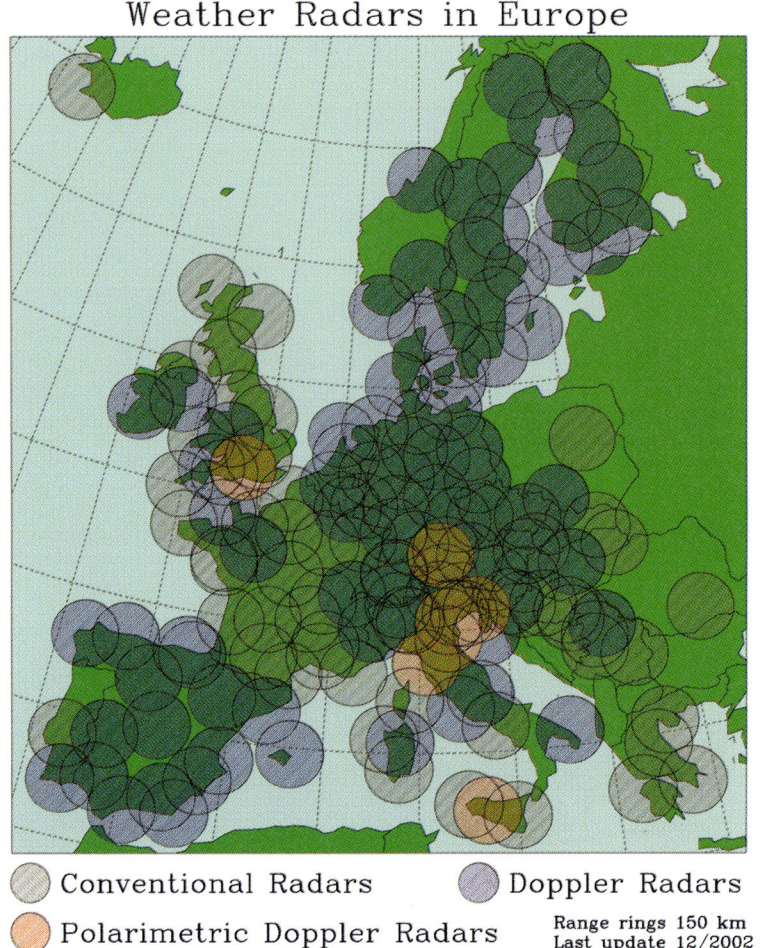

Fig. 2. Coverage of Europe with different types of weather radars. The circles are of 150 km radius

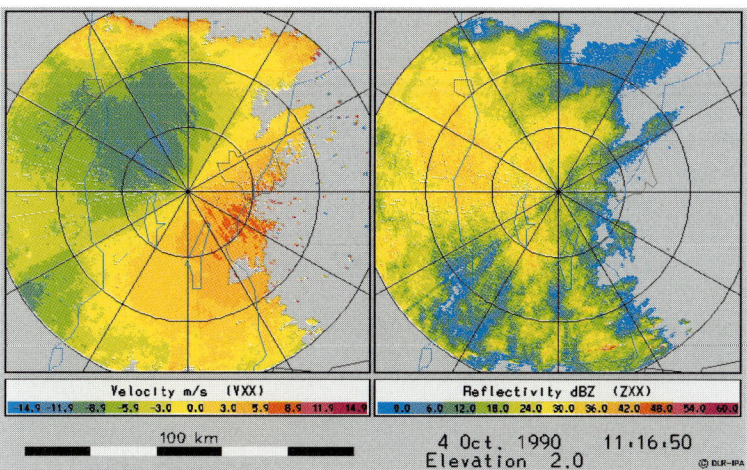

Fig. 3. PPI of Doppler velocity and reflectivity of a rather homogeneous northwesterly wind situation in southern Germany around Oberpfaffenhofen

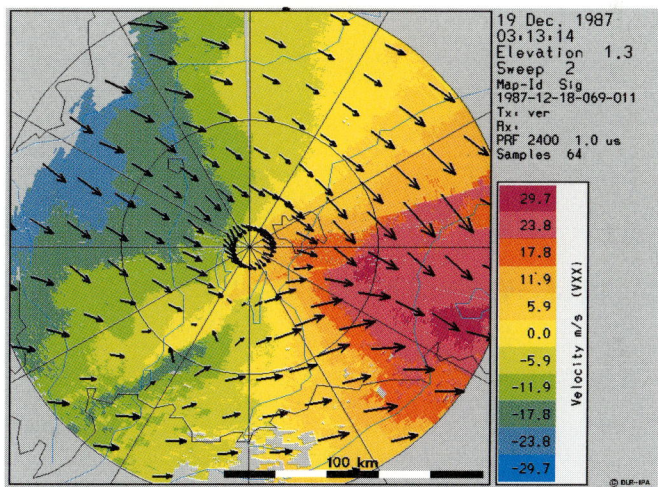

Fig. 4. Horizontal wind field around Oberpfaffenhofen, as analysed with the Uniform Wind method and overlaid on the Doppler velocity measurements used

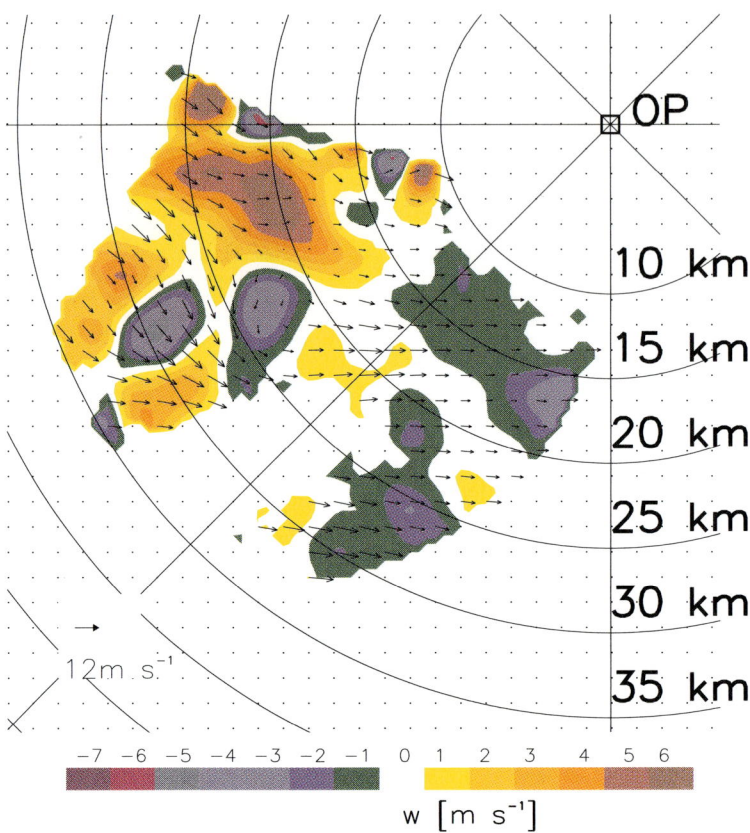

Fig. 5. Horizontal wind vector field underlaid by vertical motion at 4.35 km MSL within a moderately active convective cloud estimated by a bistatic Doppler system (Friedrich, 2002). Yellow and red is upward motion, green and blue are downward motion

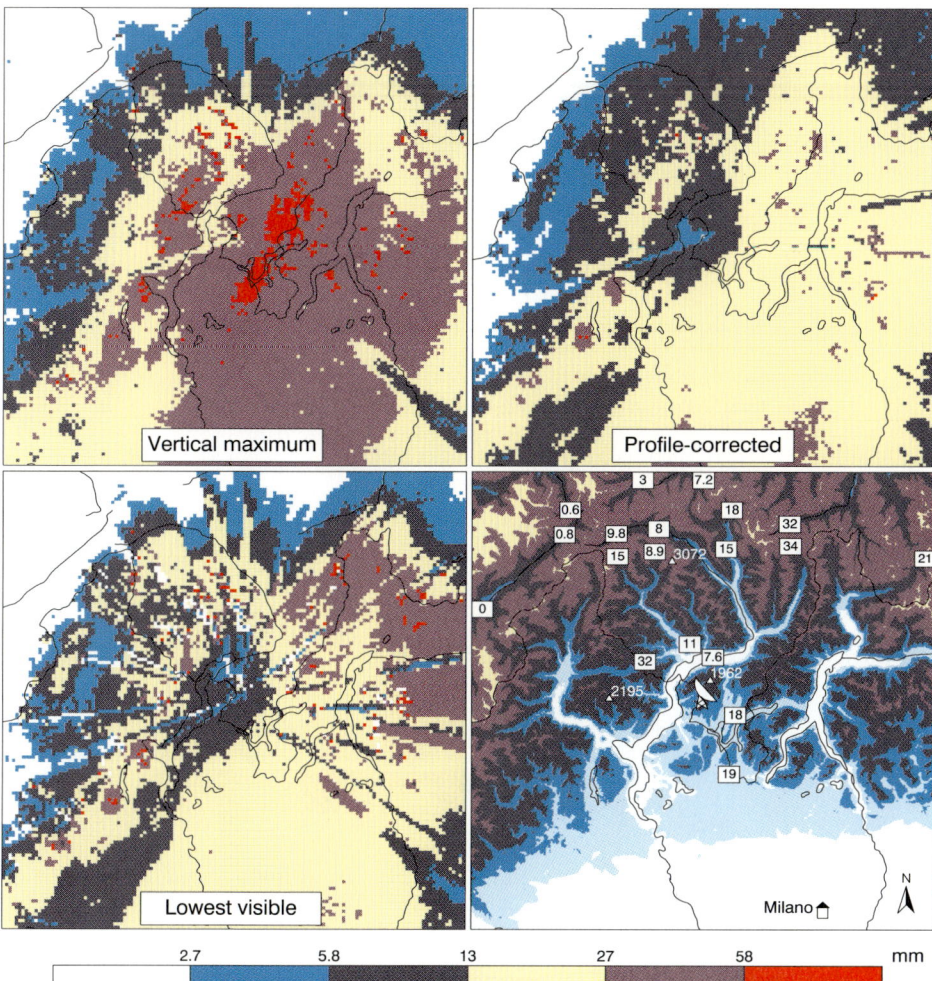

Fig. 6. Comparison of three methods for estimating rainfall amounts by radar in a mountainous region. The three estimates are based on the vertical maximum echo (*upper left*), the lowest visible echo (*lower left*), and estimates corrected using the meso-beta profile (*upper right*). Partial shielding has been corrected for the two products 'lowest visible' and 'profile-corrected,' but not for the 'vertical maximum.' See text. Rainfall amounts are daily accumulations of 3 August 1998. Topography and gauge totals of the same period in millimeters are depicted (*lower right*). Levels of shading correspond to terrain height between 0–250 m (*white*), 250–500 m, 500–1000 m, 1000–2000 m, 2000–3000 m, and above 3000 m, respectively. The frame is a 140 km × 140 km area centred over the radar site Monte Lema (1625 m). The height of three mountains is also indicated: Pizzo Campo Tencia (3072 m), Cima della Laurasca (2195 m) and Monte Tamaro (1962 m)

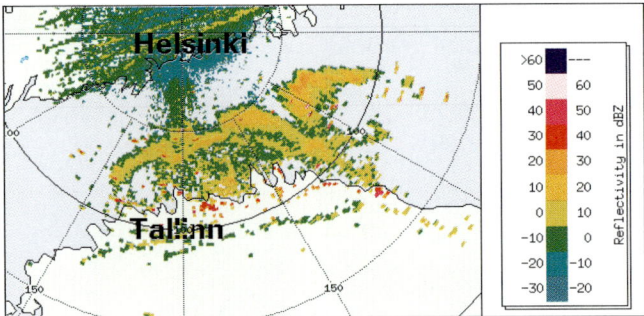

(a) Radar reflectivity factor in the 500 m pseudo-CAPPI product. Sea clutter is located in the Gulf of Finland, weak echoes in continental Finland are from insects and a few stronger spots in Estonia represent remaining coastal clutter, which has passed the Doppler filtering and quality thresholding in signal processing

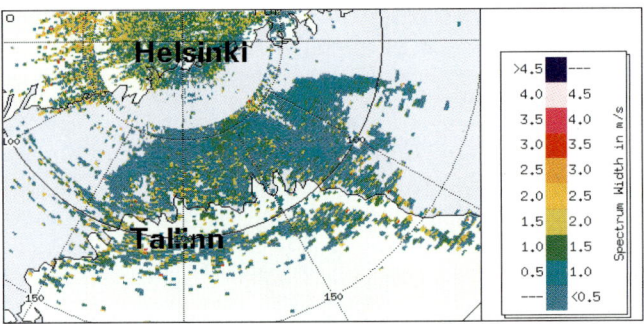

(b) Doppler spectrum width (m/s)

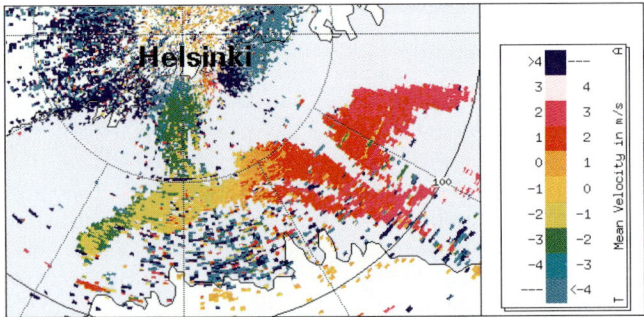

(c) Doppler velocity (m/s), PRF=570 Hz, at the lowest elevation PPI (0.4°). Dots off the Estonian coast represent ships and large flocks of migrating birds

Fig. 7. Examples of sea and ship clutter. From Vantaa on May 27, 2000 at 13:15 UTC

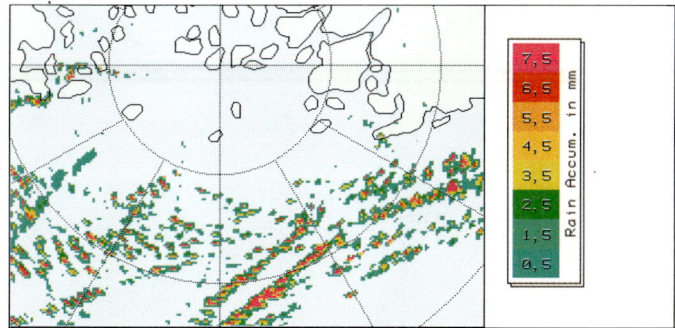

Fig. 8. Accumulated 12 hourly "precipitation" (mm) from ships, as seen from Korpo on June 21, 1997 at 06 UTC. Range ring interval is 50 km

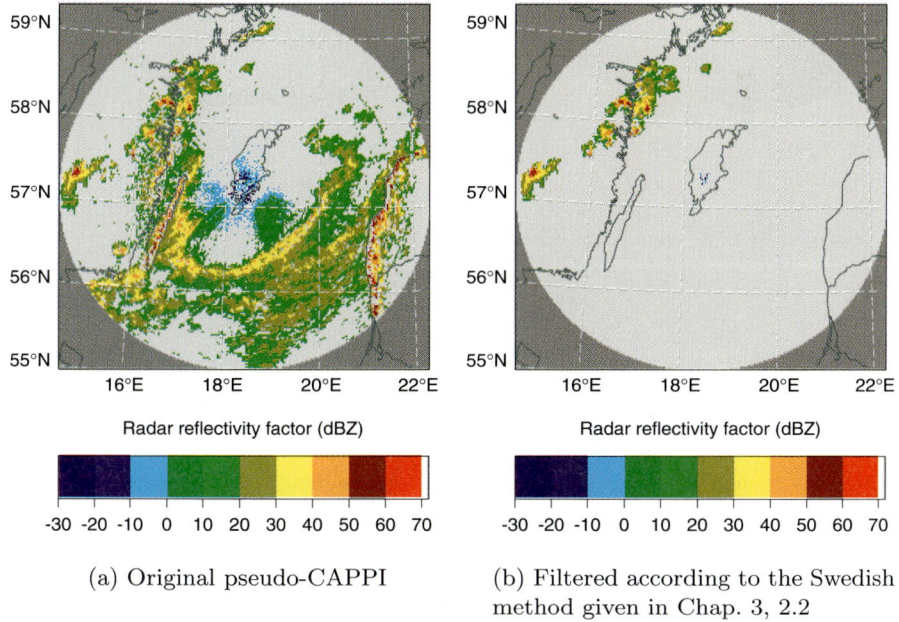

(a) Original pseudo-CAPPI

(b) Filtered according to the Swedish method given in Chap. 3, 2.2

Fig. 9. Severe AP conditions, as seen from Hemse on August 25, 1997 at 14:09 UTC. A line of convective cells stretches from west to north over the Swedish mainland. Strong AP echoes from land are seen covering the Swedish island of Öland and the Latvian and Lithuanian coasts. Severe sea clutter is organized as more or less concentric half-rings at different ranges, with the closest ring being the most discernible

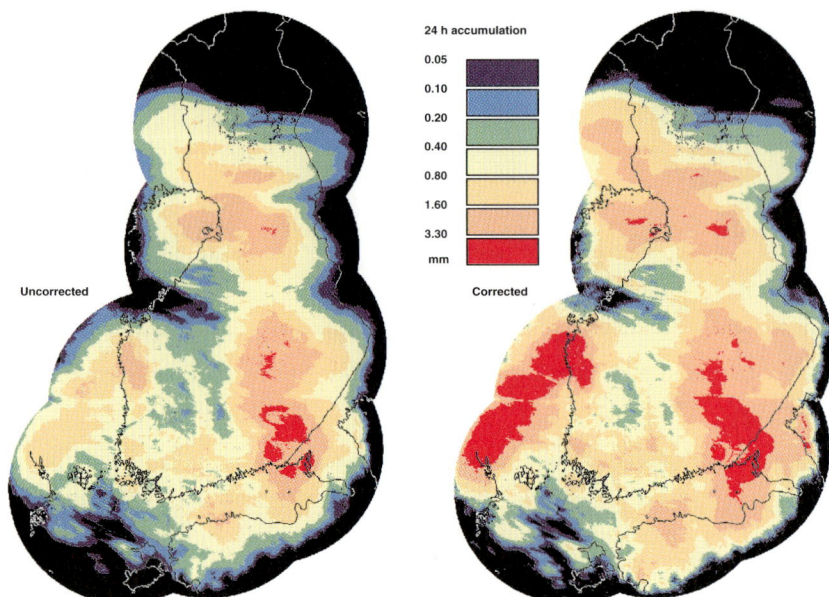

Fig. 10. Accumulated 24-h precipitation in the Finnish radar network on December 20, 2002 at 8 UTC. The outer range of the measurement range is 250 km at each radar. Left: accumulation without any VPR corrections. Right: accumulation after application of the composite VPR correction scheme

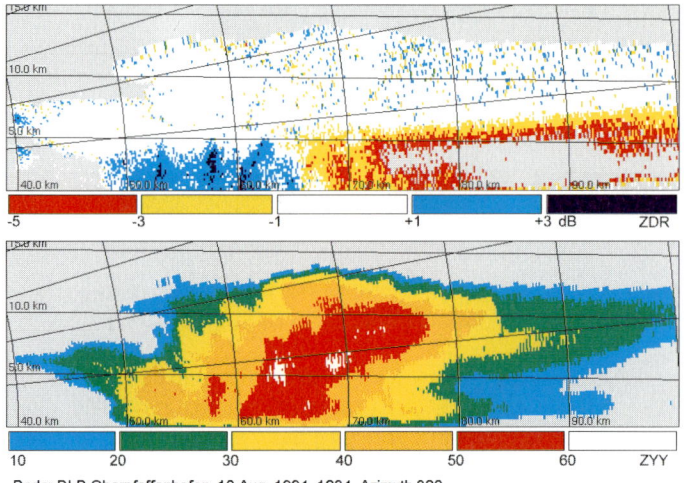

Fig. 11. RHI radar images of a severe thunderstorm. The area of high reflectivity causes severe attenuation of the radar beam such that the signal strength is much reduced beyond the area of high reflectivity. The lower image is reflectivity, and the upper image is differential reflectivity (courtesy DLR Germany). The differential reflectivity becoming smaller than about -5 dB is caused by more intense attenuation of horizontally polarised waves than vertically polarised waves

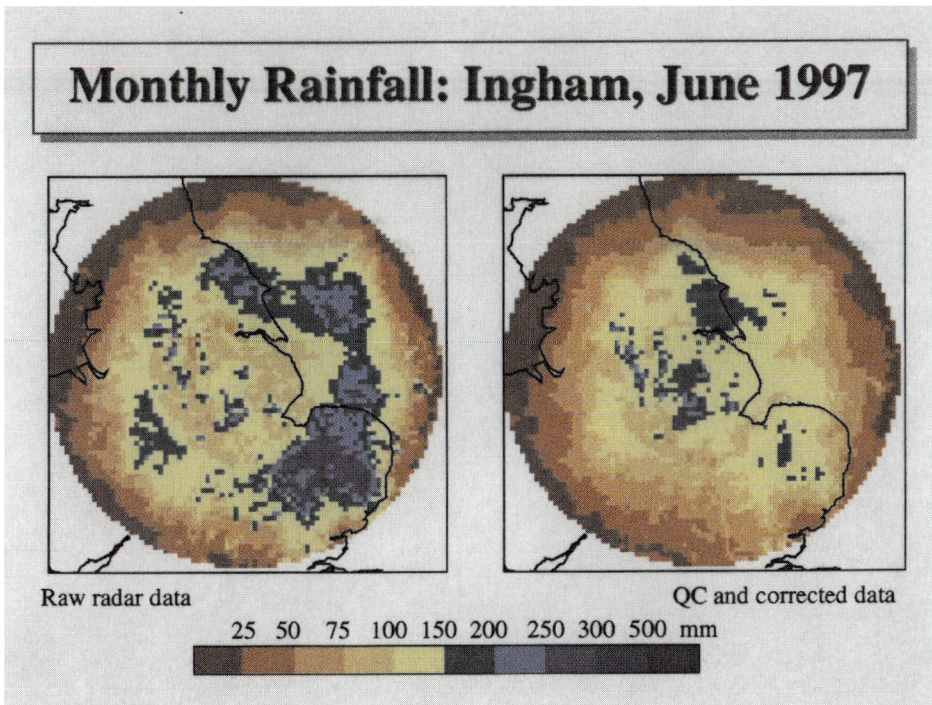

Fig. 12. Monthly integration in June 1997 before and after VPR adjustment for the Ingham radar in the UK (courtesy Met. Office, UK). The coastline of England is shown by the black line. The color scale gives the monthly rainfall measured with the rain radar data (*left panel*) and the quality controlled radar data (*right panel*)

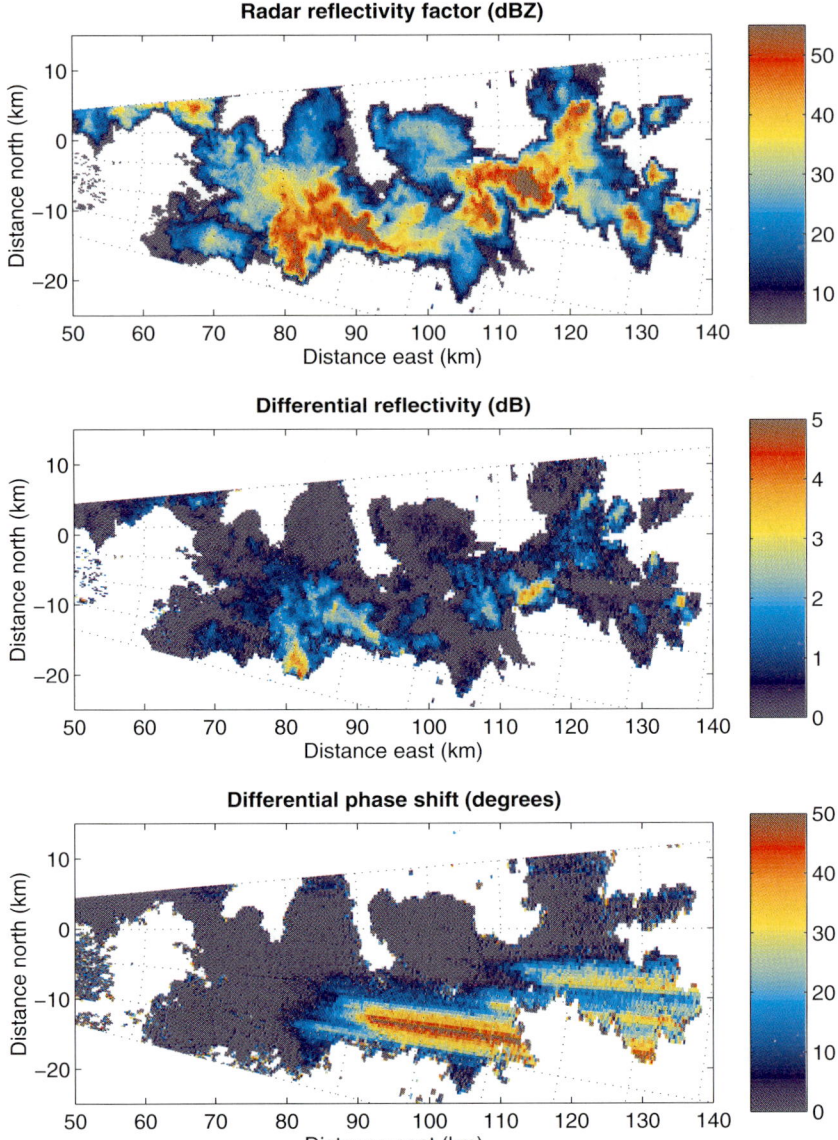

Fig. 13. An example of Z, $Z_{\rm DR}$ and $\phi_{\rm DP}$ observed with the narrow beam S-band Chilbolton radar in the UK on 28 July 2000. The data are from a low elevation (0.5°) PPI scan, so the beam is dwelling in the rain. Even for these large values of $\phi_{\rm DP}$ the derived values of $K_{\rm DP}$ are rather noisy with much poorer range resolution than Z and $Z_{\rm DR}$. Figure courtesy of R.J.Hogan (U. of Reading). Radar data kindly supplied by RCRU, Rutherford Appleton Laboratory

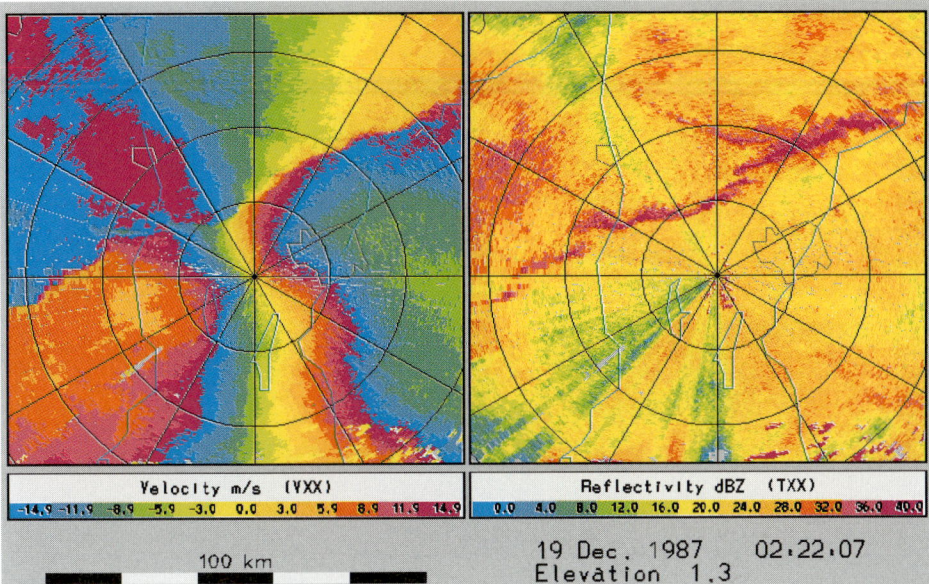

Fig. 14. Radar picture (PPI) of the front at 0121 UTC on 19 December 1987. Doppler velocity on the left, reflectivity on the right. Range circles are every 20 km. Areas surrounded by black lines are (from left to right) city of Augsburg, Ammersee, Starnberger See and the city of Munich

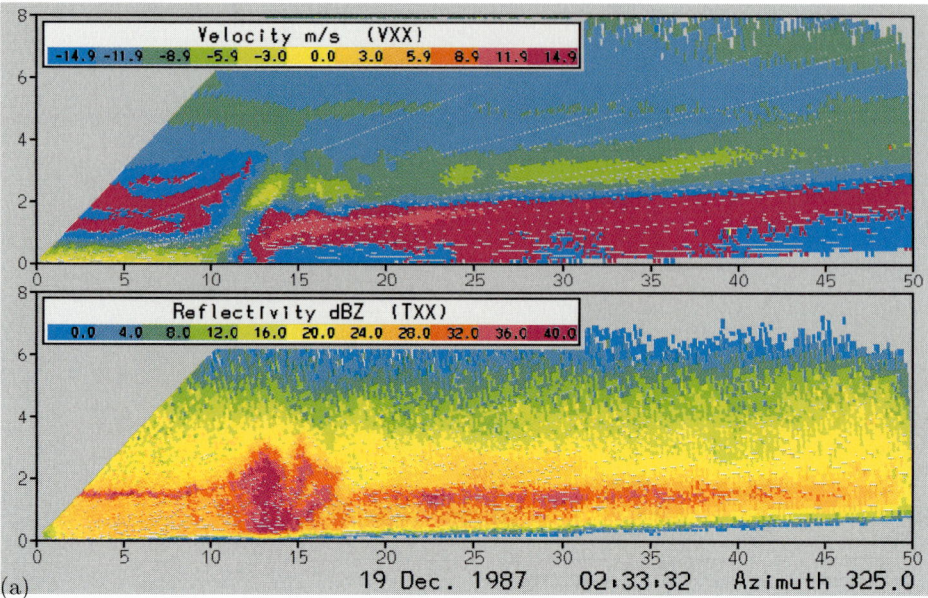

Fig. 15. (a) Radar picture (RHI) showing Doppler velocity (top) and reflectivity factor (*bottom*) perpendicular to the cold front at 0133 UTC, moving to the left. The reflectivity pattern clearly shows the narrow rainband and the difference in the melting layer height, as represented by the bright band in front and behind that rainband. The Doppler velocity pattern indicates the updraft, initiating the intense precipitation formation

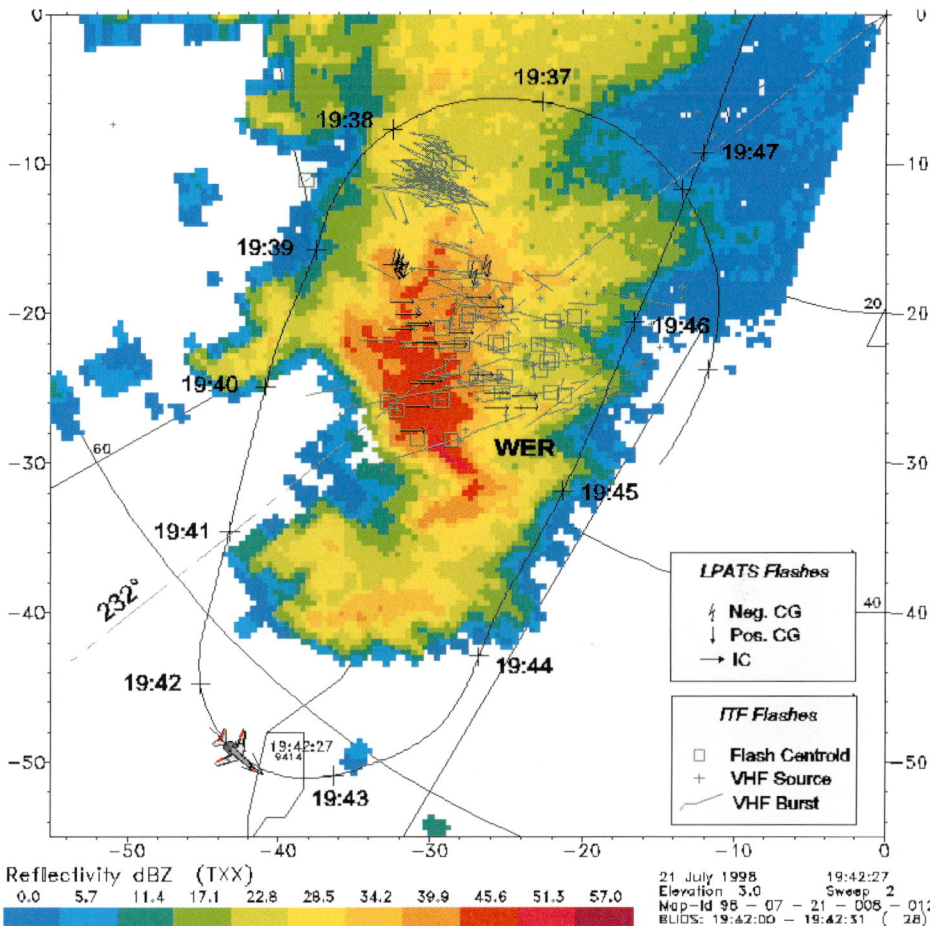

Fig. 16. The supercell of 21 July 1998. POLDIRAD radar reflectivity factor at 1742 UTC taken at 3° elevation. FALCON flight track with the position of the FALCON aircraft indicated at the time the radar scan was taken. Flashes from a 45-s period are shown. Negative, positive and intracloud flashes as measured with the Lightning Position And Tracking Systems LPATS are indicated as well as flashes indicated by VHF interferometric measurements (ITF); see Höller and Schumann (2000)

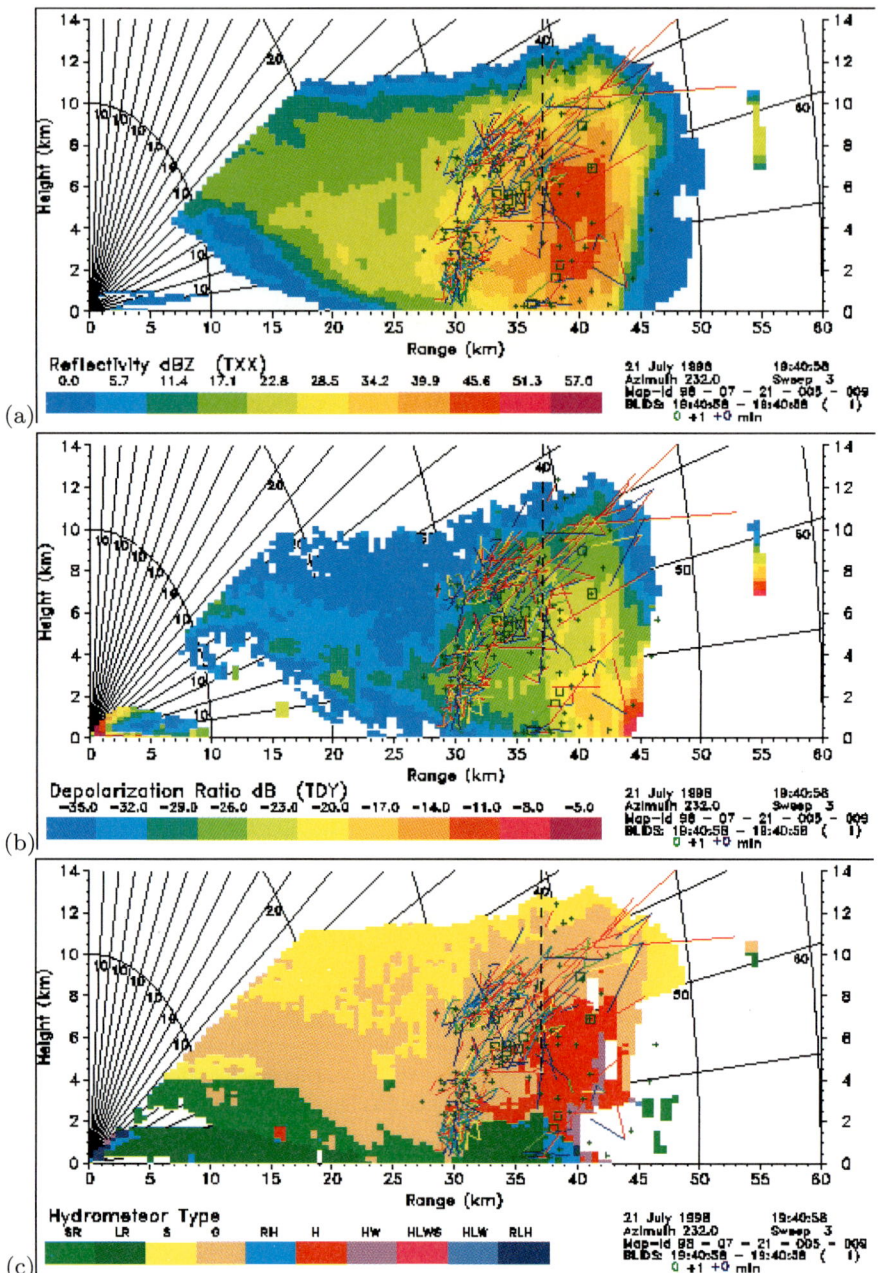

Fig. 17. Vertical section of the supercell storm of 21 July 1998 at 1741 UTC along the 232° azimuth from POLDIRAD. ITF flashes are plotted for a 30-s interval around the nominal scanning time from an azimuthal interval of ±10°. The FALCON aircraft is visible in all parameters shown at a range of 55 km and 9 km height SW of the storm. (a) reflectivity factor, (b) linear depolarisation ratio, (c) hydrometeor or particle type (with color code for Small Rain, Large Rain, Snow, Graupel, Rain–Hail mixture, Hail, Hail Wet, Hail Large Wet Spongy, Hail Large Wet, Rain–Large Hail mixture)

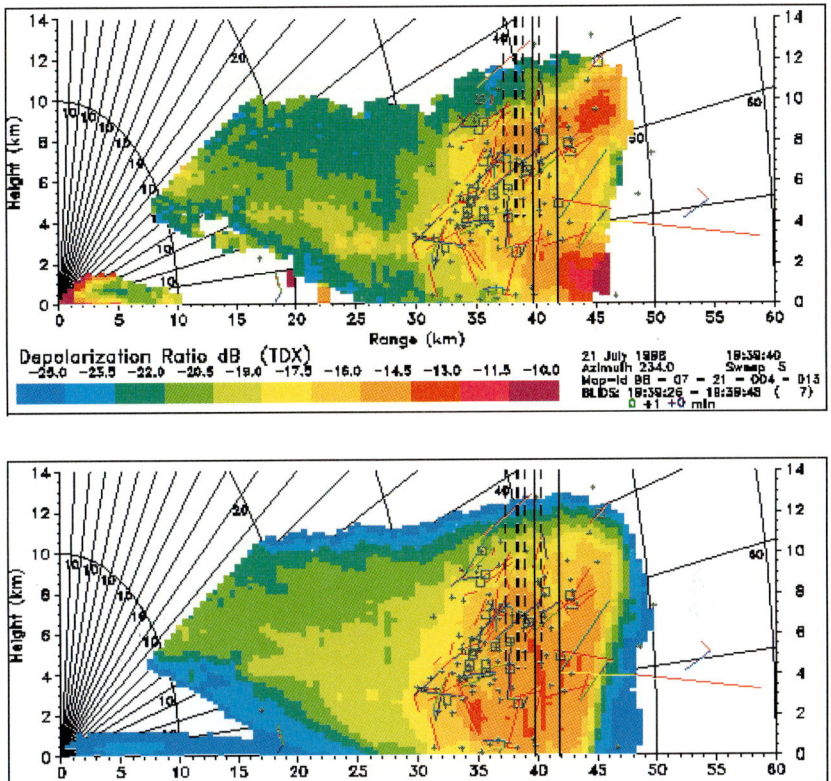

Fig. 18. Vertical section of the supercell storm of 21 July 1998 at 1739 UTC along the 234° azimuth from POLDIRAD. ITF flashes are plotted for a 30-s interval around the nominal scanning time from an azimuthal interval of ±10°. Top: Linear depolarisation ratio; bottom: reflectivity factor

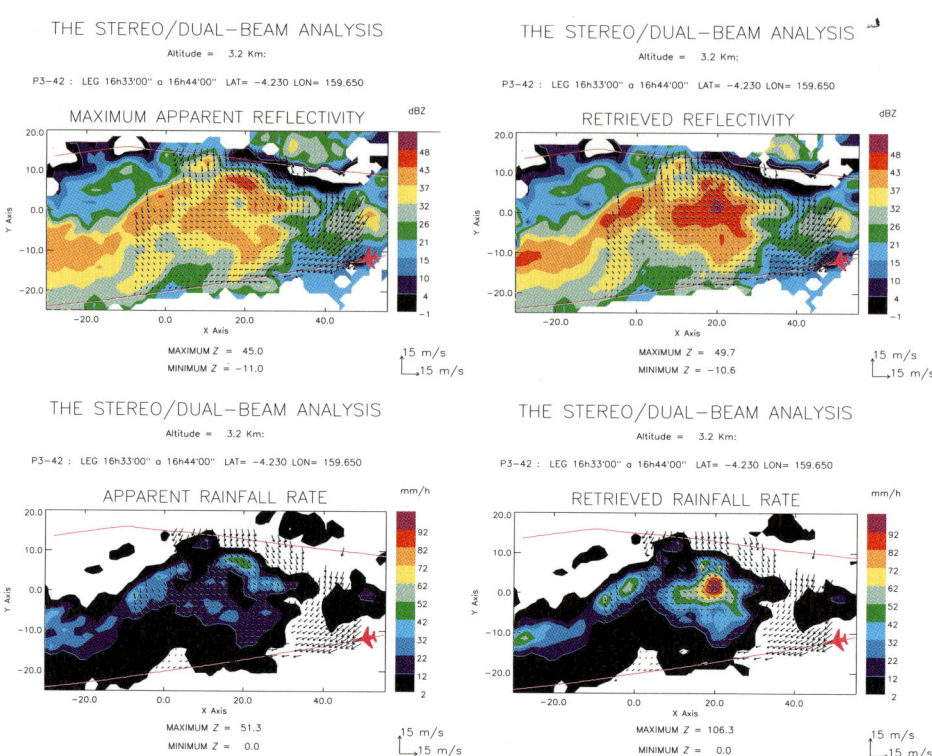

Fig. 19. Horizontal sections of the precipitation field at 3.2 km altitude in the 9 February 1993 Toga-Coare Squall line observed by the two NOAA-P3 aircraft each equipped with a dual beam (or pseudo-dual beam) X-band radar (*aircraft trajectories thin lines*). Top left shows the maximum reflectivity field observed from the four available beams, and bottom left shows the corresponding "apparent" rain rate. Top right shows the corrected reflectivity determined from the hybrid technique and bottom right the corresponding "corrected" rain rate is shown. The wind field in background corresponds to the 0.8 km altitude

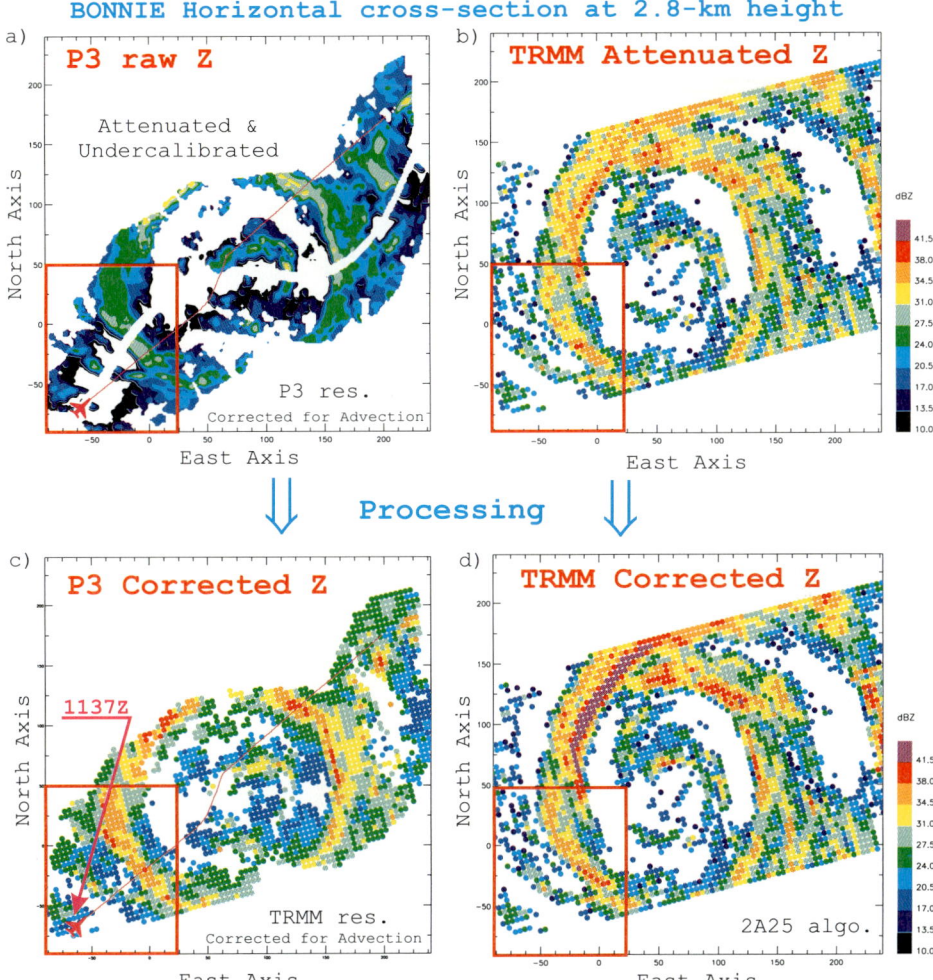

Fig. 20. Quasi-simultaneous sampling of hurricane Bonnie by the TRMM precipitation radar and by the NOAA P3-42 aircraft. Top diagrams: raw reflectivity field (max between "fore" and "aft" beam) of the P3 compared with the "attenuated" reflectivity observed by the TRMM PR. Bottom diagrams: Retrieved reflectivity from the aircraft using the hybrid technique compared with the "corrected" reflectivity of the TRMM-PR using the "rain profiling algorithm." Note the particularly good agreement in the SW corner where the time coincidence of the samplings is within ± 5min

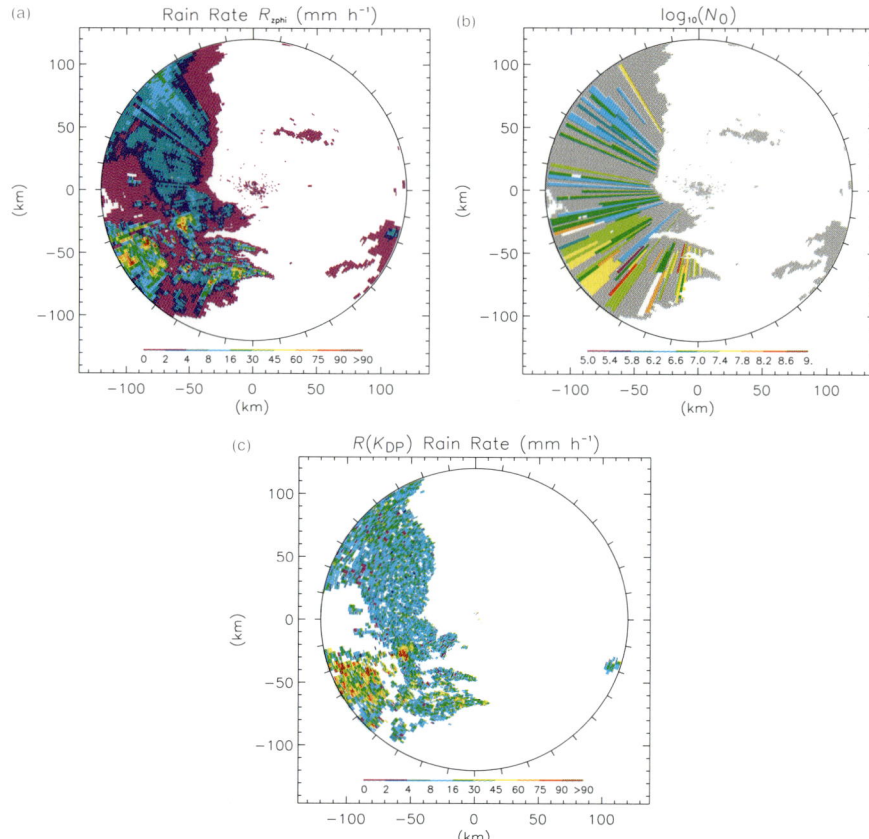

Fig. 21. PPI of the Darwin radar (at 1.6° elevation, 19 January 1998 around 2240 UTC). Top left shows the rainfall rate delivered by ZPHI. Top right gives the corresponding field of N_0^* (each ray corresponds to a segmentation; in the gray areas, ZPHI does not operate, and the classical estimate is used instead). The bottom panel displays a standard estimate of the rain rate based on K_{DP}

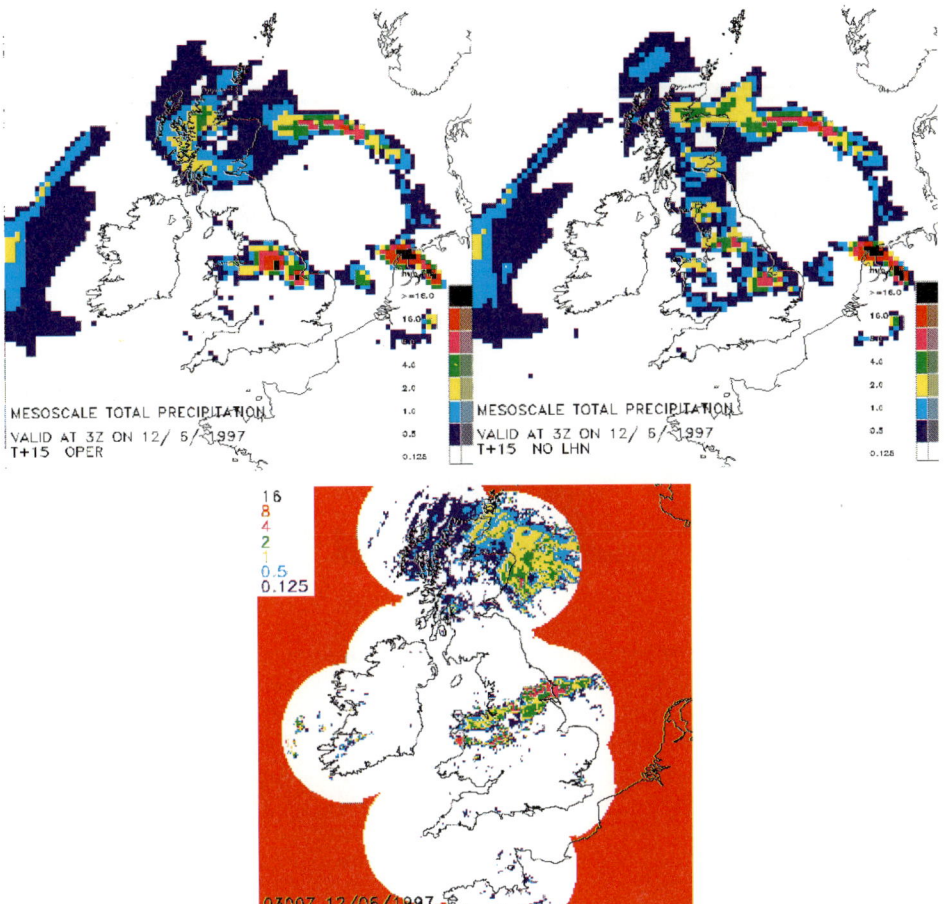

Fig. 22. An unusually long lasting forecast impact of radar rain rate assimilation in the Met Office (UK) mesoscale model, taken from Macpherson (2001). Bottom frame shows radar picture for 03 UTC, 12 June 1997, with northward moving rain band over central England. Top left frame shows operational precipitation rate forecast at $t+15$ hours. Top right frame is from an experiment with NO assimilation of radar data

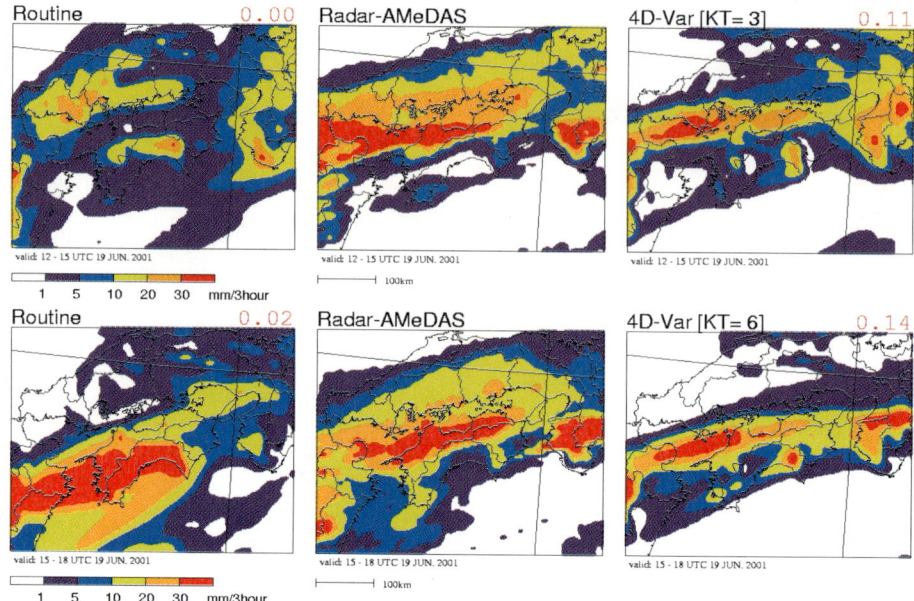

Fig. 23. Forecasts and observations of 3-hour precipitation amount. Centre column shows the Radar-AMeDAS observations of the Japan Meteorological Agency, right column shows forecasts starting from a 4D-Var analysis and left column shows the forecasts from a 'routine' analysis using PI for assimilation of rainfall data. Initial time of forecasts is 12UTC 19 June 2001. Top row is for forecast period 0–3 hours; bottom row is for period 3–6 hours. Taken from Ishikawa (2002)

Index

Z–R relation   143
– dynamic determination of   91
– uncertainties in   91
Z–R relationship   203
3D-Var assimilation of wind data   271
4D-Var assimilation of wind data   270

airborne precipitation radar   200
airborne radar systems   200
Alps   169
analog-to-digital converters   8
anomalous propagation   144, 160
– characterization of   81
– illustration of   81
anomalous propagation echoes
– illustration of   88
antenna   3, 6
anticyclonic rotation   182
assimilation of radar precipitation data in NWP models   259
assimilation of radar wind data   268
asynchronous digital subscriber lines   19
attenuation   34, 65, 117, 118, 121, 158, 162, 199, 203, 204, 320

BALTEX   98
Barnes analysis   98
baseband   7
beam-broadening   65, 71, 73
bistatic Doppler radar   37
bistatic multiple Doppler radar   42
bistatic receivers   44
bright band   34, 96, 128, 171
– diagnosis of   101
bright-band
– contamination by   69
brightband   157

calibrated radar targets   20

calibration   73, 151, 161
– absolute   71
– hardware   71
– role of   71
calibration procedures   19
calibration uncertainty   21
CLARA   244
clusters   175
clutter   79, 137
– avoiding   59
– frequency of   62
– residual   60, 69
– wavelength and   59
– with nonzero velocity   62
clutter cancellation   116
clutter elimination   59
– decision tree   61
– effectiveness   62
– neighbour test   60
– noncoherent statistical tests   59
– speckle filter   63
– using short pulses   59
clutter filter   33
clutter filtering   9
clutter map   33
– dynamic   59, 62
– static   59
coherent air scatter   237
coherent on receive operation   14
coherent particle scatter   236
coherent processing interval   9
coherent weather radar   6
cold climates
– anomalous propagation conditions in   81
– effect of network inhomogeneities in   79
– precipitation measurement in   78
– spurious echoes in   79

cold front 169
cold pool 178
composite
– heterogeneity of 78
composite products 17
convergence lines 177
copolar correlation 134
coupled atmospheric and hydrologic numerical models 123
crystals 160
cyclonic rotation 182

data assimilation concepts 256
data assimilation techniques 257
data center 17
data quality 19
deep convective systems 172
differential attenuation 134, 162
differential phase 135, 148, 155, 161
differential reflectivity 132, 145, 161
disaggregation 129
distributed hydrological models 115, 119
Doppler 116
Doppler radars 3
Doppler signature 39
Doppler spectral width 44
downconversion 6
drop size distribution 204
dual beam Doppler radar 218
dual channel dual polarisation 14
dual frequency radar 217
dual polarisation radar 14
ducting 80
duplexer 3
dwell time 9
dynamic range 6
dynamical models 123

eddy dissipation rate 46
encoders 19
ensemble prediction 121, 129
error feedback 121
error structure 52, 75
errors bias 121, 122
errors distribution 121
errors feedback 119, 121
errors random 121
exciter 6

extrapolate measurements from above 67
extrapolation 55, 74

feeder cells 173, 175
flashes 188
Flood Estimation Handbook (FEH) 125, 128, 129
flood forecasting 115, 118, 127
flood peak 124
Flood Studies Report (FSR) 129
Forward Scattering Spectrometer Probe FSSP 241
frequency analysis 125, 127
frontal systems 168

gauge adjustment 71, 98
– in real-time? 72
– representativeness of point observation 72
gauge observations
– correction of 99
gauge–radar comparison 93, 94, 99, 100
gauge-adjustment technique
– evaluation of 99
gauge-to-radar ratio 98
georegression techniques 126
Global Precipitation Mission 110, 232
ground clutter 144, 160
gust front 178

hail 157
hailswath 173, 175
heterodyne 8
homodyne receiver 8
hybrid algorithm 221
hybrid transmission 137
hybrid type storm 180
hybrid type thunderstorm 180
hydrid storms 175
hydrograph 121, 122
hydrological models 118, 123, 124, 129
hydrometeor identification 159, 162
hydrophobic coatings 14

ice water content 160
image analysis 83
– methods
– – three-dimensional 87
– – two-dimensional 83
in-phase component 7

incoherent particle scatter 236
integration time 9
intermediate frequency 8
interpolation 55
inverse model 204, 212

Kolmogorov spectrum 46

Lagrangian advection 74
large droplet issue 237
Latent Heat Nudging 261
lightning 187
line oriented storms 176
linear depolarisation ratio 138
liquid water content 244
liquid water path 245
local area network (LAN) 17
low noise amplifier 7
lowest visible echo 69

macroscale of the turbulence spectrum 236
magnetron transmitters 14
mean radial velocity 11
measurement errors 19
MESAN 89
meso–γ scale models 266
mesocyclonic rotation 194
mesocyclonic vortex 194
mesoscale 175
Meteosat 89
microwave power meter 20
minisupercell 193
MM5 model 184
moments of the DSD 212
motion vector 104
– derivation of 105
motion vectors
– illustration of 106
mountainous terrain
– precipitation measurement in 52
– radar network in 58
multicell storms 173
multicellular storms 172
multiple Doppler methods 41
multiple Doppler radars 37
multiple scattering 246

narrow rain-band 169
network products

– accuracy of 78
NORDRAD 78
normalised DSD 205, 206
nudging 258
nugget variance 55

obstacles 64
optical extinction 237
Optimum Interpolation 269
orography 73
– radar and 52
orthogonal correction schemes 71

partial beam filling 65
path integrated attenuation 216
phase stability 19
Physical Initialisation 260
point observation
– representativeness of 55
polarimetric radar 13, 226, 227
polarisation agile radar 14
polarisation diversity 117, 127
precipitation
– advection method for nowcasting of 105
– fraction of water in 92
– nowcasting of 104, 105
– – verification of 108
– point forecasts of 106
– shallow 94
precipitation measurement
– in cold climates 78
– in mountainous terrain 52
– lowest visible echo 69
– profile correction 68, 317
– qualitative overview 74
– sources of error 52
– vertical maximum echo 69
– weighted multiple regression 72
precipitation nowcasting
– general requirements 104
– illustration of 108
precipitation phase 90, 91
precipitation process 171
precipitation type
– observations of 91
probabilistic information 75
probability density functions 126
Probable Maximum Flood (PMF) 125

Probable Maximum Precipitation   125, 128
profile correction   67, 68, 317
– operational   70
propagation conditions   81
pulse repetition frequency   3

quadrature component   7
qualitative overview of precipitation   74
quality control   115, 116, 118, 125, 127
quality control of radar data for NWP   273

radar   202
radar data errors
– treatment in assimilation   274
radar network   17, 22
radar product generator   3
radar signal processor   8
radar systems
– sensitivity of   78
radome   14
rain profiling algorithm   204, 232
rain-band   168
raindrop shape   139, 147, 153, 161
raindrop size spectra   142
rainfall rate   150, 157
rainfall rates   161
range   3
range bias
– vs. Z–R relation   79
range gate   8
range sampling   8
range time clock   8
receiver   3
reflectivity   3
reflectivity gradients   133, 137
representativeness of point observation   55, 72
reproducibility   71
return period   125
RHI   320

sampling   125
sea clutter   79, 80
– illustration of   82, 88
sensor synergy   235
shielding   53, 54, 63
– partial   54, 63
– underestimation caused by   69

ship clutter   81
– illustration of   82
side lobes   14, 137
single cell storm   172
single cells   172
sleet
– measurement of   90
snow
– measurement of   90
– operational measurement of   79
snowfall   90
– accuracy of measurements   92
– socioeconomic impacts   90
spaceborne   202
spaceborne radar   215, 217
spatial continuity of precipitation   55
specific attenuation   214
spectral width   12
spurious echoes
– computer vision techniques   84
– detection algorithms   85
– identification and correction of   83, 87
– mitigation of   89
– multisource methods   89
– use of fuzzy logic   85
squall lines   176, 177
state estimators   123
stereoradar algorithm   221
storm classification   172
streamflow predictions   123
successive correction method   269
supercell storms   172, 174
superrefraction   81
surfaces of constant delay time   43
switched dual polarisation   14
synoptical observations   91

test transmitter   20
thermodynamic retrievals   269
thunderstorm   172
thunderstorm - flow structure   183
time to independence   132
tornadic storms   188
tornadoes   188
transmitter   3, 6
transponder   20
triple scattering echoes   133
turbulence   44

upconversion   8

Upper Rhine valley   188

validation   222, 230
variational assimilation   258
variational assimilation of precipitation data   263
variogram   55, 75
– of Alpine precipitation   56
verification measures   90
– limitation of   108
vertical maximum echo   69
vertical profile of reflectivity   67, 94, 100, 117
– correction for   67, 100
– representativeness   70
– variability of   69, 70
vertical sampling difference   95
– illustration of   96
visibility   73

– geometrical simulation   63, 64, 66
– observed   66
visibility correction   63
visibility map   53, 63
VPR   321
VPR correction
– illustration of   103

wave number   6
waveguide switch   14
weak echo region   174, 180
weighted multiple regression   72
wide area network (WAN)   17
wind field   219
wind fields   37

X-band polarisation   149

ZPHI technique   155, 161

Druck: Strauss Offsetdruck, Mörlenbach
Verarbeitung: Schäffer, Grünstadt